THE OTHER SIDE OF NORMAL

THE OTHER SIDE OF

NORMAL

HOW BIOLOGY IS PROVIDING THE CLUES TO UNLOCK THE SECRETS OF NORMAL AND ABNORMAL BEHAVIOR

JORDAN SMOLLER

wm
WILLIAM MORROW
An Imprint of HarperCollins*Publishers*

The geography of the Big 5 personality traits (page 61): Peter J. Rentfrow, Samuel D. Gosling, and Jeff Potter, *Perspectives on Psychological Science* (Vol. 3, no. 5), p. 31 © 2008 by SAGE Publications. Reprinted by Permission of SAGE Publications.

Mike and Judy Bornstein with one of their daughters (page 125): Used with permission of the Bornstein Family.

Examples from the "Before They Were Famous" test (page 140): Reprinted with permission of Richard Russell, Brad Duchaine, and Ken Nakayama.

Hazda faces (page 248): Reprinted with permission of Antony Little and Coren Lee Apicella.

HarperCollins books may be purchased for educational, business, or sales promotional use. For information please write: Special Markets Department, HarperCollins Publishers, 10 East 53rd Street, New York, NY 10022.

A hardcover edition of this book was published in 2012 by William Morrow, an imprint of HarperCollins Publishers.

FIRST WILLIAM MORROW PAPERBACK EDITION PUBLISHED 2013.

Designed by Jamie Lynn Kerner

Library of Congress Cataloging-in-Publication Data

Smoller, Jordan W., 1961–

The other side of normal : how biology is providing the clues to unlock the secrets of normal and abnormal behavior / Jordan Smoller.—1st ed.

p. cm.

Includes bibliographical references.

ISBN 978-0-06-149219-8 (hardback)

1. Behavior genetics. 2. Psychobiology. 3. Biological psychiatry. 4. Norm (Philosophy) I. Title.

QH457.S66 2012

591.5—dc23

2011040827

ISBN 978-0-06-149220-4

13 14 15 16 17 OV/RRD 10 9 8 7 6 5 4 3 2 1

For Ava

Contents

THE OTHER SIDE OF NORMAL

PROLOGUE

. . . [The] intellect suffers to pass unnoticed those considerations
which are too obtrusively and too palpably self-evident.
—EDGAR ALLAN POE, "THE PURLOINED LETTER"

E VERY DAY YOU LOOK AT SOMETHING RIGHT IN FRONT OF
your eyes and yet you never see it. Your retina, the part
of your eye that picks up visual information, sits behind
a network of blood vessels at the back of the eye. This vascular
window shade is so obvious and ever-present that the brain has had
to make it invisible, creating a latticelike blind spot that is filled in
by the mind. Like the purloined letter of Poe's tale, many of the
most fundamental features of the normal mind have been hidden
in plain sight. They are so basic to what we do, think, and feel, that
we hardly give them a thought. And, until recently, they have been
relatively invisible even to scientists who study the mind for a living.

This book is about the obscure and the obvious. It is about phenomena that are so complex they may seem indecipherable, even though they are so familiar we live inside them every day. It is a book about how the brain gives rise to the mind and how the mind, in turn, gives rise to everything we care about. It is also about the universal and the unique. We all enjoy the experiences and products of a mental life—thoughts, feelings, desires, relationships. And yet each of us has a unique and private life of the mind—a singular configuration of cognition, emotion, and social functioning that reflects an unprecedented combination of genes, experience, and environmental contingencies.

Self-consciousness, one of the human mind's universal features, has also given us an eternal curiosity about how and why we do what we do. Few things carry the same compelling quality. A former editor of a major news weekly once told me there are two cover subjects that could always be counted on as big sellers: stories about the brain and stories about Jesus. "If only we could find a way to do a cover on Jesus's brain," she told me.

Philosophers and scientists since antiquity have tried to fathom the human mind. In some cases, ancient theories seem surprisingly modern—the Greeks believed that imbalances of four bodily humors were responsible for temperament and mental illness, a notion that resonates with modern views about "chemical imbalances." And over the past century, we've seen the rise and fall and the powerful impact of debates about nature vs. nurture, psychoanalysis, and behaviorism.

Until recently, two obstacles—one technological and the other psychological—have limited our quest to understand the mind. First, we lacked the tools. The brain is the organ of interest for understanding how we think, feel, and behave, but before the late twentieth century, scientists wanting to study how the brain gives rise to the mind had to make do with a pretty crude set of options. They could study animals—observing their behavior or dissecting their brains—and they could ask people questions or observe their actions. In the early twentieth century, with the introduction of the electroencephalogram, or EEG, they could begin to study the brain in action by measuring electrical currents flowing though neurons by putting electrodes on the scalp. However, if they wanted to study the structure of the brain in living people, to see how the brain's circuits are wired, and to watch those circuits function in real time, they were out of luck. But not anymore.

In the last two decades, the technological floodgates have opened. A combination of applied physics and high-power computing has created a breathtaking array of machines for looking at the brain. The field of neuroimaging, which had its first breakthrough in the early 1970s with the introduction of CAT scans, now offers an alphabet soup of sophisticated techniques to study brain structure (CT, MRI, DTI), its function (fMRI, PET, ASL, MEG, SPECT, NIRS), and even its chemistry (MRS). And the new field of molecular neuroscience has introduced methods for studying the nanoworld of the brain—synapses and the signals sent between and inside cells.

We've also known for a long time that mental traits run in families. The nature/nurture debate was preoccupying philosophers, theologians, and scientists for centuries before we had any conception of genes, let alone the tools to study them. And even after researchers understood that genes influence how the brain develops and functions, they didn't have the resources to study how genes work. Not anymore.

Today we know the sequence of all human genes and can determine if variations anywhere in the genome are linked to neural or behavioral traits. We can study how nurture turns genes on and off. And we can begin to link these discoveries to the information we get from molecular neuroscience and brain imaging. We still have a long way to go, but at least we have a road map.

The second obstacle to demystifying the brain, the psychological one, has to do with what I'll call the "purloined letter effect." Many of the crucial questions about the normal brain are those that, until recently, we hadn't thought to ask. They have to do with aspects of the mind that are so self-evident we easily overlook them. How do we understand other people's thoughts and feelings? Why do we fear some things and not others? Where do we get our ability to trust? Why are we attracted to one person and not another? How does emotion color our memories? How does experience change the brain? Some of these questions are not really new—but now scientists are able to ask them with the tools of neuroscience and genetics in hand. And the answers have begun to reveal an unseen biology behind the familiar mind—a biology of normal.

This book emerged from my own experiences in psychiatric research. Over the past fifteen years, I have been studying the genetic and brain basis of psychiatric disorders such as depression, anxiety disorders, bipolar disorder, schizophrenia, substance dependence, and personality disorders. But the more I learned about these disorders, the more I came to appreciate that the only way to really understand how the brain and the mind go awry is to understand how they were designed to function in the first place. Mental dysfunctions exist because there are functions that can be disturbed. Anxiety disorders exist because we have brain mechanisms designed to detect and respond to threat. When these mechanisms are distorted or exaggerated, fear and anxiety can overwhelm our lives. But these mechanisms are evident in the earliest expressions of normal childhood temperament—the tendency of children to avoid or approach unfamiliar situations and people. My own effort to find genes that make people susceptible to anxiety disorders has evolved into finding genes that influence temperament and the activity of brain circuits that mediate normal fear.

And despite the popular critique that psychiatry pathologizes everything, we only recognize certain varieties of human behavior as disordered. That's because there are a finite number of things our minds have to handle for us to survive and reproduce: avoid harm, form relationships, assess risks, choose mates, acquire resources—to name a few of the most important. When these go wrong, it matters. There is no "athletic skill disorder" because being a skilled athlete is not on the list (luckily for me). There are certain domains where the idea of normal matters, and many more where it doesn't. Charting that territory of normal is a crucial project for psychiatry and all of the other fields, from psychology to economics, that are concerned with making sense of human behavior. The point is that many disorders can best be understood as perturbations of normal systems and mechanisms. As this theme began to guide my own research, I discovered that a surprisingly coherent picture of the normal human mind was emerging at the intersection of the social, behavioral, and biological sciences. It is by no means complete, but it provides a fascinating look at what makes us tick.

In the chapters that follow, I describe this emerging field: the biology of normal. Along the way, I draw on the latest research from a range of disciplines—psychology and psychiatry, developmental and cognitive neuroscience, genetics, molecular biology, economics, epidemiology, ethology, and evolutionary biology—to shed light on how the brain works. My hope is that by the end, you will begin to see how the complex features of the mind fit together, which in turn will give you a new way of looking at how we adapt to life's challenges.

Let me also say up front what the book is not about. First, my aim is not to provide a comprehensive review of what we know about normal brain function—that would be an encyclopedic project, and, trust me, you wouldn't want to read it. Instead, I will focus on how genes, experiences, and even chance shape our emotional and social natures. This is a book about what and whom we care about. Second, I don't intend to convince you that everything important about the mind can be reduced to biology. It would be absurd to claim that we can adequately explain or describe every mental phenomenon in material terms. The mind does emerge from the brain, but that does not mean there is a one-to-one correspondence between the firing of nerve cells and what we mean by the mind. A purely biological account of love, empathy, and other human experiences will never be a fully adequate account in the same way that a detailed account of the wavelengths of light reflected by each pigment in Picasso's *Guernica* would not capture the painting's power.

Also, I do not use *normal* to mean "right"—in either its older or newer sense. Until about the 1820s, *normal* was a term of geometry, meaning "at right angles" or "perpendicular." As the philosopher Ian Hacking has written, it later acquired another connotation of "right"—that is, the "standard" or the way things "ought to be."[1] Neither of those is what I mean. Instead, I intend something closer to the meaning coined by the eighteenth-century French physiologist François-Joseph-Victor Broussais, who was the first to conceive of normal as a spectrum of variability. As Hacking explains, Broussais believed that "pathology is not different in kind from the normal; 'nature makes no jumps' but passes from the normal to the pathological continuously."

So, just to be clear, I'm using the phrase "biology of normal" as a shorthand to refer to the underlying architecture of the brain and the mind. A full account of that architecture requires multiple perspectives and languages, depending on what we're trying to explain: neuroscience, psychology, evolutionary biology, cultural anthropology, and social experience. And we will draw on all of these languages in the chapters that follow.

HUMAN NATURE, HUMAN DIVERSITY, AND TRAJECTORIES

THREE THEMES ARE WOVEN THROUGH THE CHAPTERS THAT FOLLOW. The first provides the scope for our exploration of the biology of normal: each of us is a product of human nature as well as individual difference. These two strands—the universal and the particular— are scientific cousins, and in an almost poetic irony of history, their intellectual champions were themselves cousins.

Charles Darwin's theory of natural selection made the heretical claim that our human nature derives not from God's image but from the outcome of our ancestors' "struggle for existence." Rather than being a blank slate, the human mind comes preloaded with neural circuits that were shaped by the adaptive challenges we faced in our evolutionary past. As a result of this common heritage, our brains have mechanisms for solving the problems that determined how successful our ancestors were at leaving descendants. In the aggregate, this legacy creates the boundaries of our "human nature"—the shared functions that our minds use to navigate the challenges of life.[2]

Where Darwin laid the groundwork for understanding the universal components of human nature, his half-cousin Francis Galton pioneered the study of individual differences.* He coined the phrase "nature versus nurture" and invented the use of twin studies to tease them apart. In the course of asking questions about the causes of human variation, he derived fundamental statistical tools and principles, including the concept of statistical correlation and the field of biometry, for which he is still widely known. The modern discipline of behavioral genetics, devoted to the study of how variations in genes (and environment) cause individual differences in human and animal behavior, is descended from Galton's pioneering work.[3,4] These individual differences and their genetic basis are the other axis of normal. They contribute to the diversity of human temperament, personality, and intelligence.

The second and related theme is the unfolding of what I call *trajectories*. Our minds reflect the influence of both our shared evolutionary endowment and the particular set of genetic variations we inherit. But each of us is unique. Our singular trajectory through life is the result of two additional forces: the unprecedented set of environmental circumstances that we encounter and the stochastic nature of biological systems—in other words, experience and chance. And here, the element of time enters the equation. Within

---

*Despite some scientific differences with his famous relative, Galton was a strong supporter of Darwin's theory: "The slowness with which Darwin's fundamental idea of natural selection became assimilated by scientists generally is a striking example of the density of human wits."

the terrain of human possibility, each of us inhabits a developmental rivulet whose trajectory depends on the sequential accidents of our unique personal histories. We each walk the stage with a particular cast of fellow actors: a distant mother, a bullying brother, a special teacher, a first love. Each of us acquires a particular portfolio of experiences: the moment of birth, the first day of school, windfalls, humiliations, and traumas. And our lives depend not only on *what* happens, but *when*. As we will see, the developing brain passes through sensitive periods when experiences can set or redirect the course of our lives. For example, whether we are nurtured or neglected in the first years of life may set us on a trajectory of resilience or vulnerability.

The third recurring theme will explore how an understanding of the biology of normal informs our understanding of mental illness. Many of the mysteries of psychiatric illness begin to make sense against the backdrop of how the mind and brain do what they were designed to do.* In each chapter, we will consider not only what the mind does normally but what it would look like if these normal functions went awry.

WHAT LIES AHEAD

WE WILL BEGIN WITH A QUESTION THAT IS INESCAPABLE FOR A book with the word *normal* in its title: "What do you mean by *normal*?" In the first chapter we'll see that this is a complicated question indeed—one that has been tackled mainly by attempts to define what is *not* normal. Psychiatry in particular has struggled with this issue, often with unsatisfying results. The line between normal and abnormal is hard to draw, and sometimes cultural bias rather than scientific evidence has been used to draw it. Our search

---

* By *designed* I don't mean to imply an intelligent designer. I am simply referring to the ways that mental and neural systems developed under the influence of natural selection, genetic variation, and environmental factors.

for the boundaries of normal will take us through epidemics of multiple personalities and shrinking penises to the controversial history of psychiatric classification and the evolutionary psychology of mental dysfunction.

Having considered the definition of *normal*, we will move on to what science is teaching us about its biology. After exploring the genetic roots of temperament and personality (Chapter 2), we'll dive into the debate about the formative influence of early experience (Chapter 3). In the chapters that follow, we will discuss the development of key mental functions in childhood and adulthood including social cognition and empathy (Chapter 4), the biology of attachment and trust (Chapter 5), the roots of sexual attraction (Chapter 6), and how emotion and fear shape learning and memory (Chapter 7). Along the way, we look at how discoveries in these areas can shed light on what we call mental disorders. And, finally, in Chapter 8 we return to the question of what the "biology of normal" can teach us about our shared humanity, the singular trajectories of our lives, and how we can understand mental suffering.

In case you were wondering, though, I won't have anything to say about Jesus's brain.

A final note: Throughout this book, I use several case stories based on my experiences as a clinician to illustrate some of the ways in which the biology and psychology of the normal mind can go awry. To protect patient privacy, these stories represent fictionalized composites and do not refer to any single individual.

CHAPTER ONE

"WE'RE ALL MAD HERE"

B Y THE LATEST ACCOUNTING, MORE THAN HALF OF ALL Americans meet criteria for a psychiatric disorder at some time in their lives.[1] The current system of diagnosing mental disorders contains hundreds of labels, ranging from well-known standards like schizophrenia to less familiar ones like hypoactive sexual desire disorder. But what is a psychiatric disorder? Does *normal* become meaningless if most of us have an abnormality of mind? Where do we draw the line between normal and abnormal?

In 2007 two reports were released documenting alarming increases in the diagnosis of childhood psychiatric disorders that were previously thought to be rare. Both reports triggered a public outcry. But the nature of the outcry was quite different.

The first report, from the U.S. Centers for Disease Control, examined the prevalence of autism among eight-year-old children in the year 2002. Based on data from fourteen sites, the CDC found that 1 in 150 children (0.66 percent) had an autism spectrum disorder. That number was more than ten times higher than prevalence estimates of autism in the 1980s and seemed to validate a growing concern that the nation was in the midst of an epidemic.

The response among families, advocacy groups, and the media was, understandably, one of unmitigated alarm. Alison Singer, spokeswoman for the advocacy organization Autism Speaks, captured the sense of urgency felt by many: "This data today shows we're going to need more early-intervention services and more therapists, and we're going to need federal and state legislators to stand up for these families."[2] Singer and others called for a vast increase in research funding "so we can find a cause and understand what is fueling this high prevalence."[3]

Some families, and certain celebrities, insisted that vaccines were to blame; others weren't so sure but worried that some kind of environmental toxin might be contributing to the rise in prevalence. Many scientists and educators cautioned that the apparent epidemic might simply be a product of greater awareness and a broadening of the definition of autism (to include a larger "autism spectrum"). But few doubted the urgent need to help affected children and their families.

The outcry over the second report was equally strong but dramatically different in its tone. The study, published in the *Archives of General Psychiatry*, examined trends in the diagnosis of child and adolescent bipolar disorder using data from a large survey conducted by the National Center for Health Statistics. The authors found that between 1994 and 2003 the rate of bipolar disorder diagnoses in children up to age nineteen increased fortyfold, from 0.025 percent to 1.0 percent of the population (approximately half the rate of bipolar disorder among adults).[4] This time, the jump in prevalence was widely interpreted not as a public health emergency but a scandal. For many, the findings confirmed the suspicion that psychiatry itself was deeply flawed. The blogosphere lit up with critics who claimed that psychiatry was pathologizing normal behavior, medicalizing childhood, and even colluding with pharmaceutical companies to create a market opportunity for drugging children. Many in the medical community were also suspicious that a lot of misdiagnosis was going on.

Two numbers, two very different reactions. Considered side by side, these two episodes dramatize the charged and complicated nature of defining psychiatric disorders. There are some remarkable parallels: in the same year, the public learned that two often disabling childhood disorders, once thought to be rare, were now being diagnosed in about 1 percent of children. In both cases, part of the story seemed to be an increasing public awareness of the condition and an expansion of diagnostic labels. The new autism estimates captured the broader autism spectrum including Asperger syndrome. And the bipolar estimates reflected a broadened spectrum as well. Since the mid-1990s, some researchers and clinicians argued for expanding the diagnosis beyond the classic symptoms of manic highs and depression to include children who exhibited chronic and explosive anger and irritability.

But there were important differences. Autism had always been a disorder of childhood while, prior to the 1990s, many psychiatrists believed that bipolar disorder did not exist in children. The broadening of the autism spectrum may have been less controversial because it had a longer history. But there was another key difference. At the time the reports were published, there were few if any established drug treatments for autism. On the other hand, medications are a cornerstone of treating bipolar disorder. And many of these medicines—lithium, valproate, and antipsychotics—can have serious side effects. The idea that such powerful drugs would be increasingly used to treat bipolar disorder in young children was clearly part of what was alarming to many people. Some saw the expansion of the diagnosis as psychiatric imperialism and "disease-mongering." Scientists who collaborated with pharmaceutical companies were accused of nefarious conflicts of interest, with the implication that psychiatric research was motivated by financial self-interest.

We still don't know exactly why the prevalence of autism and bipolar disorder has been growing, but the controversy forces us to confront an important question: How do we draw the line between normal and disorder when it comes to how the mind functions?

At what point are we just pathologizing normal as some critics of psychiatry charge? Answering those questions requires that we first answer another question: What do we mean by *normal*?

Determining what is normal is a surprisingly difficult task, and that may explain why academic science has rarely tried to address it. But the definition of *abnormal* has been investigated and debated over and over again—perhaps in part because of a notion articulated a century ago by the great American psychologist William James, who believed that "the best way to understanding the normal is to study the abnormal."[5]

Modern psychiatry has largely tried to define the abnormal without much reference to the normal. And as we'll see, that's created some problems. For the most part, we have described disorders by starting at the edges of human experience—identifying syndromes from the most striking and dramatic symptoms that people express. Working our way inward from those edges, normal becomes something of an afterthought—the ill-defined residual.

But without a basic map of how the mind and brain function, our definitions of abnormal and normal depend heavily on what behaviors we decide are unusual, bizarre, or problematic. And those decisions can easily be influenced by cultural trends, historical tradition, or the opinions of "authorities."

A REVOLUTION IN PSYCHIATRY

SEVERAL YEARS AGO ONE OF MY COLLEAGUES POSED A QUESTION during a staff luncheon in our Department of Psychiatry: "Who do you think was the most influential psychiatrist of the last fifty years?"

The answer seemed obvious: Robert Spitzer. Robert Spitzer? Probably an unfamiliar name to most people; but the revolution he led transformed the way we view mental illness.

As recently as the 1970s, psychiatrists had no reliable criteria for making a diagnosis. A patient who reported hallucinations and bizarre behavior might receive a diagnosis of schizophrenia from one psychiatrist, borderline personality from another, or manic-depressive illness from a third. At the same time, the field began to acknowledge that its disorders were sometimes based on archaic views of human behavior. In 1973 the Board of Trustees of the American Psychiatric Association voted to remove homosexuality from its official manual of psychiatric disorders.

That same year, *Science,* a top-tier scientific journal, published an article challenging the foundations of "sane" and "insane."[6] The author, psychologist David Rosenhan, asked seven confederates to join him in a deception. They were each to present themselves to psychiatric hospitals with the complaint that they had been hearing voices. All eight of these "pseudopatients" were admitted to psychiatric hospitals and held for weeks. Their mission was to get discharged. "Each was told that he would have to get out by his own devices," Rosenhan explained, "essentially by convincing the staff that he was sane" (p. 252). This turned out to be very difficult, and it took nearly three weeks for the pseudopatients to be discharged. Even though they exhibited no psychiatric symptoms during their hospital stays, all eight were initially diagnosed with schizophrenia and their "normal" behavior was interpreted as evidence of illness.

In the early 1970s, another indictment of psychiatric diagnosis highlighted the need to change the way psychiatrists practiced. A study of hospital admission records revealed that a patient was much more likely to be diagnosed with schizophrenia (rather than an affective disorder, such as manic-depressive illness and depression) if he were admitted to a hospital in New York than if he were admitted to a hospital in London.[7] Could mental illness in America really be so different from mental illness in the UK?

One obvious way to answer this is to show the same set of patients to psychiatrists in both countries and see if they agree on diagnosis. As part of the U.S./UK Cross-national Project, researchers showed videotapes of patient interviews to groups of psychiatrists in the United States, United Kingdom, and Canada.[7] The results clearly showed that it was the psychiatrists not the patients that explained the transatlantic differences in diagnoses. When faced with the same patients, American psychiatrists were far more likely to make a diagnosis of schizophrenia than were the British psychiatrists. If small cultural differences among psychiatrists could have such big effects on the way they labeled symptoms, what hope was there of defining the boundaries of normal and abnormal?

The unreliability of psychiatric diagnosis led Robert Spitzer and his colleagues to overhaul the system. In 1980 they rolled out the third edition of the *Diagnostic and Statistical Manual of Mental Disorders*, or DSM-III, as it is better known. The two previous editions of the DSM (published before 1970) had been heavily influenced by Freudian concepts of psychopathology and offered few specifics about the definition of mental illnesses.

The third edition provided the field, for the first time, with an explicit set of criteria for diagnosing disorders. DSM-III also debuted a raft of conditions that are now familiar fixtures of popular culture: attention deficit disorder, panic disorder, posttraumatic stress disorder, borderline personality disorder, and others. Over successive editions of the manual, psychiatry has engaged in a cycle of lumping and splitting its diagnoses. Between the publication of DSM-I in 1952 and the latest major revision, DSM-IV, in 1994, the number of diagnostic labels in the book has swelled from just over 100 to more than 350.

Today, the DSM is the most influential book in psychiatry. It is the reference manual every psychiatrist-in-training must learn to use before being considered competent to practice. Among other things, it provides the definitions of mental disorders that insurance companies use to determine whether psychiatric treatment is reimbursable. In many ways, DSM-III and its successors also fueled the modern era of medication treatment of mental illness. With clearly defined disorders to study, researchers and pharmaceutical companies could test whether new compounds were effective treatments for these conditions. Indeed, before a pharmaceutical company launches a psychiatric drug, they usually must demonstrate its effectiveness for a "DSM-defined" disorder. More than any other psychiatrist, Spitzer (and his colleagues) shaped the way we talk about mental illness.

But it's no secret that the DSM has its limitations. Right up front, the manual acknowledges "that no definition adequately specifies precise boundaries for the concept of 'mental disorder.'"[8]

The primary goal of the DSM, since 1980, has been to provide a practical and useful set of criteria—a common language—for diagnosing mental disorders in clinical practice and research. In essence, it presents a description of syndromes—agreed upon by a consensus of experts—that are associated with distress, disability, or "a significantly increased risk of suffering death, pain, disability, or an important loss of freedom." Still, despite the claims of some critics, the DSM was never intended to be an authoritative statement about what's normal and what isn't. As Robert Spitzer himself noted, "It does not pretend to offer precise boundaries between 'disorder' and 'normality.'"[9]

By design, the DSM also doesn't attempt to tie disorders to the basic functioning of the mind and the brain. And so, as useful as it's been in providing a common language for drawing a line between mental health and illness, the application of DSM's categories can be subject to the vagaries of cultural trends in how we label behavior. That's something I witnessed in the course of my own training as a psychiatrist.

FROM EPIDEMIC TO ODDITY

"YOUR NEXT ADMISSION'S IN 314."

I stopped by the nurses' station on the way to room 314 and picked up a copy of Sarah Crane's chart. It was 2:30 in the morning, and she would be my fourth admission of the night—I needed a quick summary of her history. I glanced at the note from the resident

who had admitted her last month and skimmed a story that was by
now a familiar one.

"Hello, Ms. Crane, I'm Dr. Smoller."

A woman in her late twenties sat, with a blank stare, in the corner
of the interview room, wrapped in a powder-blue wool blanket. She
didn't make eye contact.

"Can you tell me what brings you in tonight?"

"One of my alters tried to kill me," she answered, matter-of-factly.

"Tried to kill you?"

"Yes."

"Who tried to kill you?"

She didn't respond.

"Ms. Crane, who tried to kill you?"

We sat there in silence for two or three minutes.

Then her eyes narrowed, and her face took on a stern scowl; she
spoke in a voice that was low and gruff. "I did."

In the late 1980s an alarming but previously obscure mental
illness began to reach epidemic proportions in the United States.
To accommodate the victims, psychiatric hospitals were driven to
divert their inpatient resources by opening "units" specializing in
the treatment of this disorder. The disorder was called "multiple
personality disorder" (MPD) and was believed to be due to early
traumatic sexual abuse, which itself was being recognized as vastly
more common than previously suspected.

Even more striking, MPD was becoming epidemic not only on
a national scale but also, one might say, on an individual level. Eve
may have had three faces, but the modern MPD victim could have
more than a hundred "alters," each with its own personality, name,
vocal inflection, and set of memories. Prior to 1970 fewer than two
hundred cases had ever been reported, but between the mid-1980s
and 1990s, more than twenty thousand cases were diagnosed.[10] And
then, in 1994, MPD was removed as a label in the DSM manual.

In its place, the diagnosis of dissociative identity disorder (DID)
appeared. The diagnostic criteria for MPD and DID are almost

identical, but the name change signaled a retreat from the almost supernatural notion of coexisting multiple personalities. By the time MPD was stricken from psychiatry's official list of diagnoses it had become the focus of a controversy that engaged feminists, victims' rights advocates, litigators, and mental health professionals.

The concept of "recovered memory" played a key role in disorders like MPD and posttraumatic stress disorder. This seemed to cultivate a cottage industry of therapists who elicited and helped "recover" memories of childhood sexual and physical abuse in patients with a variety of symptoms.

Exemplified (and perhaps inspired) by the 1970s story of Sybil, who reportedly developed multiple personalities after suffering horrific abuse during her childhood, the prevailing explanation for MPD was that victims of overwhelming abuse develop separate personalities to handle their unbearable memories and seal them off from consciousness. Therapists were trained to draw these memories into awareness, sometimes through hypnosis or by interviewing patients while they were under the influence of Amytal (a barbiturate touted as a "truth serum"). Suddenly patients who had never known they were abused were discovering they had been horribly victimized.

Families were torn apart, and in a growing number of cases, patients sued the alleged perpetrators (usually a family member). The recovered memory phenomenon fueled a growing cultural panic about child abuse in the 1980s, reaching a peak with prosecutions of staff members of several preschool and day care centers. Responding to accusations of abuse, law enforcement officials and therapists elicited increasingly bizarre tales of ritual and satanic abuse that should have defied credulity. In the Little Rascals Day Care Center case, the center's director was sentenced to twelve consecutive life terms based on the testimony of young children who described abuse that included the ritual killing of babies aboard a spaceship.

As research emerged demonstrating that recovered memory of severe trauma is a rare (if not implausible) phenomenon, a backlash ensued, including a new wave of litigation targeting therapists who encouraged or even induced false memories of sexual and ritual abuse. Paul McHugh, then chairman of psychiatry at Johns Hopkins University, likened the frenzied assertions about repressed and recovered memory and multiple personality disorder to the social hysteria of the late 1600s that produced the Salem witch trials.[10] Skeptical psychiatric researchers challenged the advocates of repressed memory to verify their claims. Harvard psychiatrist Harrison "Skip" Pope and his colleagues at McLean Hospital asked why it was so difficult to find documented examples of repressed memory prior to the twentieth century if it is a natural or innate capacity of the brain in the face of severe trauma.

They scoured historical works of literature and nonfiction and were unable to find any descriptions of repressed traumatic memory. And then they did something unusual for a group of academics: they offered a $1,000 reward "to the first person who could produce an example of dissociative amnesia for a traumatic event in any work of fiction or nonfiction in any language, prior to 1800." They posted their challenge in print and on websites and discussion groups all over the Internet and in multiple languages. It was an extraordinary approach to solving a highly contentious, politicized, and seemingly unending debate.

I spoke with Pope about the repressed memory challenge as we sat in his office at McLean Hospital, where he directs the Biological Psychiatry Laboratory. He is a man who speaks in paragraphs with a boyish enthusiasm that is uncommon for a Harvard professor and an erudition fitting for a descendant of Alexander Pope (another scholar with an interest in memory and forgetting: "Of all affliction taught a lover yet, 'Tis sure the hardest science to forget!").

"I had always been struck," he said, "by the fact that there did not seem to be any cases of repressed memory in Shakespeare or in Aeschylus or Euripedes or Sophocles or the *Aeneid* or *The Odyssey* or the Bible or other things and wondered if maybe it's just that I didn't have a sufficiently comprehensive knowledge of literature or whether maybe this was an indication of the fact that this was not a natural human phenomenon."

In the 1990s he had asked members of a university English Department to see if they could come up with any instance of repressed memory in literature before the nineteenth century, and they were unable to. Intriguing but hardly definitive. But a decade later, he realized that advances in technology had created an unprecedented opportunity. The reach of the Internet and the resources that are available online meant that he could launch a comprehensive test of his hypothesis. "My study is a study that right up front seeks to prove a negative and claims to have done so because it uses a technology that has not existed until the last ten years of humankind. Namely the power to ask a question of every single person in the world and then if nobody can answer the question to be able to say that there is no answer to the question."

In 2006 Pope and his colleagues issued their repressed memory challenge on more than thirty high-volume websites across the Internet, from broad-interest sites like Google Answers to more specialized sites like "Great Books Forums." They translated the challenge into French and German and posted it on websites hosted in those countries, and readers spread the challenge to other websites.

In 2007 Pope and his colleagues published the results of their quest in a medical journal— not a single case of repressed memory was uncovered by them or anyone else.[11] They concluded that "dissociative amnesia" (thought to be a core component of multiple personality disorder) was best described as what psychiatry calls a "culture-bound syndrome"—that is, an entity constructed by and limited to a particular historical culture—in this case, twentieth-century western societies.

There is a postscript to this story. Shortly after their paper was published, a response to the challenge was submitted that did appear to describe an instance of repressed memory published before 1800. It concerned a scene from *Nina*, a one-act French opera by Nicolas Dalayrac, which premiered in 1786. In the opera, Nina faints after seeing her true love, Germeuil, in a pool of blood, apparently murdered by a rival whom her father wishes her to marry. When her father presents her to the murderer for marriage, she becomes delirious. She is sent to recuperate at her father's country estate, where she develops amnesia for Germeuil's murder, believing he is on a trip from which he will soon return. When Germeuil finally does reappear, miraculously having survived, Nina gradually regains her memory of him. Strictly speaking, even this case does not meet Pope's challenge, because Nina's forgetting seems to have involved amnesia due to delirium, and there's no indication that she recovered a memory of the traumatic event. Nevertheless, Pope and colleagues awarded the prize for this entry, which moved the origin of "repressed memory" only fourteen years before the cutoff date of 1800.

Repressed memories are considered to be central to the etiology of MPD; given that the concept of repressed memory didn't appear before 1786, it's perhaps not surprising that the first case of a dual personality wasn't reported until 1791. Eberhard Gmelin, a German physician, described a local woman who, while recovering from an infectious disease, developed attacks of nodding head movements followed by a sudden shift into the identity of a vivacious French woman who described herself (in fluent French) as a refugee of the Revolution who had fled to Germany.[12] In these states, she had no memory of her German family, but just as suddenly, she would return to her true identity, with no recollection of her French alter ego. The second, and more famous case of a "multiple personality" was that of Mary Reynolds, reported in 1816 by the New York physician S. L. Mitchell. Like the earlier case, she had a second, more bubbly personality emerge following an illness that apparently included severe seizures.[13] It seems likely that these cases actually represented a neurologic alteration of personality that can occur following seizures or a variety of brain insults. The more modern concept of MPD, with its emphasis on repressed memory, did not appear until the late nineteenth or early twentieth centuries.

The story of multiple personality disorder is one of several examples in which psychiatric diagnoses have risen and fallen from favor as notions of what is normal and what is an illness have shifted. Like "hysteria" and "fugue" before it, multiple personality disorder made the journey from epidemic to oddity.[14] The point here is that our definitions of disorder can change, even within a generation, sometimes owing more to cultural preoccupation than scientific insight. And, I would argue, that's more likely to happen when we construct descriptions of syndromes without a grounding in how the mind and the brain work—that is, the psychology and biology of normal.

CULTURE-BOUND

SOME CRITICS OF THE DSM'S APPROACH TO DISTINGUISHING "DISorder" from "normal" have pointed to the influence of such cultural preoccupations as evidence that the whole enterprise is socially constructed and based on a largely Western, medical model of mental illness. For some disorders, that seems like an overstatement. For example, the psychotic disorder that Western psychiatry calls schizophrenia is recognized around the world, and rates of schizophrenia have been consistent across cultures, in the range of 2 to 5 per 1,000 population.[15]

But it's undoubtedly true that social and cultural factors affect how people express distress, experience symptoms, and engage in healing. It's also true that all definitions of mental illness involve some kind of value judgment about the bounds of normal. That is, defining the realm of abnormal or disorder depends on how a group (e.g., a society or a professional establishment) judges the boundaries of normal and the concept of deviance.

In his insightful book *Crazy Like Us*, the journalist Ethan

Watters makes a compelling case that the Western mental health establishment has exported its concepts of mental illness and psychiatric disorder around the world, in essence infecting non-Western cultures and creating epidemics of DSM-defined mental illnesses where they never existed before. He documents examples of disorders—anorexia, PTSD, depression—that have taken hold in cultures around the world as a result of the West's cultural hubris, well-intentioned naïveté, or even the marketing machinery of the pharmaceutical industry.

At the same time, cultures around the world have constructed their own conceptions of mental illness, and some behaviors that we might consider abnormal don't neatly fit into any of the DSM's categories.

SHRINKING PENISES

CONSIDER THE FOLLOWING CASE DESCRIBED IN 1965 BY A TAI-wanese psychiatrist:

> T. H. Yang, a thirty-two-year-old single Chinese cook from Hankow, in Central China, came to the psychiatric clinic in August 1957, complaining of panic attacks and various somatic symptoms such as palpitation, breathlessness, numbness of limbs, and dizziness. During the months just prior to his first visit, he had seen several herb doctors, who diagnosed his disease as shenn-kuei or "deficiency in vitality" and prescribed the drinking of boy's urine and eating human placenta to supply chih (energy or vital essence) and shiueh (blood). At this time the patient began to notice also that his penis was shrinking and withdrawing into his abdomen, usually a day or two after sexual intercourse with a prostitute. He would become anxious about the condition of his penis and ate excessively to re-

lieve sudden intolerable hunger pangs. Almost irresistible sexual desire seized him whenever he felt slightly better; yet he experienced strange "empty" feelings in his abdomen when he had sexual intercourse. He reported that he often found his penis shrinking into his abdomen, at which time he would become very anxious and hold on to his penis in terror. Holding his penis, he would faint, with severe vertigo and pounding of his heart. For four months he drank a cup of torn-biann (boy's urine) each morning, and this helped him a great deal. He also thought that his anus was withdrawing into his abdomen every other day or so. At night he would find his penis had shrunk until it was only one centimeter long, and he would pull it out and then be able to relax and go to sleep.[16]

Most people would agree that this case describes a condition that is not "normal"—but what is it? If this man were to walk into the office of a Western psychiatrist steeped in the language of DSM-IV, he might receive any of several diagnoses: panic disorder (an anxiety disorder), major depression with psychotic features (a mood disorder), delusional disorder, somatic type (a psychotic disorder), hypochondriasis (a somatoform disorder), or any number of others, although the details of the case would make an awkward fit for the DSM categories. In fact, the diagnosis given to this patient is older than any of the DSM labels I just mentioned. He is a victim of *koro.*

Koro has been recognized for centuries in China[17] but didn't appear in the Western literature until the late nineteenth century.[18] The classic presentation of *koro* is an acute state of panic in males caused by the belief that the penis is shrinking or even disappearing and that complete retraction will lead to death.[18] Not surprisingly, early Western psychiatric accounts interpreted cases of *koro* in Freudian terms as a manifestation of "castration anxiety." But there is another part of the *koro* story that makes it unlike your standard neurosis—it often occurs in epidemics.

In October and November 1967, an outbreak of *koro* occurred primarily among the Chinese population of Singapore. Rumors spread that *koro* was caused by eating pork from pigs that had been vaccinated against swine flu.[19, 20] Fanned by media reports, the rumors triggered an epidemic of *koro* that ultimately sent hundreds of victims to emergency rooms and clinics fearing that they were about to die from genital retraction.[20] The epidemic occurred when the Chinese, for whom pork was a dietary staple, felt threatened by Muslim Malays, who do not eat pork.[18] An even larger epidemic, affecting more than two thousand men, women, and children, struck Thailand in 1976 following rumors that Vietnamese immigrants had poisoned Thai food and cigarettes with a powder capable of causing genital retraction.[18, 20] Again, ethnic tensions seemed to be at the root of the outbreak because fears of invasion by the Communist Vietnamese were widespread.

Although *koro* has been commonly considered an Asian culture-bound syndrome, similar cases have been reported in Europe, Africa, and the United States.

An outbreak in Khartoum followed rumors that foreigners were roaming the city and, by handshakes, causing men's penises to disappear.[21] The panic appears to have begun in Nigeria or Cameroon in 1996 but spread to involve numerous countries over a several-year period.[22] In the Western literature, a growing number of cases have been reported in which genital retraction fears have figured prominently. In some instances, the syndrome appears to be a complication of underlying medical or neuropsychiatric diseases, a phenomenon dubbed "secondary *koro*." Thus, *koro*-like illness has been reported as a symptom of diseases ranging from brain tumors, epilepsy, and stroke to urologic disease, HIV infection, and even drug abuse.[20]

There is a debate in the ethnopsychiatric literature about how to categorize the various forms of *koro*. Is sporadic *koro*, affecting isolated individuals and resembling an anxiety or psychotic syndrome, really the same as the epidemic form that is often ignited by folk beliefs or ethnic tensions? Should "secondary *koro*" and "chronic *koro*" be considered separate subtypes? How do genital retraction syndromes differ from other cultural syndromes like *dhat* (an Indian syndrome) and *shen k'uei* (a Chinese syndrome), which involve anxiety and panic about "semen loss"?[23] We can imagine, and experts have proposed, an elaborate classification of genital retraction syndromes and their causes. Now, chances are, a few minutes ago you didn't know *koro* existed. But already you can see the complexities of trying to define the boundaries of disorder. When we're classifying disorders based largely on descriptive syndromes instead of a road map of how the mind works, it's easy to get into the kind of lumping and splitting of categories that many have criticized in the growth of the DSM.

A LINE IN THE SAND?

WE'VE SEEN THAT DEFINITIONS OF NORMAL AND ABNORMAL CAN BE highly contingent on time and place. They can rise and fall depending on the historical moment or cultural setting. Is there no way to ground the relationship between normal and abnormal functioning? One of the arguments I will make in this book is that there is. But doing so requires that we reverse the strategy that psychiatry has pursued for most of the past century. Rather than constructing disorders by labeling the extremes—the troubled mind and the broken brain—we must start with an understanding of the normal. What were the mind and the brain built to do? How do mental and neural functions develop? How are they organized? By understanding the basic architecture of the mind and the brain and how they make sense of the environment and experiences they encounter, we can begin to see where the dysfunctions are likely to occur and how they emerge from the normal spectrum of human experience. Our definitions of mental illness become less arbitrary. That doesn't mean that cultural influences will no longer matter. Indeed, as we learn more about the fundamental structure of the mind, we can see more clearly how culture shapes our experience and judgments about behavior.

One of the most influential attempts to grapple with the basic organization of the mind has turned to evolution for answers. The functioning of our brains, like the rest of our bodies, evolved in response to the challenges that ancestral humans faced in their struggle to survive and reproduce. Our most fundamental mental processes are organized around the most important of these challenges: avoiding harm, making plans and decisions, selecting mates, negotiating social dominance hierarchies, and so on. Jerome Wakefield, a professor of social work and psychiatry at NYU, has proposed a simple but powerful definition of mental disorder: a disorder is a "harmful dysfunction."[24] The line between mental health and mental disorder is crossed when a behavioral or psychological condition causes harm to an individual and represents a dysfunction of some naturally selected mental mechanism. Wakefield's solution nicely struck a compromise in a long-standing and contentious debate that spanned the last several decades.

On the one side were those who claimed that psychiatric diagnoses and the distinctions drawn between normal and abnormal behavior are inherently value judgments. The Rosenhan pseudopatient experiment and the American Psychiatric Association's vote to depathologize homosexuality were certainly examples where the line between normal and abnormal seemed to be drawn based on cultural value judgments. The extreme version of this critique was exemplified by Thomas Szasz and the so-called antipsychiatry movement that arose in the late 1960s. Szasz, whose 1961 book *The Myth of Mental Illness* was probably the most influential statement of this position, claimed that psychiatry's diagnostic labels were merely tools used for the exclusion and subordination of individuals. A less radical view is that psychiatric diagnoses may be useful, but they are ultimately just social constructions. On the other side of the debate were those who claimed that mental illnesses are biomedical disorders that can be defined just as objectively as diabetes or cirrhosis of the liver.

But both the strict "values" and the strict "biomedical" positions are ultimately incomplete. For one thing, the idea that psychiatric disorders are simply myths or social constructions ignores a vast body of evidence about the biological basis of mental illness. Our biological understanding of psychiatric disorders is admittedly limited, but decades of scientific research have established that people who meet the criteria for these disorders do have profiles of genetic risk and brain structure and function that differ from those who do not meet criteria—although the differences are usually matters of degree. What's more, as a psychiatrist, I have seen the pain and desperation that individuals and their families have to bear when psychosis, mania, depression, or panic overtake the mind. I have seen people so overwhelmed by this pain they wanted to end their lives rather than face a future filled with these symptoms. I have also seen medication and psychotherapies transform suffering and save lives. And the notion that defining these conditions as illnesses is merely an exercise in mythmaking trivializes the suffering of those who must bear them.

At the same time, it is hard to argue against the claim that the definition of psychiatric disorders involves some normative judgment about behavior. Severe shyness and social inhibition can be diagnosed as a disorder (social phobia) when they impair functioning (e.g., by inhibiting someone from advancing in their career). But that impairment occurs in part because social inhibition is devalued by employers and the larger culture.

Wakefield's notion that mental disorder is a "harmful dysfunction" accommodates both values and biology.[25] The first necessary condition for a mental disorder is that it involves mental states or behaviors that are harmful to an individual according to social norms. The syndromes we call schizophrenia, bipolar disorder, depression, and so on clearly fulfill this criterion. But "harm" is not sufficient to define a mental disorder. Plenty of behaviors are harmful but we would not call them disorders—procrastination or illiteracy, for example.

The other requirement is that the mental states or behaviors result from failure of a biologically designed function. Our brains exist to perform certain functions. Natural selection has sculpted the contours of those functions by enhancing the reproductive success of early humans whose brains best met the challenges that life threw at them. Some of these are obvious—detecting and avoiding danger, mating and reproducing. Others are more subtle—not being cuckolded, recognizing the intentions of others, cooperating and competing effectively, and maximizing available resources. In modern times we have given these functions names like trust, attraction, empathy, selfishness, and so on.

THE NORMAL SIDE OF DEPRESSION

AN IMPORTANT IMPLICATION OF THE "HARMFUL DYSFUNCTION" model is that psychiatry's DSM system may be diagnosing mental illness when no disorder is present. Modern psychiatric diagnoses are based almost entirely on clusters of symptoms, with little attention to the circumstances in which those symptoms occur. Take the example of depression. The DSM-IV diagnosis of depression (officially known as "major depressive disorder") requires two weeks or more of at least five symptoms, including persistent depressed mood and/or loss of interest or pleasure in activities most of the day, nearly every day. The other symptoms are significant weight loss or gain, sleeping too much or too little, physical agitation or slowing, loss of energy, feelings of worthlessness or excessive guilt, impaired concentration or indecisiveness, and recurrent thoughts of death or suicidality.

To reach the level of a diagnosis, the symptoms must cause significant distress or impaired functioning, and they can't be due to the effects of a drug or another medical illness. And there's one more thing—the symptoms can't be due to bereavement. That's a key exclusion, because the grieving process normally involves most of the symptoms of depression. Imagine a mother whose child has just died of leukemia. For a month, she cries nearly every day, loses interest in sex, has trouble falling asleep, and can't muster the energy to go back to work for three weeks. Should this woman be given a diagnosis of depression? Of course not. She is experiencing a normal grief reaction in the face of a devastating loss.

But should bereavement be the only situation where depressive symptoms are considered normal? What about other painful losses, traumas, and stresses that many of us experience over the course of a lifetime?

A man pulls me aside at a dinner party to seek my advice: "I'm worried about a friend of mine. Howard's been with our firm for twenty-five years, and three weeks ago, the company downsized and Howard got axed. He's fifty-nine years old, and his work was his life. I saw him last week and I was really alarmed. He's devastated—he just had this blank stare, he's lost weight, and he looked like he hadn't slept in a week. I tried to get him to come out golfing this weekend—something he's always loved to do—but he just said, 'No, some other time.' He looks so lost and his wife says he just mopes around the house. I think he's depressed. Is there some kind of medicine that could help?"

Does Howard need treatment for depression? He certainly seems to have symptoms of a major depressive episode. And we know that episodes of depression are often triggered by major stresses in vulnerable people. Here's a man whose whole adult life was organized around his work, and now the core of his self-concept is gone. He clearly has suffered a terrible loss. Had his wife died, we would ascribe his symptoms to bereavement and his friend would probably not even have asked me about the need for medication. So why is one traumatic loss so different from another? How clear is the line between normal sadness and depression?

Wakefield and his colleagues asked this question using data from a large study of the prevalence of psychiatric disorders in the United States.[26] They looked at people who met the criteria for a depressive episode and who said that their symptoms were triggered by either the death of someone close to them ("bereavement-triggered") or by some other type of loss ("other loss–triggered"); they then divided these groups into "complicated" or "uncomplicated" cases. "Complicated bereavement" is a term used in the DSM to describe genuine cases of depression that are triggered by bereavement. According to the DSM, bereavement crosses the line from uncomplicated (normal) to complicated (true depression) when it is prolonged and accompanied by serious symptoms like impaired functioning, suicidal thoughts, or morbid preoccupation with worthlessness.

When Wakefield compared the uncomplicated bereavement group to the uncomplicated "other loss" group on nine indicators of major depressive disorder (things like the number of depression symptoms, suicide attempts, functional impairment, and treatment for depression), he found essentially no differences between them. On the other hand, complicated cases were significantly more severe for all of the indicators, whether they were triggered by bereavement or some other type of loss. In other words, there was no evidence that bereavement was a special kind of loss in terms of its connection to depression.

So what? Well, right now, if you experience two weeks of intense sadness, trouble sleeping, loss of interest, and trouble concentrating after losing your job, getting divorced, or some other major loss, you would qualify for a diagnosis of major depression. Wakefield and his colleagues estimate that if psychiatry treated these other losses the same way it treats bereavement and categorized uncomplicated cases as normal sadness, the prevalence of depression in the United States would drop by nearly 25 percent. Wakefield doesn't claim that psychiatric diagnosis is inherently flawed—he's just suggesting that it can be improved by adopting a framework that places it in the context of normal mental function and the situations that people find themselves in. We can't define a line between normal and disorder by simply declaring a set of extreme behaviors as symptoms. Context matters. And we need to start by asking where these behaviors come from and how they fit into the full spectrum of human experience.

In other words, if we want to understand mental illness, we first need to understand how and why the mind functions the way it does. Perhaps that seems self-evident, but most attempts to define mental dysfunction—including the DSM—have not started with an account of normal function. So let's look at one example of how understanding normal function can tell us something about disorder.

STEP ON A CRACK?

OUR MENTAL CAPACITY TO SENSE RISK AND AVOID HARM WAS clearly developed during our evolutionary past. An animal without this ability would not have survived long enough to reproduce. Natural selection promoted those mental mechanisms that could anticipate and avoid danger. What if our normal harm-avoidance mechanisms went awry? What would it look like if we saw danger where none exists?

In fact, many of the syndromes we refer to as anxiety disorders are exaggerated and inappropriate forms of detecting and responding to threats. For example, psychiatry defines obsessive-compulsive disorder (OCD) as an anxiety disorder in which individuals suffer from recurrent, anxiety-provoking, intrusive thoughts (obsessions) or repetitive behaviors aimed at preventing harm or relieving anxiety (compulsions). But the content of these obsessions and compulsions is not random; they tend to fall into certain domains.

Four groups of symptoms account for the majority of obsessions and compulsions: (1) contamination obsessions and washing compulsions; (2) aggressive obsessions and checking compulsions; (3) symmetry obsessions and ordering compulsions; and (4) hoarding compulsions.[27] Each of these tap into fears and rituals we all may experience from time to time and each likely reflects a dysfunction of a mental system that evolved to avoid danger and stay safe.

What evidence do we have that these harm-avoidance and pre-cautionary systems exist in all of us? For one thing, we see them bubble up to the surface during certain moments in our lives. As little children, dependent on our parents and with little experience to distinguish what is safe from what is harmful, we are particu-larly vulnerable. Not surprisingly, early childhood offers a showcase for fears and rituals. Think bedtime—that fearsome and dreaded moment when parents leave their children at the mercy of monsters lurking under the bed.

Bedtime fears are common, and many young children develop elaborate rituals to quell their fears: repeatedly checking under the bed or reciting safety scripts like little shamans warding off evil spir-its. And then there is the awesome responsibility children often feel to prevent harm to themselves or their caregivers—fears that some-times fuel perfectionistic compulsions to avoid making mistakes or to get things "just right" ("Step on a crack, break your mother's back"). As one group of scientists put it, "These rituals may re-semble pathology when taken to an extreme, but within their appro-priate ontogenetic context, they are crucial in teaching children to manage their anxiety about the outside world"(p. 858).[28]

There's another life stage when intrusive fears and compulsive behaviors normally flare: pregnancy and the postpartum period, a time whose importance is hard to trump from an evolutionary per-spective. Natural selection is fundamentally a race for reproductive fitness—that is, maximizing the transmission of an individual's ge-netic makeup to subsequent generations. Preoccupations with the safety of the fetus and newborn are understandably common during pregnancy and early parenthood when our reproductive fitness is most directly at stake.

James Leckman and his colleagues at the Yale Child Study Center have been studying the biological basis of OCD for more than twenty years. Several years ago, they decided to explore the hypothesis that the preoccupations of early parenthood could be thought of as a normal variant of OCD. They interviewed parents during the eighth month of pregnancy and within the first three months after childbirth and found some intriguing parallels.[29] Just before the baby was born, more than 80 percent of mothers and fathers experienced worries about "something bad happening to the baby" and more than a third had thoughts about doing harm to the baby.

When they were interviewed at two weeks and three months postpartum, more than 70 percent of parents continued to have preoccupations with their babies' vulnerability or safety. In some cases these fears had a key feature seen in the obsessions of OCD: intrusive worries that an individual recognizes are irrational. Nearly 25 to 40 percent of parents had thoughts about doing harm to the baby. Parents reported graphic images of dropping or throwing the baby, scratching the baby with their fingernails, injuring the baby in a car accident—despite being sure they would never do something like that.[28]

More than 75 percent of parents also reported that they felt a compulsive need to check on the baby, even though they knew "everything was okay," and, at two weeks postpartum, about 20 to 30 percent recalled "telling themselves that such compulsive checking was unnecessary or silly."[29] When new parents were played recordings of their infant's cries while undergoing brain scans, fear centers lit up and correlated with OCD-like intrusive fears and compulsive harm avoidant behaviors.[28] The anxieties and preoccupations of early parenthood were greatest just before and just after the birth of the baby, and then began to decline.

So the perinatal period is a time of a normal increased sensitivity to avoiding harm and errors. It's not that new parents suffer from a psychiatric disorder. Most parents who experience intrusive anxieties and compulsive safety behaviors report that they are brief and do not cause marked distress or interfere with functioning—that is, they don't cross the threshold necessary for a diagnosis of OCD. But the perinatal period seems to tap into the same mental mechanisms that overtake the minds of those suffering from OCD.

DIRTY THOUGHTS

CONTAMINATION FEARS PROVIDE ANOTHER EXAMPLE OF THE CON-tinuum between normal and pathological obsessions and compulsions. At the extreme, OCD sufferers can wear their hands raw from excessive washing, or become housebound from obsessive contamination fears about the outside world. But the same fears are triggered when we avoid shaking hands with the sniffling, sneezing person in the next cubicle at work. The irrational fears of AIDS victims that swept the United States in the 1980s demonstrated the powerful and sometimes violent shape that engaging these harm avoidance responses can take.

More recently, fears about deadly flu epidemics and other germs have created a massive market for hand sanitizers: in a one-year period (2004–2005), sales increased by more than 50 percent,[30] creating a community of Purell-soaked germophobes that has been dubbed "hand-sanitation nation."[31]

The emotional counterpart of contamination sensitivity is the feeling of disgust, certainly a universal and familiar experience. Typically, disgust (literally, "bad taste") is triggered most potently by the thought or act of oral contact with objects or fluids derived from animals or other humans (feces and decaying meat are two of the most universal triggers of disgust). Disgust most likely evolved as a mechanism for avoiding disease.[32]

But even those of us without OCD experience irrational disgust and contamination fears. In a series of intriguing studies, Paul Rozin and his colleagues at the University of Pennsylvania found that people's feelings about food contamination often involve a degree of "magical thinking."

What if I asked you to eat a bowl of your favorite soup but told you that it had been stirred by a washed-but-used flyswatter? Would you eat it? When Rozin asked a group of healthy adults, most said no. But 50 percent said they still wouldn't eat the soup if it had been stirred by a brand-new flyswatter. In another test, subjects were offered two pieces of fudge that differed only in shape. They were happy to eat the fudge that was shaped like a muffin but rejected the fudge that was in the shape of dog feces, even though they knew it was just fudge.[33]

Neuroimaging studies have even pinpointed some of the brain regions that specialize in handling this function and appear to be overactive in people with OCD. When shown pictures of objects like public phones, toilets, or ashtrays and told to imagine coming into contact with them without washing afterward, individuals with OCD and contamination fears activate a system of brain regions involved in the processing of emotions, especially disgust.[34] Interestingly, similar regions light up in healthy individuals given the same task, though to a lesser degree,[35] suggesting again that OCD is an exaggeration of normal brain mechanisms.

A little ridge of cortex in the brain known as the insula is a key player in the biology of disgust.[36] Among its other responsibilities, the insula is the brain's clearinghouse for gut "feelings" and heart "aches": it keeps tabs on bodily sensations and connects them to emotional responses.[37] It is also the primary taste cortex, the region where our experience of taste is registered and integrated with our sense of smell.[38, 39] Electrical stimulation of this area in the brain triggers nausea and stomach churning.[40] So the insula is perfectly suited to handle disgust, an emotional response to bad tastes and smells. And indeed, brain imaging studies have confirmed that the experience of disgust activates the insula.[36] When healthy volunteers are presented with disgusting tastes, odors, or pictures (spoiled food, mutilated bodies, etc.), the insula goes into overdrive.[41]

So disgust seems to be hardwired and we are evolutionarily prepared to find some things disgusting—things like feces and putrid food. Overcoming our contamination sensitivity takes effort or self-deception. Think about the "five-second rule": food that falls on the floor is safe to eat if you retrieve it within five seconds. (Sadly, this turns out to be a myth because most of the transfer of bacteria from the floor to a piece of bologna happens within the first five seconds.)[42]

But wait—any parent knows that two-year-olds will put anything in their mouths. Where's the hardwired disgust? They have to be taught that a dead cockroach is totally gross. That's true. A child's sense of disgust and contamination sensitivity emerges gradually as demonstrated in a study of three- to twelve-year-old children who were given cookies and juice under a progressively more disgusting set of conditions.[43] The children were offered a glass of apple juice. But before pouring the juice, the experimenter pulled a comb out of her purse, combed her hair, and returned the comb to her purse.

She then produced another comb, telling the child, "This is a brand-new comb that I bought yesterday, all washed and cleaned. I am going to stir your juice with this comb."

After stirring the juice, she asked, "Will you drink some juice?"

If the child drank the juice, the experimenter produced another comb from her attaché case and said it was the one she used to comb her hair every day, but it was washed and clean. If the child was willing to drink juice stirred with this comb, the experimenter pulled a comb from her purse and said it was the one the child had seen her use to comb her hair (it was actually a clean duplicate of the original comb). Would the child drink the juice after she stirred it with a comb they'd just seen her use on her hair?

The answer depended on how old the children were: 77 percent of children ages three to six years would drink the juice compared to only 9 percent of the nine- to twelve-year-olds. In another version of the experiment, the experimenter brought forth a real (sterilized) grasshopper and dropped it in the juice. She asked the children if they would drink some juice from the bottom of the glass using a straw. Sixty-three percent of the youngest children were perfectly happy to oblige compared to only 19 percent of the older children.

So what happened to make a ten-year-old disgusted by the thought of drinking bug juice? One possibility is that the brain of a three-year-old simply doesn't have the capacity to think of juice being contaminated by a floating bug.[44] In other words, the development of disgust sensitivity has to wait until certain cognitive abilities come online. But social learning is another likely contributor: older kids have seen other people express disgust about contamination. Entomophagy (the practice of eating insects) is actually common in many parts of the world, but Americans find the idea revolting, and children's disgust reactions are stronger to things their parents find disgusting.[45]

And that social learning seems to have a neural basis: the same brain structure that activates when we experience disgust also lights up when we see facial expressions of disgust in others. French neuroscientist Bruno Wicker and his colleagues performed functional MRI scans of subjects in two conditions. First, the subjects watched movies of actors who smelled the contents of a glass that contained either water, perfume, or a disgusting-smelling liquid (the contents of a toy with the pungent name "stinking balls"). The actors made facial expressions appropriate to the contents of the liquid they smelled (neutral, pleasure, or disgust, respectively).

In the second experiment, the subjects were asked to inhale a series of pleasant smells (passion fruit, lavender, and so on) and a series of disgusting smells (including ethyl-mercaptan, once dubbed the "smelliest substance in existence" by the *Guinness Book of World Records*). Both the sight of others expressing disgust and the direct experience of disgust lit up the anterior insula. In other words, watching others react with disgust triggers our own disgust center. Perhaps the ten-year-old learns that bugs are gross by seeing those around him react with disgust. A broader implication of this work, and one we will return to in Chapter 4, is that "we perceive emotions in others by activating the same emotion in ourselves" (p. 660).[46]

Contamination-related disgust is central to some forms of OCD, and it occurs in a less harmful form in daily life, suggesting that there is a normal system for experiencing disgust and when that goes awry, mental illness can result. While brain-imaging studies have found that the insula and other emotion-processing regions may contribute to obsessive-compulsive symptoms, they also point strongly to dysregulation of a circuit connecting the frontal cortex to deeper structures like the basal ganglia, which are involved in avoiding errors and adjusting our behavior to threats and rewards. The point is that we can begin to understand OCD not as some mysterious affliction but as a dysfunctional expression of safety mechanisms that we all have.

THE DISTRIBUTION OF NORMAL

IN 1754 A FRENCH MATHEMATICAL GENIUS NAMED ABRAHAM DE Moivre died in poverty and relative obscurity in London. Two years after his death, the third edition of his great work *The Doctrine of Chances* appeared, containing a discovery that has become an iconic symbol in scientific and popular culture. De Moivre was concerned with describing the outcomes of random events—for example, if you flip a coin one hundred times, what's the probability that you'd observe thirty tails? He noticed that as the number of trials increased, the probability of its outcomes (e.g., heads or tails) formed a predictable pattern. Most trials of a fair coin toss will result in an equal number of heads and tails, so the most likely outcome is that we will observe fifty tails. For numbers much less or much more than fifty, the probability trails off. Using these simple observations, de Moivre derived a formula that produced an intriguing result. Graphing the probabilities of each number of tails produces a curve shaped like a bell. As it turns out, this bell-shaped curve can describe the distribution of a remarkable range of physical, biological, and even social phenomena; it has clearly earned its other familiar name: the "normal distribution."

I bring up the normal curve to address the question I posed at the beginning of this book: What is normal? If you look up the word *normal* in most dictionaries, the first definition is usually one with a statistical basis—something like: "conforming to the usual standard, type, or custom"—that is, normal is the most common or perhaps the average. But the metaphor of a "normal distribution" usefully goes beyond this.

Normal distributions are entirely defined by two numbers: one is the mean (the average), and the other is the variance (or its square root, the standard deviation). In other words, in statistical terms, a normal distribution encompasses both the average and deviations from the average: variance is an essential part of normality. By analogy, we'll see in this book that the biology of normal human functioning encompasses variations in how the brain processes the conditions of the physical and social environment it encounters. The result is a broad range of normal when it comes to temperament, empathy, trust, sexual attraction, and social cognition.

The recurring story of this book is that each of us finds our place in this great distribution by the intersection of three major players: evolution, genetic variation, and the particular environment and experiences we've encountered. The first—our shared evolutionary heritage—begins long before we're born. The countless trials and errors of natural selection have compiled a basic text of biological instructions spelled out in the human genome. The overwhelming majority of letters in that text are shared by all humans and provide a common set of possibilities and constraints within which our minds develop, function, and interact with each other and our world.

But the other two players—genetic variation and experience—shape the unique trajectory we travel within the broad distribution of the possible.

NIGHT AND DAY

IF WE ACCEPT THAT NORMAL IS NOT ONE STATE—THE MOST COMMON, the average, or the ideal—but rather a distribution or a spectrum of human possibility, how are we supposed to draw the line between normal and abnormal? A distribution may have a bulge in the middle and tails on the end, but there are no dividing lines in between.

If you've been waiting for me to give you my answer to the question of where the line between normal and abnormal is, here it comes. I don't think there is one. Sorry. It's not that I'm dodging the question, it's that I think it's not the right question to ask. There are no bright lines.

If that's the case, why write a book about the biology of normal?

Actually, there are two reasons. When I talk about the biology of normal, I'm referring to an understanding of what the brain and the mind are designed to do and how they function across the spectrum of human endeavor. We are now beginning to build that understanding through an unprecedented convergence of anthropology, genomics, psychology, and neuroscience. The story that's emerging is worth telling because it sheds light on how we become who we are. That's the first reason.

The second is that characterizing the biology of normal can ground our understanding of how things can go awry and contribute to what we recognize as mental disorders. But, you might be asking, if I'm claiming there is no sharp line between normal and abnormal, how can we even say what a mental disorder is?

Wakefield's harmful dysfunction model gives us one answer, but useful definitions of psychiatric disorders don't depend on identifying a single "true" line between normal and abnormal. We draw lines to create useful and "real" distinctions all the time, despite the fact that such lines are at some level not really there. The practice of medicine has many examples. Hypertension is defined as a blood pressure greater than 140/90, but no one thinks that there's a qualitative difference between a blood pressure of 141/90 and 139/90. And yet, high blood pressure can be deadly; hypertension has been a useful concept for research and clinical medicine.

Normal and abnormal are like night and day. That is, both are meaningful descriptions of two states that we recognize as different. But the line between them is impossible to draw. When exactly does day become night? We might decide to draw the line at sunset—a specific moment in time that we've constructed to separate the two.

But that's clearly somewhat arbitrary. Nevertheless, we'd all agree that day and night are meaningfully real. We schedule our lives around them; we make plans based on them. But we rarely worry about the moment that one becomes the other. We're comfortable with the fuzziness of twilight.

The same principle applies to the distinction between normal and abnormal or between disorder and nondisorder. Any specific line we draw to define disorder will require a judgment. But that doesn't mean that these disorders are simply fictions. There is clearly value in identifying syndromes that cause people harm and suffering: they allow us to develop treatments, to predict prognoses, and perhaps even formulate strategies for prevention.

TOWARD A BIOLOGY OF NORMAL

WAKEFIELD'S HARMFUL DYSFUNCTION MODEL PROVIDES A FRAMEwork for classifying disorder—the abnormal. But it also says something central to the subject of this book: there is much to be learned by understanding normal. To be on solid ground in defining and studying mental dysfunction, we first need to understand what functions are being "dys-ed." We need to grasp what the brain and the mind are designed to do—How do they function? What problems are they designed to solve? The answers to these questions are what I refer to as the biology of normal.*

In the chapters that follow, we'll see how research has begun to answer these questions that define who we are and what makes us tick. And along the way, we'll see that unpacking the science of normal can help demystify the nature of mental illness. Indeed, with a century of science at our back, it's time to turn William James's maxim on its head: the best way to understand the abnormal is to study the normal.

---

* As I explained in the prologue, I'm using the phrase "biology of normal" as a shorthand for describing underlying architecture of the brain and the mind. It involves multiple perspectives including evolutionary biology, neuroscience, genetics, and psychology.

CHAPTER TWO

HOW GENES TUNE THE BRAIN: THE BIOLOGY OF TEMPERAMENT

I 'VE ALWAYS BEEN THIS WAY." TIM CORNING WAS TRYING TO
describe the roots of his social anxiety in our first meeting. He
had come to see me because, after working for years as a computer programmer, he had decided to return to school for a master's degree in education. But now his dream of becoming a science
teacher was being hijacked by his anxiety around other people, and
he wanted to reclaim it.

"As long as I can remember, I was shy." He recalled his first day
of school, feeling frozen amid the overwhelming buzz of new faces. "I
don't think I spoke all day. I remember asking the teacher if I could
stay inside while the other kids went out for recess. But she said I
had to go . . . so I went. And I stayed by the door while the other kids
played.

"But I think I would have been okay if it wasn't for that day a
couple of months later when we went on a school trip to a museum.
Before we got on the bus to go back to school, the teacher told all the
kids to go to the bathroom so we wouldn't have to stop on the way
home." As he recounted the story, his face began to flush. "We were
lining up at the urinals and when my turn came, I couldn't go. I felt

like everyone was watching me. I stood there, waiting and praying that something would happen. One of the kids behind me laughed—I don't even know if he was laughing at me, but I felt humiliated." Ever since then, he was unable to urinate in a public bathroom, a condition known as paruresis, *or "shy bladder."*

Things got worse for Tim. His father left the family when Tim was seven years old. Midway through second grade, he refused to go to school unless his mother chaperoned him to the classroom. Unfortunately for Tim, his mother went along with this. One of the fundamental principles about anxiety is that avoiding what causes it is the surest way to turn a fear into a phobia.

Tim recalled his mother's own struggles with anxiety. "I don't think she ever went out when I was a kid—she was too worried about me. I guess she was the classic overprotective mother."

Over the years, Tim turned inward and focused on his studies. Schoolwork became one of the few areas that gave him a feeling of competence. In high school, he developed a fascination with science and engineering and managed to find a circle of friends who shared his interests. With a tentative self-confidence, he went off to college and studied computer science. But his social inhibition kept intruding. He recalled a job interview for a teaching position after college. He really wanted the job but, on the day of the interview, he started to imagine getting up in front of a class every day. Before the interviewer came out to greet him, he was gone. He took a job as a computer programmer, working mostly from home.

And now, ten years later, he was sitting in my office telling me he had come to a realization. He had set off in life with a sense of where he was heading, but so many times, he had taken slight turns to accommodate his shyness: declining an invitation to present his work at a scientific meeting, avoiding another party, or not quite feeling comfortable enough to call the woman who had given him her number. And now, suddenly, he looked up and realized he was miles from where he thought he would be.

How did Tim Corning end up where he did? How do any of us

end up with the emotional and social lives we do? The answer has much to do with where we begin—the genes we inherit and the temperament we are born with. As every parent knows, children begin to signal their approach to life well before they can verbalize it. Walk into any preschool classroom and within minutes you can pick out the shy, inhibited kids who are wary of unfamiliar people. You can also spot the bold, uninhibited kids who are talking and playing with anyone who will oblige. What makes one child fearful and another gregarious or even aggressive?

As we'll see, the foundations of our personalities can be traced to the genes and brain circuits that subtly shape temperament—the basic patterns of how we react to the world around us. We've known for some time that personality traits are influenced by genetic differences among people. But only recently have we begun to see how specific genes contribute to the development of these traits and how it all plays out in the brain.

The evolutionary history of our ancestors has selected suites of genes that "worked" for navigating the challenges of life. But we each inherit particular versions of these genes from the parents we happened to have. Our genetic differences bias our brains to be more or less sensitive to our environment, more or less emotional, more or less prone to behave in specific ways. They set our internal thresholds for reacting to our earliest experiences. Are we more likely to approach or avoid new situations and unfamiliar people? Are we prone to negative or positive emotions? Are we open to new experiences or wary of change? These subtle biases orient us to the world in different ways. We start life's journey pointed in slightly different directions, and as we interact with our families, our social environments, and the stresses and opportunities that life throws our way, these differences are amplified and elaborated into distinctive personalities. Our genes can even influence what kind of experiences we have—nudging us to seek out risky situations or perhaps shy away from social connections. These temperamental styles are all well within the distribution of normal. But sometimes, when our innate biases collide with the demands of the world around us, we suffer. That seems to have been the story that played out for Tim Corning, who began his journey with a tendency to be wary and shy and ended up with a life constricted by what psychiatrists have called social phobia.

In this chapter and the next, we'll explore the emerging picture of how nature and nurture interact to shape the trajectories of our life stories. The emphasis in this chapter is on the genetic roots of temperament and personality, how temperament is encoded in the brain, and how early differences in how our brains are tuned can have long-lasting effects on our emotional and social lives.

Your Mind: Day One

The most dramatic transition any of us makes is also one we all share. And none of us can remember a thing about it. I'm talking about the moment we travel from the secure and self-contained world of the womb into a new world of light, noise, and discomfort. Suddenly we have needs—and they are not being met. We've been thrust into a world of unforeseeable challenges and threats, and we have precious few resources of our own to draw on. Fortunately, we are not totally unprepared. Thanks to the trials and errors suffered by our evolutionary ancestors, we enter life with a rudimentary set of capacities that will help us negotiate the demands of this new world.

Imagine you are responsible for designing a mental survival kit for the newborn brain—a set of mental functions to ensure that this new visitor to our world will make it through the first months of life and meet the challenges of child development. Here are some constraints: the newborn can't walk or talk and has had no experience of the outside world. And he hasn't yet developed self-awareness or the concept of other people. What capacities would you pack into that little brain?

The most parsimonious answer would include at least three elements. First, you'd want to have built-in drives to satisfy immediate needs that are essential for survival—food, water, air. Next, you'd want a basic set of tools that could guide the infant brain to seek out helpful parts of the environment and avoid harmful ones, along with the capacity to control behavioral and emotional responses to the environment. This second part of the survival kit is the foundation of what we call temperament.

But of course, that's not enough. You would also want to somehow equip the brain to respond to the infinite specific environmental contingencies that it may encounter. So the third element you'd want to include is what neuroscientists call plasticity. As experience presents novel challenges to the human brain, neural connections or synapses are formed and strengthened so that we can adapt and respond. (We'll explore the biology of neural plasticity in more detail in the next chapter.) At the start of life, the nervous system is mainly a collection of possibilities. Only later will experience carve the detailed architecture of personality, desires, values, knowledge, and memories that make each of us unique. The adult mind reflects a record of the particular experiences that we encounter over a lifetime.

CONSTITUTIONAL CONVENTIONS

BUT LET'S RETURN TO THE BEGINNING. WE ENTER THE WORLD with a set of cognitive, behavioral, and emotional biases that allow us to respond to general features of our physical and social environments. We call these biases "temperament." The psychologist Jerome Kagan defines temperament as a profile of "stable behavioral and emotional reactions that appear early and are influenced in part by genetic constitution"(p. 40).[1] The notion of temperament has a long and storied history dating at least to the Greek view that differences in behavior, rationality, and emotionality were due to a balance of the four essential humors: yellow bile, black bile, blood, and phlegm.

Well before modern psychiatrists spoke of psychiatric disorders as "chemical imbalances," the humoral theory was the prevailing view of mental health and disease from the time of Hippocrates to the Enlightenment.[2] To the Greeks, disease occurred when there was an imbalance of the humors, but an excess of specific humors was also responsible for individual differences in mental traits. The Greek physician Hippocrates, and, later, the Roman physician Galen, recognized four temperaments corresponding to the effects of each humor. An excess of black bile (literally "melan" "cholic") produced a depressive temperament. The choleric temperament, owing to an excess of yellow bile, was irascible, ambitious, and passionate; while the phlegmatic type was apathetic and calm, and the sanguine type, reflecting an excess of blood, was optimistic and hopeful.[3]

The modern study of childhood temperament began in 1956 when Stella Chess, Alexander Thomas, and their colleagues attempted something that no one had done before: instead of deducing the nature of temperament from preconceived theories, they decided to study it systematically. They began a long-term study of 133 infants. By interviewing parents, observing their children, and evaluating childcare practices and parenting attitudes, they identified three temperamental styles that characterized most children—the "easy child," the "difficult child," and the "slow-to-warm-up child."[4]

About 40 percent of children could be classified as "the easy child." These were kids who were regular in their bodily functions (sleeping, feeding), generally happy and smiling, easily approached new people and new situations, and adapted to change quickly.

Another 10 percent of children were at the other end of the spectrum: "the difficult child" was loud, moody, prone to tantrums, and didn't take well to new situations or change. And the third temperamental category, dubbed "the slow-to-warm-up child" applied to about 15 percent of the children. These children were initially reserved and uncomfortable in new situations but would gradually adapt and engage.

By following the children over time, the researchers found that these early tendencies remained relatively stable into adulthood, and that temperament at age three was a significant predictor of behavioral traits in adulthood. But, crucially, Chess and Thomas also discovered that a child's successful development depends not only on how the child responds to the world around it (temperament) but also on how that world responds to the child. They called this concept "goodness of fit"—or how well the child's capacities and behaviors align with the expectations and demands of those around it.

For example, the "slow-to-warm-up child" whose parents express disappointment or anger at his shyness or difficulty making friends may have a troubled development. The same child born to parents who are accepting of his shyness is likely to do just fine.

In 2011 Yale law professor Amy Chua published a memoir that ignited a firestorm of controversy over parenting styles. *The Battle Hymn of the Tiger Mother* told the story of how Chua raised her two daughters, Sophia and Lulu, in the same way that her own Chinese immigrant parents had raised her: with love but also with a fierce commitment to excellence and hard work. In the popular press, the debate became a culture war: the uncompromising Chinese "tiger mother" versus the coddling Western style of indiscriminate praise and parental indulgence. In reality, the book was a brutally honest and often self-deprecating tale about the importance of knowing your child. And, in many ways, *Tiger Mother* was a book about "goodness of fit": what happens when one style of parenting meets two very different temperaments. Of her elder daughter, Chua writes, "From the moment Sophia was born, she displayed a rational temperament and exceptional powers of concentration. . . . As an infant Sophia quickly slept through the night, and cried only if it achieved a purpose." She was, from infancy, "calm and contemplative." In short, Sophia was the paradigm of what Chess and Thomas called "the easy child." Chua's Chinese parenting fit seamlessly with Sophia's easy temperament. She met her mother's high standards and intense work ethic with equanimity and her own drive to exceed expectations. In an open letter to her mother that appeared in the *New York Post*, Sophia wrote, "Early on, I decided to be an easy child to raise." By age fourteen, she was a model student and piano virtuoso with a Carnegie Hall debut under her belt.

But things were a little different with Chua's younger daughter, Lulu: "From the day she was born, Lulu had a discriminating palate. She didn't like the infant formula I fed her, and she was so outraged by the soy milk alternative suggested by our pediatrician that she went on a hunger strike. But unlike Mahatma Gandhi, who was selfless and meditative while he starved himself, Lulu had colic and screamed and clawed violently for hours every night." From the start, Chua writes, Lulu was willful and hot-tempered—the type Chess and Thomas might have called "difficult." Having known both Sophia and Lulu all their lives (my wife and I are their godparents), I think *formidable* would be a better word. Amy Chua describes how she was forced to adapt her cherished principles of child rearing to the vibrant reality of her younger daughter's temperament. In the end, both daughters have grown to become remarkable young women with a deep love for their parents. Amy Chua's ability to "fit" her parenting to the unique characters of her children was an example that Chess and Thomas would have advocated for mothers (and fathers) of all stripes.

REDISCOVERING OUR SENSE OF HUMORS

FOLLOWING CHESS AND THOMAS'S LANDMARK STUDIES, RESEARCHers have highlighted different temperamental traits, but almost all of them have agreed that children differ from birth in how reactive (both physically and emotionally) they are to the environment. The temperamental difference between children who are predisposed to approach unfamiliar situations and those who tend to avoid the unfamiliar is commonly called "boldness vs. shyness," and it has long-lasting effects on everything from our social relationships to our willingness to have unprotected sex.

Arguably, no one has taught us more about this temperamental difference than the psychologist Jerome Kagan, now an emeritus professor at Harvard, where he has been for more than forty years. In 2002 Kagan was ranked as one of the twenty-five most eminent psychologists of the twentieth century, just ahead of Carl Jung and Ivan Pavlov.[5] But half a century ago, Kagan was a freshly minted PhD psychologist from Yale when he was offered a job at the Fels Research Institute, in Ohio. His Ivy League mentor told him, "Don't take it, you're going to isolate yourself on an island—you'll never be heard from again."

Kagan didn't take that advice. At the Fels Institute, he was shown a room filled with piles and piles of notebooks containing the observations of children followed from birth through adolescence. Kagan reinterviewed the children as young adults, and when he and his colleague Howard Moss put the data together, they were struck by the fact that from early in life, a group of these children were passive and inhibited, and this trait seemed to follow them into adulthood. But when they wrote up their findings, they downplayed the possibility that biology played a role. This was an era when the two dominant strands of American psychology, Freudian psychoanalysis and behaviorism, had established the orthodoxy that child development was all about nurture.

"I was trained to believe in environment," Kagan said, "that was my politics. I was against biology. So I didn't pursue it—I didn't pursue temperament."

But by the late 1970s, having observed children across cultures and reading the latest research on the neurochemistry of behavior, Kagan was becoming increasingly convinced that temperament was rooted in our "constitution"—that is, our biological endowment. Modern psychology now seemed to be rediscovering the wisdom of the Greeks. "How amazing it is that Hippocrates and Galen got closer than Freud," Kagan marveled. "Blood, bile, phlegm—those are neurotransmitters—how did they do that? How did they guess right? I think that is just extraordinary."

ON THE SHY SIDE

KAGAN AND HIS COLLEAGUES PIONEERED THE STUDY OF SHYNESS and boldness by developing a method for picking up temperamental differences in the laboratory. In research that has spanned decades, they've found that differences in how children react to unfamiliar people and situations are evident as early as sixteen weeks of age. Kagan brought mothers and their four-month-old infants into the laboratory and observed them while the infants were exposed to a prespecified battery of events. The events were unfamiliar but not overtly threatening: mother stares at baby, a recording of an unfamiliar voice is played, a set of colorful mobiles are dangled in front of the baby, and so on.

Twenty percent of the infants exhibited behavior that Kagan called "high reactive"—they cried frequently and thrashed about, tensing their bodies and arching their backs when presented with the unfamiliar stimuli. Another 40 percent were low reactive: they seemed serene in the face of all these odd events. And that simple distinction turns out to tell you a lot about the developmental trajectory that these kids will take over the next twenty years. When the children were brought back to the laboratory at fourteen months and twenty-one months of age, they were again exposed to a series of unfamiliar events and people and objects. At one point, a woman dressed in a red clown costume and mask entered the room, talking and bearing toys, and invited the child to play with her. Next, the examiner brought in a radio-controlled metal robot. After a minute of silence, the robot began making noises, emitting lights, and moving, and the examiner invited the child to approach and touch the robot. Notice that these little challenges are subtle. Having a clown walk into the room is mildly stressful. If you had the clown burst into the room and yell "BOO!" you'd get little or no information about individual differences because every kid would probably react the same way: scared out of their pants. The point is to elicit differences in how children react in unfamiliar situations by gently challenging their approach/avoidance systems.

The kids who had been high reactive at four months were much more likely to be behaviorally inhibited, that is, fearful and avoidant of these unfamiliar challenges. By age four, they were much more likely to be shy, quiet, and timid around unfamiliar peers.[6] By age seven, children who were high reactive at four months or extremely inhibited at age two were more likely to be anxious, cautious, and socially avoidant,[7] whereas those who were low reactive or had been uninhibited at age two were much more sociable, smiling and talking spontaneously with strangers. These differences were also seen when the children were studied at age eleven and age fifteen.[8] Children who are inhibited in both infancy and later childhood are at increased risk for anxiety problems later in life. About a third of these children have significant problems with social anxiety in adolescence and adulthood.[9] They followed the trajectory that Tim Corning had so poignantly described to me.

It may come as little surprise that people differ in how prone they are to approach or avoid life's challenges or that inhibited children are more likely to become shy adults. The question is why? What determines where infants and young children fall on the shyness/ boldness spectrum? The answer seems to involve subtle differences in how our brains are tuned to the world around us.

IT'S WRITTEN ALL OVER YOUR FACE

THE VARIATION IN HOW LIKELY WE ARE TO FEARFULLY AVOID NEW experiences or boldly seek them out is rooted in deep and evolutionarily older parts of the brain. The amygdala, an almond-shaped collection of brain cells, plays a key role in putting an emotional stamp on our experiences ("watch out, this guy is dangerous!") and recognizing emotions in other people ("uh-oh, she's angry").

One of the major jobs of the amygdala seems to be evaluating the emotional expressions of other people—not a trivial assignment, since the face is the window through which we judge one another's intentions and emotions. Facial expressions are a kind of social vocabulary. In 1872 Darwin wrote that our expressions "reveal the thoughts and intentions of others more truly than do words, which may be falsified" (p. 364).[10] (I'll have much more to say about this in Chapter 4.) The amygdala also plays a major role in our response to novelty, triggering behavioral responses that make us either approach or avoid unfamiliar people and situations. Emotional faces and unfamiliar faces have something important in common: they tell us that we're in a situation that may be good or bad for us. Unfamiliar faces, like angry and frightened faces, signal potential threat.

Neuroimaging research has shown that one of the most reliable ways to fire up the amygdala is to show someone pictures of emotive or unfamiliar faces.[11–13] My colleague Carl Schwartz and others at Harvard conducted a twenty-year follow-up study of children whose temperament previously had been observed in Kagan's laboratory when they were only fourteen months old.[14] The children had been exposed to unfamiliar people and situations, and while some responded to novel stimuli with fear and avoidance, others were quite uninhibited and unafraid of new people or surroundings. Twenty years later, these same subjects returned to the laboratory to undergo functional MRI (fMRI) studies.

Even though the subjects were all now healthy adult volunteers in their early twenties, their brain scans revealed a hidden signature of distinct childhood temperaments. When shown a series of unfamiliar faces, the adults who had been inhibited as infants had a much stronger amygdala response than those who had been uninhibited. This work has been confirmed in other studies that have shown that adolescents who were inhibited as infants have a heightened amygdala response to faces that evoke uncertainty or are expressing emotions.[15]

In another study, Schwartz and his team studied eighteen-year-olds who had been classified in Kagan's lab as high or low reactive at four months of age.[16] When he ran them through an MRI, he found something striking: their temperament as infants predicted differences in the actual structure of their brains at eighteen years of age. Those who had been high reactive had significantly thicker brain tissue in the right ventromedial prefrontal cortex, a region known to play an important role in regulating brain regions involved in fear and avoidance. On the other hand, adults who had been calm, low reactive infants had greater thickness in the left orbitofrontal cortex, a region involved in inhibiting fear reactions and suppressing unpleasant feelings. And when they were shown pictures of unfamiliar faces, the eighteen-year-olds who had been high reactive at four months had stronger amygdala responses compared to those who had a low reactive temperament in infancy.[17]

Infant temperament, it seems, leaves a "footprint" that can be seen in the brains of adults decades later—visible in the structure and sensitivity of emotion centers like the amygdala and prefrontal cortex.

Based on studies like Schwartz's, Kagan and others have concluded that high reactivity in infancy and shyness/inhibition in childhood reflect an innate difference in how the brain reacts to novelty and threat. In shy, inhibited children, the emotional circuits of the brain (the "limbic system") seem to have a lower threshold for detecting and responding to uncertainty and potential harm. The system is more excitable, more vigilant, like an amplifier with the gain turned up. Once activated, the amygdala, a key node in the limbic circuitry, sends signals to other centers that activate stress responses. The sympathetic nervous system prepares for "fight or flight" and the stress hormone axis releases cortisol, triggering a broad range of defensive reactions in the brain and body.

The evolutionary roots of shy and bold temperaments run deep: fear behavior in response to novelty are seen across the evolutionary tree of life. And in mammals from mice to monkeys, the biology of shy, inhibited temperament involves many of the same brain and hormone systems that we see in the human version.[18, 19]

Temperament Grown Up

So these subtle biases in how we approach the world have an underlying biology that follows us from infancy into adulthood. And they can leave more visible traces in our adult lives: our relationships, our work, even our mental health. Children who were temperamentally shy in early childhood are more likely to have smaller social networks as young adults[20] and a greater risk of developing anxiety disorders,[21] especially social phobia,[9, 22, 23] in which fear of social and performance situations can be debilitating. That was the path that led Tim Corning to my door.

"It was like an out-of-body experience." In our second meeting, Tim was describing what lunchtime was like at a software company where he'd worked after college. Every day, his team would go to the cafeteria for lunch. As they sat down to eat and began to banter, Tim's mind went into overdrive. Instead of enjoying a casual lunch, Tim felt like he was onstage without a script, under a big spotlight, and playing to a crowd that was scrutinizing his every word and gesture. He was sure they could hear his heart pounding and his voice cracking, they could see his hand shake as he brought his fork to his mouth. After a few weeks of this, he stopped joining his colleagues for lunch, explaining that he needed to stay at his desk to catch up on work.

Tim's mind seemed to be tuned with an exquisite sensitivity to social judgments, the core feature of social phobia. Indeed, the biology of social phobia seems to be an extension of the biology of normal shyness. Brain-imaging studies have found that people with social phobia have exaggerated responses of the amygdala or medial prefrontal cortex when they're asked to get up in front of a group and give a speech,[24] to think about embarrassing situations,[25] or even just to look at faces of people expressing contempt.[26, 27]

What about children at the other end of the shyness/boldness spectrum—those who are temperamentally disinhibited early in life? These are the children who boldly approach unfamiliar situations. They tend to be impulsive and risk-taking. The trajectory for these children looks quite different from those who are temperamentally inhibited. One study that followed nearly one thousand children into adulthood found that those who were "undercontrolled" at age three were more likely to engage in risky or dangerous behaviors as adults—violent crime, alcoholism, unprotected sex, and drunk driving.[28] They had difficulty forming intimate and trusting relationships and were more likely to be unemployed and to have been fired from a job. Temperamentally disinhibited children are also more prone to behavior problems including aggressive and antisocial behavior and to what child psychiatrists call "disruptive behavior disorders," including attention deficit hyperactivity disorder (ADHD).[29, 30]

Just to be clear: when we talk about the spectrum of infant temperament, we're talking about normal variation in how children approach the world from the first months of life. But for some children the extremes of shyness and boldness can set the stage for vulnerability to disorder.

So if you're born with a nervous system that biases you toward high reactivity or inhibition, or toward disinhibition and boldness, has the story of your life been written? Of course not. It goes without saying that the world around you can shape the trajectory you take. Inhibited children whose mothers tend to be overprotective and intrusively controlling are more likely to remain inhibited and socially reticent. Children with less protective mothers and children placed in day care within the first two years of life appear to be less likely to remain inhibited by age four.[31, 32]

The interaction between temperament and the environment can

be complex and subtle. For example, inhibited children are more likely to be bullied,[33] and bullied children are more likely to become shy and withdrawn. Temperament is only the raw material. The family we are born into, the experiences we have, and the unique disappointments and windfalls that life brings sculpt the undifferentiated stuff of temperament into the textured contours of our adult personalities.

THE BIG FIVE

WHEN I TALK ABOUT PERSONALITY, I'M REFERRING TO THE ABIDING traits that give us our characteristic styles of operating in the world. They jell over time as our temperamental predispositions meet the specific conditions of life that we encounter. These are traits that allow us to make judgments about each other: "She's a really friendly person," or "He's so hard to get to know," or "He's so conscientious." We have an enormous variety of words like these that we use to characterize ourselves and others: selfish, gregarious, wimpy, cold, upbeat, and so on. The online dating site eHarmony markets their "29 Dimensions of Compatibility that are scientifically proven to predict happier, healthier relationships." But how many varieties of personality traits or dimensions are there?

In 1936 psychologists Gordon Allport and H. S. Odbert[34] set out to answer that question in a systematic way. They started with the premise that if a trait is recognizable enough to represent something real, there ought to be a word for it. Next, they undertook a project that one can only marvel at. They took the 1925 edition of Webster's *New Unabridged International Dictionary* and looked for every word that described individual differences in human behavior. From more than half a million words in the dictionary, they identified 17,953 descriptors of human behavior. They whittled that number down to about 4,500 words that describe "real" personality traits.

But people who do personality research for a living will tell you that there are only a handful of stable personality domains that describe individual differences in our behavior. Based on massive amounts of data gathered over the past several decades, these researchers have shown that our personalities can be boiled down to how each of us varies along just five overarching domains: the "Big Five" as they are commonly called: neuroticism, extraversion, openness to experience, agreeableness, and conscientiousness. I'm going to bet that if I had asked you to name the five dimensions of personality, that's not the list you would have come up with. But there they are—virtually all personality measures can be encompassed by these five factors, and they seem to be universal. The same domains emerge from studies in countries as diverse as Finland, Israel, Korea, Japan, China, Germany, and Portugal.[35]

Neuroticism refers to a tendency to experience worry, unstable moods, and negative emotions as opposed to being calm and emotionally stable. The extraversion factor captures the tendency to be active, enthusiastic, and to seek stimulation and the company of others. Those low in extraversion (i.e., high in introversion) tend to be quiet, reserved, shy, and withdrawn. Openness is a dimension that involves a tendency to be curious, creative, and open to new ideas and experiences. It correlates with aesthetic appreciation as well. Those low in openness are close-minded, conservative, and conventional in their tastes. Agreeableness indexes the tendency to be compassionate, empathic, and cooperative as opposed to being antagonistic, suspicious, and unfriendly. And, finally, conscientiousness refers to being self-disciplined and achievement-oriented as opposed to being disorganized and irresponsible.

Though these traits capture variation in personality across countries and cultures, there are still interesting differences in the personality profiles of people in different regions. In fact, the results of a survey of more than six hundred thousand Americans[36] were oddly consistent with stereotypes we have about the personality profiles of different regions of the United States. Neuroticism tended to be high on the East Coast, while the outgoing positivity of extraversion was concentrated in the Midwest. Where are the friendly folks high in agreeableness? You guessed it: the Midwest and the South. But some states stand out. North Dakotans seem to be the most outgoing, friendly bunch of traditionalists you'd ever want to know: they topped the list of all states in agreeableness and extraversion but came in last on openness. On the other hand, Alaska scored at or near the bottom on all five traits, suggesting that the typical Alaskan is a calm but disagreeable and introverted slacker who doesn't like unconventional ideas. If you're looking for open-minded, enthusiastic, friendly neighbors who are emotionally stable and conscientious, your best bet is to move to Utah.

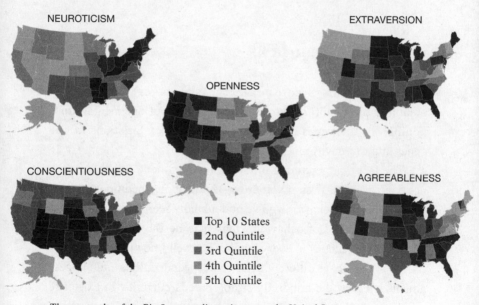

The geography of the Big 5 personality traits across the United States.

Studies of cultures around the world suggest that nations also differ in their personality profiles. In a study of data from fifty-one cultures across the globe, Brazilians reported the highest levels of neuroticism, the Northern Irish were the most extraverted, the

German Swiss scored highest on openness, Czechs reported the highest levels of agreeableness, and Filipinos and German Swiss were tied for first place on conscientiousness.[37]

The Big 5 model is not without its critics. Some have pointed out that these traits are based on questionnaires that ask people how they typically behave, without addressing the fact that what's "typical" in one situation may not apply in another. Nevertheless, these personality dimensions can even be found across the animal kingdom. The same dimensions have been observed in animals as diverse as guppies, octopus, cats, dogs, pigs, monkeys, and chimps. Even donkeys have a personality trait dubbed "vivacity" that closely resembles extraversion (though I must admit the notion of a vivacious donkey is a little disturbing).[38]

DIFFERENT STROKES

SO WHAT ACCOUNTS FOR INDIVIDUAL DIFFERENCES IN TEMPERA-ment and personality? Like everything else about our mental functioning, the answer is that it's a combination of variations in our genes and our environments.*

The fact that genes can influence temperament and behavior is not controversial. Even if you are a skeptic about the importance of genes in human behavior, you've undoubtedly seen the phenomenon of genetic control of behavior. Dogs may be the "poster species" for behavior genetics in everyday life. It's well known that different breeds of dogs have different behavioral specializations and temperaments. In fact, the hundreds of dog breeds currently in existence are basically the result of breeders using genetic selection to create animals that share specific temperaments and physical characteris-

---

* By the environment, I mean everything that isn't encoded in the sequence of our DNA. That includes the natural environment (the availability of food, weather patterns), the social environment (interactions with other people, the size of our families and communities), and even the gestational environment (the womb).

tics. The most popular dog breed in the United States year after year is the Labrador Retriever, and that's not because most dog owners have unmet retrieval needs. The appeal of the dog is its temperament, which the American Kennel Club describes as "a kindly, outgoing, tractable nature; eager to please and nonaggressive toward man or animal."[39]

But people aren't bred for behavior, and so it may be less obvious that genes affect temperament in humans. Though the phrase "the apple doesn't fall far from the tree" is a well-worn truism, the fact that traits run in families could in principle have little to do with genetics. Children observe their parents' behavior throughout development. It's entirely possible that any behavioral resemblance is simply a matter of this modeling effect. Siblings have many traits in common. But that could be entirely due to the fact that their family environments have been so similar.

So how can we find out whether genes affect human temperament and personality? One way to isolate the effect of genes would be to compare genetically identical clones who were raised in the same family environment to nonidentical siblings raised in the same family environment. If the clones are more similar in their behavior, you could conclude that this is due to their greater genetic similarity. This may sound like the musings of a mad scientist, but this very study has been done many times. It's called a twin study.

In twin studies, researchers compare the trait similarities of pairs of identical twins ("monozygotic," or MZ twins) to that of genetically nonidentical twins ("dizygotic," or DZ twins). If the environment is not more similar for MZ twin pairs compared to DZ twin pairs, then the greater similarity among MZ twin pairs for, say, introversion, can be attributed, at least in part, to the fact that their genes are more similar.

Behavior geneticists measure the importance of genetic variation between people in terms of a number called *heritability*. The heritability of a trait is the proportion of variation in the trait within a population that is due to differences in people's genes. It ranges from 0 percent (no influence of genetic differences) to 100 percent (completely due to genetic differences).

Because the concept of heritability will come up in many of the chapters in this book, let me clarify a couple of things.

First, heritability says nothing about the genetics of an individual. That's because heritability is about how much variation in a trait is due to genetic differences in a population. So heritability is only meaningful when we talk about populations, not individual people. If the heritability of body weight is 70 percent, that doesn't mean that 70 percent of your aunt Zelda's obesity is genetic and 30 percent is environmental. It just means that 70 percent of the variation in body weight in the population is due to genetic variation. For an individual, we can't tease apart nature and nurture in that way.

And, second, the heritability of a trait can vary depending on what population you're talking about. Take the example of body weight again. In a population where there's a lot of variation in what people eat, the environment may account for most of the differences in weight, whereas in a population where diet is more uniform, genetic differences may dominate. The heritability of weight would be smaller in the first group than in the second.

So the main value of heritability is that it gives us a measure of how strongly genetic variation influences trait differences in a population.

And twin studies have consistently shown that the heritability of the common temperamental and personality traits is in the range of 40 to 60 percent. That is, genetic differences account for about half of the variation in how shy or inhibited kids are in a population, how extraverted or neurotic adults are, and so on.[40–47]

One of the problems with heritability is that it tells us nothing about *which* genes influence temperament and personality. It also tells us nothing about how many such genes there are, how they act, and how big an effect each gene has. Until recently, scientists simply didn't have the tools to identify the specific genetic variations that contribute to complex traits like personality. All that changed over the past two decades through a remarkable series of discoveries and scientific breakthroughs that have given us the ability to decipher and analyze the human genome. In recent years, scientists have applied the tools of molecular genetics to unravel which specific genes are involved and how they exert their influence on temperament and personality. If variations in our genes account for about 50 percent of the differences in temperament and personality, we can now ask, which genetic variations are they?

"PEOPLE" PEOPLE

SOME OF THE STRONGEST EVIDENCE THAT GENES CAN AFFECT human temperament and personality comes from rare syndromes where genes go missing.

When I met him for the first time, nineteen-year-old Greg Chislon fairly bounded forward with an outstretched hand and an exuberant "Hi, how are you?" It was the kind of greeting one might expect from an eager job applicant, but we were sitting in an exam room at my hospital. Greg was always drawn to people, his mother told me. On the first day of school, Greg would go up to other children in the class, ask their names, and try to engage them in conversation. The wariness that most kids experience when they encounter new people didn't seem to register with Greg. If temperament depends in part on how the brain's approach/avoidance systems are tuned, Greg's seemed to be set all the way over to the "approach" side of the dial. Within minutes of meeting me, he was telling me about his love of music and asking me about my interests. His mother recalled that he would approach strangers in the supermarket and begin talking to them with a friendly smile. His eagerness to engage with other people was charming to his teachers and other adults, but sometimes made him the victim of teasing and practical jokes by his peers. There was a cost to his indiscriminate affability, and his mother frequently worried that he would be taken advantage of by some unscrupulous stranger. When he was seven, Greg was diagnosed with Williams syndrome. The diagnosis came as a shock, although in retrospect it made sense of a lot of Greg's behavioral and medical history.

Williams syndrome is a genetic disorder that affects about 1 in 7,500 children who are born with a chunk of DNA missing from one copy of chromosome 7. Children born with Williams syndrome are typically missing about 1.6 million bases of DNA sequence, covering about twenty-five different genes. Not surprisingly, the loss of that many genes can have widespread effects. There can be abnormalities of the major blood vessels due to the absence of one copy of the gene that makes elastin, a protein needed for the strength and flexibility of connective tissues.[48] There is increased risk of problems with blood calcium and hormonal systems, and just about everyone with Williams syndrome has a heightened sensitivity to sounds. Children with Williams syndrome are often described as having pixielike facial features, with a short upturned nose, and a wide mouth with full lips. They usually have some degree of intellectual disability, with an average IQ of 55.[49]

But the most remarkable characteristic of children with Williams syndrome is their intense interest in other people. It's an interest that's clear from infancy, when they become enraptured by faces. To meet a toddler with Williams can be intense—she may lock her gaze onto yours and hold it with a fascinated stare. As they grow, they become gregarious and exude a cheerful warmth and guileless affability.

Children with Williams are hypersociable. They're "people" people. They seem to yearn to connect and affiliate. They're often effortlessly good at small talk and unusually sensitive to other people's feelings.

Somehow, the genes deleted in Williams syndrome affect how the brain's emotional circuitry responds to the social world, resulting in nearly the opposite pattern seen in people who are shy and socially anxious. When presented with fearful or angry faces, they have a subdued amygdala response.[50] But that's only part of the story. It turns out that they have an increased amygdala response to happy faces.[51] So the chromosomal deletion that causes Williams syndrome seems to shift the bias of the amygdala toward approaching others and processing positive emotions. And that may hold a key clue to the gregarious personality style of individuals with Williams. Where anxious, inhibited children are hypersensitive to threat, the Williams child is biased to see happiness.[52] Not surprisingly, individuals with Williams syndrome seem to be relatively immune to social anxiety and social phobia.

THE LONG AND SHORT OF IT

WHEN IT COMES TO THE GENETIC BASIS OF TEMPERAMENT, THE example of Williams syndrome is an outlier. For most of us, our style of interacting with the world around us can't be traced to a single stretch of DNA. The broad spectrum of temperament and personality reflects the action of many genes, each contributing a small amount to the overall picture and interacting with our environments and life experiences. This is why when you read headlines like SCIENTISTS FIND THE GENE FOR ANXIETY! or . . . THE GENE FOR BIPOLAR DISORDER or . . . THE GENE FOR almost anything, you should roll your eyes. There is no "*the* gene" for these kinds of complex traits. There are many genes, and they act like risk factors. Having a high cholesterol increases the risk of developing heart disease, but it doesn't guarantee it. There are many people with elevated cholesterol who don't go on to develop heart disease, and many people with heart disease who don't have high cholesterol. So cholesterol is only one of many risk factors for heart disease, just like individual gene variations (alleles) can be risk factors for developing bipolar disorder.

In fact, recent studies have convincingly identified specific genetic variants that increase risk for psychiatric disorders such as schizophrenia and bipolar disorder.[53, 54] But doing so required very large studies involving thousands of subjects, because by themselves each of these genetic variants has a small effect.

So we know that variations in our DNA contribute to individual differences in behavior and risk for mental illness, but we also know that any single gene variant can have a subtle effect. Personality is "highly polygenic"—that is, there are hundreds or even thousands of individual genetic variations involved. Nevertheless, behavior geneticists have so far focused on only a few, and if they had a top ten list of favorite genes, *SLC6A4* would be on it.

The *SLC6A4* gene makes a protein called the serotonin transporter (also known as 5HTT). Serotonin is a neurotransmitter that has, for many years, been known to play a key role in the development and functioning of brain circuits involved in mood, anxiety, and aggression. Generally speaking, neurotransmitters act as chemical messengers, crossing the tiny junctions known as synapses between nerve cells. At a serotonin synapse, serotonin released from a neuron on one side of the synapse (the "presynaptic neuron") crosses the submicroscopic divide and binds to receptors on a neighboring ("postsynaptic") neuron. The neuron that released the serotonin quickly grabs any excess serotonin through its serotonin transporters, which act like little pumps that pick the serotonin up from the synapse and bring it back into the neuron. So the job of the serotonin transporter is "reuptake" of serotonin from the synapse.

This tiny molecular pump has been one of the pharmaceutical industry's biggest cash cows. Based on the idea that depression involves a dysregulation of brain serotonin, the most widely used antidepressants—Prozac and its cousins Zoloft, Paxil, Lexapro, and others—were designed to block or inhibit the serotonin transporter. That's why they're called SSRIs: selective serotonin reuptake inhibitors.

Like other genes, the *SLC6A4* gene includes a region called the promoter that regulates how active the gene is—that is, how much serotonin transporter protein it makes. In 1996 German scientists found a common variation (known as the 5HTTLPR) in the

DNA sequence of the *SLC6A4* gene promoter: the "long" variant, or allele, has an extra forty-four letters of DNA that are missing in the "short" allele. And that simple difference makes the gene carrying the "short" allele less active in making serotonin transporter. Neurons carrying one or two copies of the "short" allele make about half as much of the serotonin transporter as those without any short alleles.[55]

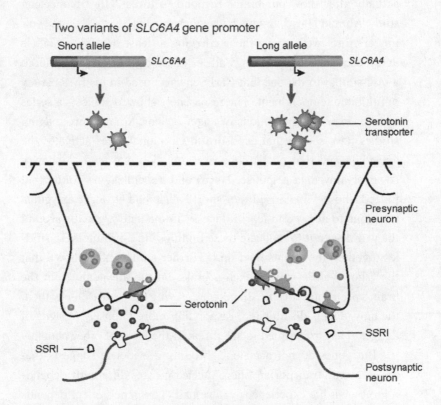

Variation in the serotonin transporter gene (SLC6A4) promoter (*top*) and its effects at a serotonin synapse (*bottom*). Genes carrying the "short" allele make fewer serotonin transporters, which are responsible for reuptake of serotonin into (presynaptic) neurons. SSRI antidepressants act by blocking serotonin transporters

As it turns out, about 75 percent of people with European-American ancestry carry at least one copy of the "short" allele and about 20 percent carry two copies. The lower levels of serotonin transporter made by the "short" allele seem to create a subtle bias in the way the brain responds to threatening or adverse experiences. Several—though not all—studies have shown that people carrying the "short" allele score higher on personality traits like neuroticism and harm avoidance that are related to anxious temperament.[56–58]

Does the DNA variation in the promoter of the *SLC6A4* gene actually affect how our brains respond to threat? In a pioneering study, Ahmad Hariri and his colleagues divided a set of healthy volunteers into two groups: those carrying at least one "short" allele and those carrying two "long" alleles. Then they had them undergo a task similar to the one that Carl Schwartz used in his fMRI study of inhibited temperament. The researchers showed subjects a series of faces expressing emotions of anger or fear. Many neuroimaging studies have shown that exposure to emotional faces activates the amygdala and that people who are anxiety-prone tend to show a stronger amygdala response. Hariri and his colleagues found that indeed, those who carried the "short" allele had stronger amygdala reactions to angry and fearful faces.[59] Immediately, groups around the world set out to replicate these findings, and putting these studies together, the results hold up.[60] Further studies have shown that the "short" allele may weaken a brake on the amygdala from the brain's prefrontal cortex.[61] In people carrying the "short" allele, then, the amygdala is disinhibited, responding more intensely to signs of danger. They are primed to see threat in the faces of other people.

But genes do not act alone. Mounting evidence suggests that the serotonin transporter "short" allele operates differently depending on what life experiences you've had. The gene's effect depends on the world you live in. For example, in one study, children carrying the "short" allele were more likely to be shy and behaviorally inhibited if their mothers had low levels of social support,[62] and the activity of brain circuits involved in perceiving threat is greater among short allele carriers who have experienced more stressful life events.[63]

OF MICE AND MEN

EVEN IF THE SEROTONIN TRANSPORTER GENE TURNS OUT TO BE A gene that influences temperament, we know that it is not *the* gene. As I said before, the heritability of temperament involves many genes, each making a small contribution to how our brains interact with the world around us. So if these genes have such subtle effects, is there any way to find them? My laboratory has been pursuing this question for more than a decade. We reasoned that we could make use of the fact that shy, inhibited temperament is seen in many animal species, including mice. Mouse models can be very useful because they allow scientists to perform genetic studies that would be impossible in humans. And genetically, we're a lot like mice. In fact, more than 99 percent of mouse genes have a counterpart in the human genome.[64]

The ability to breed and cross large numbers of mice provides a powerful way to rapidly map genes that affect behavior. And beyond that, using sophisticated "gene targeting" techniques, scientists can actually delete or insert specific genes in the genomes of mice. For example we can "knock out" a gene from a mouse embryo and see what effect deleting the gene has on the behavior of the mouse later in life. If mice missing the gene are more fearful, for example, we have evidence that the gene is somehow involved in fear behavior. And if we can find genes related to shy or fearful temperament in mice, we can see whether the same genes affect temperament in humans.

In 1995 a group of scientists at Oxford reported that a region on mouse chromosome 1 is linked to mouse fear behavior.[65] Later studies replicated this finding, but the precise gene or genes involved remained a mystery. Then in 2004 the Oxford group seemed to have an answer. Using newer methods for mouse gene hunting, they fingered a gene called *rgs2* as at least one of the culprits. Mice carrying one version of the gene were inhibited in a battery of fear behavior tests. Also, *rgs2* "knockout mice"—mice born without the gene—

are behaviorally inhibited and "anxious."[66, 67] Like temperamentally inhibited children, these mice also have an overactive sympathetic ("fight or flight") nervous system.[68]

So what does the *rgs2* gene do? Among other things, it makes a protein that controls how nerve cells respond to neurotransmitters like serotonin, norepinephrine, and dopamine—key players in the biology of temperament, anxiety, mood, and stress responses. When these neurotransmitters bind to their receptors on nerve cells, the receptors activate proteins (called G proteins) that set off a cascade of events within the cell. The *rgs2* protein gloms onto the activated G proteins and shuts them down, providing an essential brake on the neurotransmitter signal. So a malfunctioning or missing *rgs2* gene might leave brain cells vulnerable to overstimulation by neurotransmitters like serotonin. As you might expect, *rgs2* is active in many of the key brain regions known to influence temperament and emotion, including the amygdala, hippocampus, and cerebral cortex.[69-71]

We humans have an *RGS2* gene, too. If *rgs2* contributes to anxious temperament in mice, could the human version play a similar role? Researchers at my lab studied children who had previously come to Jerry Kagan's laboratory at Harvard and been exposed to unfamiliar people and situations in the battery of temperament measurements that I described earlier. Later, we had these children and their parents spit into a cup, allowing us to extract DNA from their saliva, which we then analyzed to find variations in the human *RGS2* gene. We found that children with specific variants of the *RGS2* gene were three times more likely to be shy and inhibited. And one of these variants had previously been shown to correlate with lower expression of the *RGS2* gene. In other words, it appeared that having less *RGS2* protein was associated with anxious temperament—just what we'd predict from the mouse studies.

Then, in a collaboration with my colleagues Murray Stein and Martin Paulus, at the University of California–San Diego, and Joel Gelernter, at Yale, we asked whether the *RGS2* gene might also affect introversion in adults (since introverted adults, like inhibited children, are often shy and wary of unfamiliar people). We analyzed the DNA of nearly 750 adults who had completed a personality assessment and found, indeed, that the same *RGS2* variants that predicted childhood shyness were associated with adult introversion.

The key question, though, was whether we could see the effect of this gene in the brain centers that regulate temperament and social anxiety. To answer this, we looked at the *RGS2* variants in a cohort of adults who underwent fMRI scans while being shown emotional faces. Sure enough, those carrying the variants associated with inhibited temperament and introverted personality had a substantially stronger response to emotional faces in two key areas of the brain's emotion (limbic) circuitry: the amygdala and the insula.[72]

A WINNING PERSONALITY?

THE *RGS2* STORY IS NOTABLE FOR TWO REASONS. FOR ONE, IT provided one of the first demonstrations of how a specific gene's effect on temperament and personality can be seen at both a behavioral and a neurobiological level. But secondly, the *RGS2* story provides key evidence that at least some of the genetic influences on temperament and anxiety are evolutionarily conserved. Here's an example where the same gene seems to affect anxiety-related behavior and brain function from mice to humans. The serotonin transporter short/long variation also points to our evolutionary history, though a more recent one. That variation seems to have arisen about forty million years ago because it is present in monkeys and apes but not in earlier mammals.[73] Interestingly, the frequency of the short and long alleles in rhesus monkeys is quite similar to our own. That raises an interesting question—why?

Why would natural selection maintain a common variation like the "short" allele that seems to make animals more timid and "neurotic"? There are several possible explanations, including the possibility that the variation is "invisible" to natural selection—that is, it doesn't really affect a primate's reproductive fitness and so natural selection leaves it alone. But if that were true, it's hard to explain how a mutation in the serotonin transporter gene that apparently arose as a onetime event forty million years ago became so common. It's also possible that variation in genes that affect temperament and personality is not just evolutionary "noise." It could be that natural selection might favor certain temperaments. It might seem obvious that a trait like extraversion could provide selective advantages. Individuals who are more social and outgoing might have more opportunities to find mates, for example. But even behavioral inhibition could be a good strategy from a reproductive standpoint: being wary of new situations or people could prevent someone from being preyed on or from engaging in fatal conflict.

So gene variations that contribute to these traits could easily be promoted by natural selection. But if you think about it further, there's a problem. Remember, these traits are heritable—that means that individuals differ in these traits, and we're trying to explain the genetic differences that underlie the trait differences. If one temperament or personality type is clearly advantageous, shouldn't natural selection cause it to become a fixed part of universal human nature? If being shy or avoidant protected our ancestors from harm, the shy ones should have had a reproductive advantage and gradually replaced all the risk-taking types in the human population. Alleles that promoted shy, inhibited behavior would be selected and become "fixed" at high frequencies. So, we'd expect that there would be very little, if any, genetic variation in genes that shape personality. With no individual differences or genetic variation, the heritability of temperament and personality would gradually approach zero. But we know that's not the case.

Everything in Moderation

One explanation for how natural selection might maintain variation in personality involves something called "balancing selection." The idea is that personality traits, like everything else in life, involve trade-offs. Agreeableness is great when it allows you to form alliances, but not so great when you need to fight to defend your interests or your family. One form of balancing selection, called "heterozygote advantage"—an advantage that comes with carrying one copy of a mutation—is widely known in medicine. For example, people who carry two copies of a mutation in the β-globin gene, which encodes a protein essential for making hemoglobin, develop the painful and devastating disease called sickle cell anemia. The sickle cell mutation interferes with the flexibility of red blood cells, causing them to assume a rigid "sickle" shape when oxygen levels become low. Sickled red blood cells can get stuck in capillaries, cutting off blood supply and oxygen delivery to the tissues, which in turn can cause excruciating pain and even death.

So why hasn't natural selection gotten rid of this deadly mutation? It turns out that carrying one copy of the sickle cell mutation results in a mild form of sickling that can reduce the risk of malaria, one of the world's biggest killers. Because sickle cell anemia only occurs if there are two copies of the mutation, people who carry only one copy of the mutation (heterozygotes) are protected from both sickle cell disease and malaria. That trade-off has allowed the sickle cell mutation to persist in areas where malaria is common. A similar phenomenon could balance out the effects of natural selection on alleles that contribute to personality traits.

Of course, temperamental styles and personality traits come in many flavors, each of which entails trade-offs alone and in combination with other traits. Different combinations of extraversion, neuroticism, openness, and so on may each have their own risks and benefits. So if we have a situation where each trait is affected by many genes, and the advantages of each trait vary over time and context, the alleles that influence that trait may not be eliminated by natural selection, even if they sometimes have disadvantages.[74]

OUT OF AFRICA

IN THE CASE OF TEMPERAMENT, WE CAN ACTUALLY BEGIN TO SEE the faint footprint of natural selection on human behavior. For example, there is evidence that a gene variant related to personality may have become more common as humans migrated across the globe by subtly enhancing the fitness of those who explored new environments.[75]

Many studies have pointed to the neurotransmitter dopamine as a key player in regulating how eager we are to seek out new experiences. Dopamine receptors are located in brain regions involved in motivation, exploration, and reward. Just as the amygdala and related fear circuits stamp certain stimuli as dangerous threats to be avoided, these regions mark other stimuli as rewarding opportunities to be approached.

Variations in the gene that makes one of the dopamine receptors (the dopamine receptor D4 or DRD4) have been associated with boldness, novelty-seeking, and high levels of activity and exploration in humans as well as birds, dogs, and horses.[76–81] One of these variations involves repeated DNA sequences in part of the gene that determines how the DRD4 receptor responds to dopamine. People differ in how many copies of the repeated DNA sequence they carry. Many people have four copies of this repeated sequence (the "four-repeat allele"), but there are other variants, including a seven-repeat form that appears to make less efficient DRD4 receptors. This seven-repeat allele has been linked to novelty-seeking and may be a risk factor for attention deficit hyperactivity disorder (ADHD).[82]

With detailed knowledge about the structure of the genome, researchers can now conduct evolutionary detective work (think

CSI-meets-Darwin) to track down the history of genetic variations that influence behavior. Using clues from the patterns of variation in the DRD4 gene, a team of geneticists was able to date the origin of the novelty-seeking seven-repeat allele to a mutation that occurred in Africa about fifty thousand years ago (relatively recently in the history of human evolution and around the time of the last major human exodus from Africa).[83] Somehow, this recent mutation became common in human populations throughout the world. Rather than remaining rare or disappearing, it actually flourished.

But why would natural selection preserve a genetic risk factor for ADHD? One possibility is that those carrying the seven-repeat allele had an advantage under some circumstances. In an environment where the availability of food or other resources might suddenly change or disappear, people who were able to rapidly respond, move, and seek out new resources might have done better than those who were slower to respond. These were the people whose credo was WHEN THE GOING GETS TOUGH, THE TOUGH GET GOING. As you would expect for a genetic variant that promotes novelty-seeking, the seven-repeat allele becomes more common at the farthest distances from Africa. In South America, which is the endpoint of migrations that would have had to span Africa, Europe, Asia, and North America, the seven-repeat is actually the most common form.[84] Perhaps those carrying the genetic form of DRD4 were "bold enough to go where no man had gone before."

Again, it's clear that one or two genes don't tell the whole story of how shy, extraverted, fearful, or aggressive we are. Rather, many genes contribute small amounts to the development and functioning of brain circuits that underlie how we feel about and interact with the outside world. Your 5HTTLPR genotype or DRD4 repeats don't determine what kind of person you will be. But there is growing evidence that common genetic variation between people may contribute to slight shifts or biases in brain function that influence how we respond to the world around us.

The subtlety of these effects has recently been brought home by very large studies that have used newer DNA chip technology to scan the whole genome for variants affecting the Big 5 personality traits. These "genomewide association studies" are able to examine genetic variations across the entire genome and typically use very large sample sizes. And yet these studies have been unable to find many specific variants related to personality traits, despite the fact that we know these traits are substantially heritable. In other words, we know there are variants to be found, but the vast majority have effects that are too subtle to be picked up even in powerful studies. One meta-analysis of these studies that surveyed more than 2.4 million genetic variations across the genome in more than 17,000 subjects was only able to find two genetic regions that were strongly associated with Big 5 traits—one for openness and the other for conscientiousness.[85] In another study that included thousands of people, my colleagues and I found another genetic region associated with excitement-seeking, a central feature of extraversion.[86] Finding all of the genetic differences that account for the heritability of personality traits (which you'll recall is about 50 percent of the total variation in these traits) will probably require massive studies. That's the lesson emerging from other genetic studies of "complex traits" like obesity. The heritability of obesity is similar to personality traits, but it took a genomewide study of nearly 250,000 people to pick out genetic variations that together account for about 2 percent of the individual differences in body mass index.[87] If personality is anything like body weight, you might need to study millions of people to find all of the genes involved.

SHADES OF THINGS TO COME

THERE'S ANOTHER IMPORTANT LESSON IN ALL OF THIS. BY CREATing subtle biases in how we approach life, temperament and its underlying neurocircuitry can sometimes have long-lasting and cascading effects that set us on a troubled trajectory. Children who are extremely inhibited are much more likely to develop significant problems with social anxiety later in life. A temperamental bias toward impulsivity and distractibility can evolve into attention deficits and hyperactivity. And an early tendency to refract one's experience through the lens of negative emotionality can produce a vulnerability to depression, especially when someone faces adversity. So we begin to see how our temperamental approach to life, itself the reflection of how our neural systems are calibrated, can evolve into symptoms and syndromes that appear as disorders. Not surprisingly, then, genetic variation that underlies temperament and personality seems to account for much of the genetic component of common disorders, including depression and anxiety disorders.[88, 89] The biology of normal shades into the biology of disorder.

And that may provide a clue to how treatments for depression and anxiety work.

At the end of my first meeting with Tim Corning, I suggested that we try treating his social anxiety with cognitive-behavioral therapy (CBT), a proven therapy for treating a wide range of anxiety disorders. In CBT, clients learn to recognize and overcome the cognitive biases that exaggerate their social fears. And by gradually exposing themselves to social situations, they become desensitized to the fear that they will do something embarrassing or that other people will judge them harshly. Unfortunately, after twelve weeks of therapy, Tim felt that he wasn't making enough progress, and he wanted to try something else. And so we agreed to start an SSRI. We began with 25 mg of sertraline (Zoloft) and worked our way up to 100 mg per day. Eight weeks went by with little change. As we continued to increase the dose, Tim began to notice something: he was starting to feel more comfortable eating at the local diner. After two weeks at 200 mg, he told me he'd surprised himself by accepting an invitation to go to a party thrown by his former coworkers. "I could never have done that a year ago. I thought I'd just stand around and watch, but I actually talked to people." A month later, he went on his first job interview in two years. "I'm a different person," he said.

Tim may have been on to something. Recent research suggests that SSRIs like sertraline may work in part by subtly changing personality itself. People with depression and anxiety disorders like social phobia tend to be high in neuroticism and introversion (low extraversion). In a placebo-controlled study of the SSRI paroxetine (Paxil) for depression, researchers found that the SSRI "normalized" neuroticism and introversion whether or not their depression improved. And when they controlled for changes in these personality traits, there was no independent effect of the drug on depression. In other words, the antidepressant effect of the SSRI seemed to depend on its ability to recalibrate personality.

Psychological and neuroimaging research has shown that anxiety- and depression-prone people have emotional circuitry that tends to process the world as "a glass half-empty"—a bias toward registering negative emotional and social features of the environment and away from seeing and feeling the positive side of things.

So here's the emerging picture. Variations in genes make some people more prone to shyness, anxiety, and negative emotions, in part by creating an amygdala-prefrontal cortex circuit that is oversensitive to threat and biased toward negative thoughts and emotions. SSRIs, and perhaps CBT, seem to retune that circuitry, damping down a sensitive emotional system and shifting its bias toward a more "glass half-full" approach to life.[90] When people with depression or anxiety and even healthy volunteers are given SSRIs, MRI scans show a cooling down of emotional circuits that have been linked to inhibited, anxious temperament and neuroticism.[91-93] So perhaps for Tim Corning, whose genetic endowment and life experiences had amplified his inhibited temperament into debilitating social anxiety, medication offered a way to nudge him back toward the middle of the normal distribution.

Evidence from both psychological and genetic studies points to the notion that the most common psychiatric disorders are really the extremes of normal, quantitative traits that all of us share. And like other quantitative traits—blood pressure, body weight, cholesterol levels, for example—where we draw the line between normal variation and dysfunction may be a pragmatic decision. Some experts have gone so far as to say that "what we call common disorders are, in fact, the quantitative extremes of continuous distributions of genetic risk . . . there are no common disorders—just the extremes of quantitative traits" (p. 877).[94]

We enter life with brains that have been tuned by our genes and the environment of the womb. Temperament shapes the timbre of our earliest notes and probably constrains our dynamic range throughout life. But as we will see in the next chapter, early experience can profoundly alter the path we take in life. The picture emerging from neuroscience and genetic research is one of continual dialogue between the brain and experience, each modifying the other throughout development. As Tim Corning found, the people we become, our personalities, are the product of the small adjustments we make and the imperceptible turns we take as our innate temperaments encounter our own particular world.

Chapter Three

Blind Cats and Baby Einsteins: The Biology of Nurture

READING THIS BOOK WILL LITERALLY CHANGE YOUR *brain!* That may sound like marketing hype, but as we'll see in this chapter, it's a pretty safe bet. The fact is that almost any experience we have can "change" our brains. The connections between our neurons (synapses) are continually changing and adapting as we perceive and respond to the world around us.

In the last chapter, I told you how variations in our genes may shape our temperament, emotional responses, and personalities. But no self-respecting scientist believes that genes alone are destiny. There is no dichotomy between nature and nurture because they are two sides of a single coin. The effect of any gene depends on the environment it's expressed in. It's not even a sensible question to ask what part of our behavior is genetic and what part environmental. The ability to speak a language is a universal human ability that is made possible by our genetic endowment. But in the absence of exposure to people who speak, our capacity for language wouldn't be expressed.

And early in life, certain experiences can be particularly powerful in shaping the development of our minds and behavior. How much of who we are depends on what happened to us in the first few years of life? The two major strands of psychology in the twentieth century—psychoanalytic theory and behaviorism—gave opposite answers to this question. Sigmund Freud, the father of psychoanalysis, claimed that our emotional lives and the way we approach relationships are forever shaped by what happens during our first five years. His contemporary John Watson, the father of behaviorism, claimed that we are blank slates on which experience can write and rewrite learned behavior throughout life.

Over the last twenty years, developmental neuroscience has provided a more nuanced view of how early experience affects us. As we will see in this chapter, it turns out that both Freud and Watson were on to something. Freud was right in claiming that early experience has a formative and enduring influence on our relationships and how we interact with the world around us. But fortunately, as Watson emphasized, we are lifelong learners. In fact, our brains are continually being remodeled as they encounter new information. And this ongoing plasticity is the root of the remarkable resilience of the human spirit. In this chapter, we'll explore how experience interacts with our genes to shape the trajectory of our lives.

GETTING AHEAD

DESPITE THE MEDIA HYPE SURROUNDING GENE DISCOVERIES, THE notion that early experience can shape brain development is alive and well. In fact, it spawned a multimillion-dollar industry that began in the 1990s with the debut of "Baby Einstein" videos. The marketing of educational videos for infants and toddlers later expanded to evoke the whole pantheon of genius in Western civilization: there's Baby Mozart, Baby Da Vinci, Baby Van Gogh, Baby Beethoven, Baby Shakespeare, and Baby Wordsworth. (Even those who want their infant to revere a great football team needn't waste a moment since the advent of "Baby Bama," a video that "uses officially licensed footage of Crimson Tide sports, mascot, marching band, and campus attractions to expose children to The University of Alabama in an exciting . . . and educational manner.") As we'll see, the growth of the market for baby brain enrichment products provides a fascinating example of how neuroscience can be hyped beyond the laboratory and co-opted to fuel commercial interests and even public policy agendas.

One of the seminal events in this story was the 1993 publication of a brief paper entitled "Music and Spatial Task Performance" in the prestigious scientific journal *Nature*.[1] Frances Rauscher and her colleagues at the University of California–Irvine reported a study in which thirty-six college students underwent three procedures in which they listened to ten minutes of Mozart's sonata for two pianos in D major (K. 448), a relaxation tape, or silence. Immediately after each procedure, the students were given three tests of spatial reasoning. Compared to the other two conditions, listening to Mozart was associated with a cognitive boost that corresponded to an 8- to 9-point increase in spatial reasoning IQ. The results quickly captured the media's attention. In a front-page *Los Angeles Times* article entitled "Study Finds That Mozart Music Makes You Smarter,"[2] Rauscher expressed unease about the potential for exploitation of the original findings: "You can never control what the marketers will do. It is a very scary thought."

Indeed, although the cognitive boost lasted only ten to fifteen minutes, this didn't deter others from seizing upon the results and dubbing them "the Mozart Effect." While other scientists had difficulty replicating Rauscher and her colleagues' results, the notion that classical music could enhance brain function seemed too appealing to ignore. Classical recordings for babies and toddlers began to be marketed to parents who were eager to give their kids an edge.

In 1996 Julie Aigner-Clark, the mother of a one-year-old girl, began creating homemade videos to entertain and educate her child. Within a year, she had begun selling the videos under the name "Baby Einstein," and the product quickly took off. Sales topped $100,000 in the first year and reached $1 million by the second year. Just around this time, music critic and author Don Campbell trademarked the phrase "the Mozart Effect" and published a best-selling book with that title. That was followed by *The Mozart Effect for Children*, which didn't claim that music could make your kid a genius, but "certainly it can increase the number of neuronal connections in her brain, thereby stimulating her verbal skills" (p. 4).[3]

The timing couldn't have been better for the "Baby Einstein" franchise. As Aigner-Clark told CBS's *The Early Show* in 2005,[4] "All kinds of research was done that said 'Listen to Mozart; Mozart is great for you.' There are wonderful studies showing that listening to Mozart will stimulate your mind. And I had a video called *Baby Mozart*. So I was really lucky." By 2004, three years after Disney purchased Baby Einstein, annual sales had reached $170 million, a success story that earned Aigner-Clark a special mention in the president's 2007 State of the Union address.

Meanwhile, child advocates were energized by what they saw as a scientific consensus that early environmental enrichment was essential to wiring a healthy brain.[5] And even politicians jumped on the bandwagon. Florida passed a law mandating that classical music be piped into all state-funded day care centers, and in 1998, Georgia governor Zell Miller allocated funding to ensure that every newborn would leave the hospital with a classical music CD. As he put it, "No one questions that listening to music at a very early age affects the spatial, temporal reasoning that underlies math and engineering and even chess. Having that infant listen to soothing music helps those trillions of brain connections to develop."[6]

Unfortunately, the story was getting ahead of the science. Remember, the original study had shown a fleeting effect of a Mozart piano sonata on a limited domain of reasoning in a small group of undergraduates. A spate of studies that followed had decidedly mixed results. A combined analysis of sixteen studies found no significant evidence that listening to Mozart improved IQ, even in the limited realm of spatial reasoning.[7] In fact, several studies suggested that any transient effect on cognitive function was probably due to the arousal and positive mood induced by listening to pleasurable material.[8, 9] That might explain why similar increases in cognitive functions were reported for adults listening to the Greek composer Yanni,[10] and ten- and eleven-year-olds listening to the rock band Blur (dubbed the "Blur effect").[8, 9] Kenneth Steele, one of the scientists who was unable to replicate the results of the original study, concluded that "The Mozart effect is pretty much on the wallet of the parents who are buying the CDs."[11] For her part, the original researcher, Frances Rauscher, claimed that the failures to replicate were due to methodological problems with the later studies. She wrote, "Because some people cannot get bread to rise does not negate the existence of a 'yeast effect.'"[12] Regardless of the scientific debate, the Mozart effect became fixed in the public consciousness.*

Fueled in part by this widespread belief that early stimulation is important for cognitive development, the baby video and DVD market exploded. Although the Baby Einstein folks demurred that their products "are not designed to make babies smarter," surveys suggested this is exactly the hope that motivated parents to buy children's videos and DVDs (to the tune of nearly $5 billion in 2004).[13] An analysis of top-selling DVDs for babies in 2005 found that more than 75 percent made educational claims. For instance, according to the packaging of the Brainy Baby Left Brain video, the video series was the first "that can help stimulate cognitive development." Parents of children under two years say that the most important reason for having their babies watch TV, videos, and DVDs is that it is educational or good for their child's brain.[14] Although the American Academy of Pediatrics issued recommendations in 1999 and again in 2011 discouraging media use for children under age two,[15, 16] children age six months to three years spend nearly two hours per day watching TV and other video media.[13]

---

* While the evidence for an early effect of music listening on brain development and IQ has been mixed, there is a more consistent set of findings suggesting that early musical training can enhance musical aptitude, appreciation, and performance.

Is there evidence that watching videos early on affects cognitive development? The answer is yes, but not necessarily in ways parents would hope for. In one study, greater media exposure in six-month-olds has been shown to predict lower language and cognitive development when the babies are fourteen months old.[17] In another influential study, Frederick Zimmerman and his colleagues surveyed more than one thousand parents of children who were between two months and two years of age.[18] For infants eight to sixteen months of age, every hour spent viewing baby videos and DVDs correlated with a substantial *decline* in scores on a standard measure of language development. The more they watched, the fewer words they had learned. And it didn't matter whether or not parents watched the videos with their infants.*

In contrast, reading or telling stories to the children was associated with an improvement in vocabulary learning. Though the study doesn't prove that baby videos hurt infant language development, it's clear that they don't seem to help. At best, other studies have found no correlation between TV or video viewing in infancy and later cognitive skills.[22, 23] In one study of a best-selling baby DVD designed to boost vocabulary, twelve- to eighteen-month-old children were randomly assigned to one of four conditions. The first group watched the DVD with a parent at least five times per week over four weeks; the second group watched the same amount, but on their own. The third group didn't watch the DVD but their parents were given a list of the twenty-five words featured on the video and asked to "try to teach your child as many of these words as you can in whatever way seems natural to you." And finally, the fourth group had no intervention. When the chil-

---

* Other analyses suggested that the content of TV shows that infants and toddlers watch is relevant. For example, heavy TV viewing before age three has been associated with attention problems in later childhood,[19] but the effect seems to hold only for noneducational TV.[20] Research on toddlers has shown that watching certain TV shows—*Sesame Street, Blue's Clues, Dora the Explorer*—is associated with improved literacy and language skills. These shows tend to elicit the child's participation, offer a clear story line, and avoid overstimulation.[21]

dren were tested for how many words of the twenty-five words they learned over the four weeks, only the children who were taught by their parents without watching the video did better than chance. The other groups did no better than the control group who had no intervention.[24]

In 2009 the Walt Disney Company offered a refund to parents who purchased the Baby Einstein videos, but emphasized that this was an expression of their confidence in the product.[25] Meanwhile, cofounders Julie Aigner-Clark and her husband went to court to defend the legacy of their product and to challenge the studies that suggested adverse effects of early video watching.[26] As Aigner-Clark noted, baby videos are hardly the worst thing for a child: "Welcome to the twenty-first century. Most people have televisions in their houses, and most babies are exposed to it. And most people would agree that a child is better off listening to Beethoven while watching images of a puppet than seeing any reality show that I can think of." And so the controversy continues.

NOW OR NEVER?

ANOTHER INTERESTING FINDING EMERGED FROM THE ZIMMERMAN study—baby videos were associated with language decrements in the eight- to sixteenth-month-olds but not the seventeen- to twenty-four-month-olds. Timing mattered. So while this study doesn't support the benefits of baby videos, it ironically supports the logic of why some parents buy these videos in the first place. The primary rationale people give for plunking their babies in front of a "brainy" video is the belief that there may be an early window for boosting brain power: Expose your infant to the right stimuli and you may affect the wiring of the brain in ways that last a lifetime. The problem is that before age two, children may not be able to learn from media, and time spent watching TV or videos is time away from playing or interacting with parents and siblings—activities that *can* promote cognitive growth.

No one doubts that development works on a schedule. If you don't develop a right arm in the womb, you are destined to live a one-armed life. No amount of nurturing, good diet, or physical therapy will get you a second arm later in life. Does the same thing apply to the organs of the mind?

The idea of "windows of opportunity" for brain development has a long and controversial history, but, in some domains, they clearly exist. Scientists refer to these windows as sensitive periods or critical periods—when the brain is especially sensitive to some kind of input from the environment, and may need to get that input in order to develop normally. In the case of a critical period, it's "now or never." If the developmental event doesn't happen in the critical time frame, it may be lost forever. Sensitive periods are less absolute—they represent a time of maximum sensitivity to an environmental stimulus, but the developmental changes may occur later to a lesser degree. The difference between critical and other sensitive periods can be illustrated by two familiar examples.

The first—imprinting—is one you probably remember from ninth-grade biology. The Austrian ethologist Konrad Lorenz is well-known for describing how newborn greylag geese instinctively follow their mothers around within about twenty-four hours of their birth. Lorenz observed that goslings born in an incubator will follow the first conspicuous object they are exposed to—whether it's a human being, a pair of boots, or even a wooden block. After a brief period of exposure to the object, they treat it as though it were their mom. This process, known as filial imprinting, has been considered a classic example of a critical period because there is a limited period (typically the first two days of life) when the gosling's brain is prepared to make the association between an object and the concept of mother goose. Filial imprinting seems to release a behavior (in this case, finding mom) that is waiting for its environmental trigger.

The second example is the sensitive period for learning a second language. If you've ever tried to learn a second language as an adult, you know it's harder than it is for a typical child. Immigrants who learn the language of their new country may retain an accent of their native language depending on how old they were when they emigrated. My grandmother and mother were both born in Poland and came to this country as Holocaust refugees—my mother when she was nine and my grandmother when she was thirty-nine. Both became fluent speakers of English, but my grandmother retained a thick accent of the old country.

If Nobel Prizes are any measure of the importance of a scientific question, the study of critical periods is clearly up there. Lorenz received the Nobel Prize in 1973 for his work on imprinting. And then, in 1981, Torsten Wiesel and David Hubel shared the Nobel Prize for groundbreaking insights into how critical periods actually work in the brain.*

It was known that children who are born with congenital cataracts covering the lenses of the eye continue to have problems seeing even after the cataracts are removed. This is, of course, not the case for people who develop cataracts as adults. Adults who develop cataracts have their sight restored when the cataracts are surgically removed.

Hubel and Wiesel wanted to know how this difference in timing can have such a profound effect. To tackle the question, they developed a model to study the visual system in cats.[27] They found that if you raise a kitten with one of its eyes sewn shut and then remove the sutures after several months, the animal never develops the ability to see in that eye. Somehow, depriving the eye of visual input early in life had an irrevocable effect.

---

* The work for which they received the Nobel Prize also involved the first detailed description of how and where visual information is processed in the visual cortex of the brain.

After a series of studies, they showed that the problem wasn't with the eye itself or even with the eye's connections to the visual cortex of the brain. Instead, they discovered, the brain actually changes in response to a lack of visual stimulation. Areas of the cortex that would have been committed to processing information from the occluded eye are taken over by neurons from areas that received input from the nonoccluded eye. In other words, the brain has changed to allow the "good eye" to do the work of both.

Hubel and Wiesel had uncovered a striking example of brain plasticity—changes in the brain's architecture that result from an animal's experience. But they also found that this plasticity had a critical window—if the kitten's eye was kept closed beyond about three months of age, the loss of vision in that eye would be irreversible.

Hubel and Wiesel's work provided the first detailed description of two key mechanisms by which the environment affects the brain: *critical periods* and *neuroplasticity*. These two phenomena, which have motivated a vast amount of brain research, are actually quite closely connected. In essence, critical periods could be thought of as temporally constrained periods of environmentally dependent brain plasticity. Or, put more simply, critical periods are developmental windows during which experience can powerfully shape how the brain is wired.

EARLY EXPERIENCE: HISTORY AND MYSTERIES

THE NOTION THAT EARLY EXPERIENCE CAN HAVE PROfound and perhaps irrevocable effects on the mind has a long and storied history. It was, of course, a key question that framed the nature vs. nurture debate. And the answers that have been offered have ranged from superstition and myth to widely influential scientific theories.

Until the twentieth century, there was a widespread belief, endorsed by prominent physicians, that children could be permanently harmed by emotional frights, longings, and traumas that befell their mothers during pregnancy. The idea was that the circumstances of the emotional situation would leave an impression or mark on the fetus. *Maternal impressions* they were called, and they were blamed for all manner of deformity and intellectual weakness in the offspring. In an 1870 article from the medical literature, we find the following typical case:

> *A lady in the third month of her pregnancy was very much horrified by her husband being brought home one evening with a severe wound of the face, from which the blood was streaming. The shock to her was so great that she fainted, and subsequently had an hysterical attack, during which she was under my care. Soon after her recovery she told me she was afraid her child would be affected in some way, and that even then she could not get rid of the impression the sight of her husband's bloody face had made upon her. In due time the child, a girl, was born. She had a dark red mark upon the face, corresponding in situation and extent with that which had been upon her father's face. She proved also to be idiotic. (pp. 251–52)[28]*

In other cases, the emotional distress was more modest:

> *A woman, between four and five months advanced in pregnancy, had an irresistible desire for a fine salmon which she saw in a market; this she purchased, despite her poverty, and as a result, at the end of the full term of normal gestation she was delivered of a child "the head*

*and body of which presented a peculiar and strange con-
formation, in truth it was salmon-shaped, whilst the fingers and toes were webbed, representing the fins or tail of
the salmon." (pp. 247–48)*[28]

Fortunately, salmon envy is no longer invoked as a
threat to the well-being of young children, but one of the
most dramatic and ancient narratives about the effects
of early experience has continued to capture the popular
imagination.

BABIES GONE WILD

What would happen if a child were utterly deprived of
normal human nurturing? Is there a minimal set of experiences that are crucial for normal development—and
if they don't occur, can the damage be undone? These
questions have made tales of so-called feral children eternally fascinating,

Stories of children raised by animals date back at
least to the myth of Remus and Romulus, the twin sons
of Mars who were left to drown in the river Tiber. Res-
cued and suckled by a she-wolf, they went on to found
Rome. In more modern times, wild children have been a
recurrent theme in literature and the arts, from Rudyard
Kipling's Mowgli to Edgar Rice Burroughs's Tarzan.

The legend of feral children took on new significance
after the eighteenth century as revolutionary ideas about
human nature permeated Western thought. Jean-Jacques
Rousseau's notion of the "noble savage" romanticized the
ideal of mankind's natural roots, before it was exposed to
the polluting effects of civilization. It was against this backdrop that eighteenth-century France was captivated by the
discovery of Victor of Averyon (the subject of Francois Truf-

faut's 1970 film *L'Enfant Sauvage*), who was said to have spent most of his life alone in the woods until he emerged at age twelve. Unable to speak and apparently without any capacity for emotional engagement, he was judged to be a hopeless case by the medical establishment.[29]

Victor lacked all social skills; he defecated in public, ate like a wild animal, and showed no interest in human attachments. He was taken in by a young medical student, Jean Marc Gaspard Itard, who devoted years to rehabilitating the boy. Victor made little progress. He never acquired language nor developed social connections. Through the lens of twentieth-century medicalizing, some wondered whether the boy suffered from autism or another developmental disorder. But at the time, Victor's sad life seemed to be a tale of the immovable force of early experience.

Tales of feral children continued to grab headlines over the past two hundred years. And yet, for all the romance and sensationalism that these stories have generated, the underlying story is anything but exotic. In most of these cases, including Victor's, there is a history of abuse and neglect.[29, 30] In the end, these stories illustrate the profound and lasting effects of early adversity and trauma—a theme we'll return to later in this chapter.

TIMING IS EVERYTHING

OVER THE PAST SEVERAL YEARS, NEUROSCIENTISTS HAVE BEGUN to sketch a fascinating picture of how sensitive periods shape the development of our mental and emotional lives. As Hubel and Wiesel discovered in the case of vision, a lot depends on getting the right instructions from the environment at the right time. As we'll see, the same idea turns out to apply to many of the brain's other fundamental functions—our sense of taste, our capacity to learn language, our ability to process emotions and to form attachments.

In the 1980s William Greenough and his colleagues at the University of Illinois provided new insights into how the environment shapes brain development. They suggested that there are two different developmental phases of plasticity during which the environment impacts the wiring of the brain.[31]

The first they called the *experience-expectant* phase, in which the brain makes use of "environmental information that is ubiquitous and has been so throughout much of the evolutionary history of the species" (p. 540).[31] Our brains have been prepared to await signals from the environment that have reliably been there over millennia— like the visual contours of objects in the world or the presence of a mother. Sensitive periods of development are essentially windows of experience-expectant learning. If you deprive an animal of an expectable environment during these periods—as Hubel and Wiesel did when they patched one eye of a newborn kitten and Lorenz did when he substituted himself for mother goose—you interfere with the development of fundamental brain functions. In other words, if you mess with the expectable environment, you mess with the brain in ways that can last a lifetime.

Early in brain development, experience can shape how neurons hook up. Humans, for example, are born with billions of neurons packed into our little heads. In the first twelve to eighteen months of life, these neurons undergo an explosion of connectivity, sprouting branches and forming trillions of junctions known as synapses. Over the next several years, this neuronal thicket is scaled back. During the experience-expectant phases that underlie sensitive periods, these connections are refined by a use-it-or-lose-it strategy. The expectable environment reinforces useful connections and eliminates irrelevant ones. As a result of this synaptic pruning, brain circuits are refined in ways that enable the animal to adapt to key features of the environment. Some brain regions, notably the prefrontal cortex, which is involved in higher cognitive functions and self-control, continue to be wired well into adolescence. So experience is literally remodeling the brain and experiences that occur during times of active pruning might have long-lasting effects.

As Greenough and colleagues put it: "If the normal pattern of experience occurs, a normal pattern of neural organization results. If an abnormal pattern of experience occurs, an abnormal neural organization will occur" (p. 544).[31] There is a kind of irreversibility that accompanies this process "because a set of synapses has become committed to a particular pattern of organization, while synapses that could have subserved alternative patterns have been lost" (p. 546).[31]

The experience-expectant or sensitive period of development allows the brain to develop the fundamental, species-typical skills it needs to navigate the world. But some of the most important circumstances an animal faces as its life unfolds are not expectable. They are unique to a particular geographic location, family, or social system. And this is where the second phase of plasticity—*experience-dependent* learning—comes in. This is how the brain responds to the fine-grained information that is unique to the world around it: where the nearest food sources are, the politics of the social hierarchy it lives in, and so on.

As the specific circumstances of our lives play out, the brain can adapt. Some of this adaptation involves the formation of new synapses. For example, dendrites, the projections from neurons that form the receiving end of a synapse, can sprout extra branches that create new synaptic connections to other neurons. As these synapses form and interconnect, they create new wiring of brain circuits that respond to the demands of the environment.[32] This lifelong process is one of the crucial ways that our brains shape the twists and turns along the unique path that each of us follows in life.

Turn On, Tune In, or Drop Out

The synaptic pruning that occurs with experience-expectant learning has a remarkable implication. The normal development of many of our key brain functions—including language, emotional responses, and social cognition—involves a *loss* of abilities. In other words, we actually begin life with capacities that we must lose over time so our brains can develop normally.

Take the case of language. If you really think about the task facing an infant who has to learn to speak and understand language, you begin to doubt that it could ever occur. The cognitive neuroscientist Patricia Kuhl has called it "cracking the speech code."[33] Out of the hundreds of available consonants and vowels, each language has derived a distinct set of about forty phonemes that change the meaning of a word (for example, from *take* to *lake* or from *cream* to *creep*). With a brain that's only a few weeks old, infants must begin to recognize these acoustic differences. They also have to learn that sounds are grouped into the distinct units that make up words. That's no mean feat. Acoustical analyses show that there are actually no silences between spoken words in a sentence.[33]

Imaginehowhardreadingthisbookwouldbeiftherewerenospacesbetweenthewords. Or think about what you hear when you listen to someone speaking a foreign language: you don't know the words or even where they begin and end. To your ear, it's just a stream of sounds. And at least *you* understand that there are words there and you have experience with how sounds are grouped into words in your own language. The infant has none of this experience. What's more, the same word almost never sounds the same. Every occurrence of the word *play* may sound different depending on who is saying it and in what context. An infant may hear the same word spoken by a man, woman, or child, each of whom says it at a different rate, pitch, inflection, and tone. The word may sound quite different in one context ("Do you want to play with that?") than in another ("Don't play with your food!"). And yet infants are instinctively able to categorize words and pick them out of a string of sounds. The genetic program that allows them to do this is far better than any program software engineers have been able to write, as anyone who has used speech recognition software can tell you.

What's even more amazing is that infants are actually better than adults at distinguishing speech sounds. We are born with a universal capacity to distinguish phonemes and sounds that are found in any of the world's languages.[34] An infant can tell the difference between word sounds used in any language and can hear the boundaries between words in any language. You cannot.

Japanese infants, like English infants, can tell the difference between "r" and "l" sounds, but Japanese adults, unlike English adults, don't hear that distinction.[33] That's because the Japanese language combines those sounds into a single phonetic unit while English treats them separately. Before you were about eight months old, you were prepared to understand any human language. And then you lost it. That's because the environment encouraged your brain to make a commitment to your native language. How does this happen?

As Kuhl explains,[33] the infant brain seems to use a kind of "statistical learning" for language acquisition. As infants are exposed to more and more examples of their native language, they start to register the statistical frequency of certain sounds and lexical patterns. The brain's language circuits use this statistical analysis to progressively tune into the language around it.* By committing to one language, the brain becomes better and better at learning that language. The faster and more fully the brain makes this transition, the faster and better the child is at learning its native language.[35] But like any commitment, this one comes at a cost. As the infant brain tunes into the ambient speech of native language speakers, it tunes

---

* In addition to statistical learning, social interaction plays a key role in language acquisition. Mothers in cultures around the world use a particular speech pattern when talking to their infants. This "motherese" involves exaggerating phonemes of the native language, making it easier for the baby to pick them out. In other words, motherese facilitates the baby's neural commitment to its native language.[35]

out nonnative (foreign) language sounds. After the first year of life, the brain has an increasingly difficult time understanding nonnative speech sounds. This explains in large part why learning a second language is so much harder for adults than for children (whose language circuits are not yet fully committed).

The developmental program involved in language acquisition is marvelously efficient. Within three years, a newborn morphs into a toddler who is fluent in her native language. But this story also reveals key elements that our brains use over and over to acquire the species-typical (human nature) skills needed to make sense of our environment. We begin with an open mind, primed to detect information that evolution has designed our brains to expect. During this phase, the brain casts a wide net. It is sensitive to a broad variety of relevant information because the neonate may be born into any of a range of environments. So at birth, we can distinguish phonetic information in any language because our brains have to be prepared to be born in Peoria, Kabul, or Tokyo. After a time, cues from the environment clue the brain into where it has landed and the salient details of that world. This triggers an experience-dependent phase, in which brain development depends on specific inputs from the environment. A commitment is made and other roads are not taken.

THE FIRST TIME EVER I SAW YOUR FACE

BABIES USE THIS SAME GENERAL MECHANISM TO LEARN THE LANGUAGE of social and emotional communication. Just like words are the units of spoken language, facial expressions are the units of emotional language. Before they can speak, babies use facial expressions to recognize Mommy and Daddy and figure out whether they're in trouble or safe. If you've only recently emigrated from the womb, these abilities are about as important as you can get. Not surprisingly, then, evolution has prepared the baby brain to read faces.

Like so much else, we owe this insight to Darwin himself. In

1872 he devoted an entire volume to cataloguing the expression of emotions in humans and animals[36] and postulated that facial expressions of emotion are innate, universal, and evolved tools of communication. As Darwin concluded, "That these and other gestures are inherited, we may infer from their being performed by very young children, by those born blind, and by the most widely distinct races of man" (p. 1468).

In a series of fascinating studies, the developmental neuroscientist Charles Nelson and his colleagues have shown that babies have a facial recognition system that starts out, like the infant's language system, "broadly tuned." In the first of these studies, they showed infants pairs of pictures of monkey and human faces to see if they could distinguish individual faces of either species.[37] At six months of age, infants were equally good at recognizing both human and monkey faces—that is, they could tell the difference between one person and another, but also one monkey and another. You might say they were bilingual for faces.

How can you tell if a six-month-old recognizes a face? Nelson and his colleagues used a well-established test. Babies will look longer at an object they've never seen before than they will at a familiar object. So they showed babies pairs of faces, one they'd seen before along with one new one. By nine months, babies could still tell the new human face from the familiar one, but they'd lost the ability to discriminate monkey faces. What does this mean? It appears that, like learning a native language, face recognition passes through an experience-expectant window—a sensitive period—when it's powerfully shaped by experience. As their environment provided more exposure to human faces and no exposure to monkeys, the babies' brains became committed to the human variety.

Once again, the brain, initially open to many possible worlds, makes an irrevocable commitment to the one it's given. In many ways, brain development, like life itself, is about making choices.

OKAY, NOW YOU'RE SCARING ME

EXPERIENCE DOES MORE THAN SHAPE OUR ABILITY TO RECOGNIZE the people around us—it's also crucial for tuning into their emotions. Between six and nine months of age, babies enter an experience-expectant phase in which their brains are looking for information about the important features of faces, including expressions of emotions. The first emotion we learn to recognize is fear. More than any other expression, a fearful face is a signal that our very survival may depend on. Seven-month-old infants fixate on fearful expressions (but not happy or neutral faces) even though such faces don't yet trigger fear responses in the infants themselves.[38, 39] It's as though babies recognize that a frightened expression is telling them something important, but they don't yet understand what it is. Not coincidentally, the brain's fear recognition circuitry comes on line when the baby is beginning to explore its environment and thus needs feedback about what's safe and what's dangerous. Babies who crawl toward a visual cliff, where the ground appears to fall away, will continue crawling if their mother's face looks happy but will stop in their tracks if their mom looks scared.[40]

The network of brain structures that allows babies to decode facial expressions includes two regions of the cortex that appear to be specialized for recognizing faces—the fusiform gyrus and the superior temporal sulcus (STS)—and two regions that are key players in our experience of emotion—the amygdala and the orbitofrontal cortex (OFC), a ridge of brain cells above the eyes (fig. 3.1). When a baby sees a face, the visual cortex relays information to the amygdala, which quickly uses low-level information to get a rapid read on the novelty and emotional tone of the face. The amygdala communicates with the OFC, as more detailed information on the salience of the face is processed. Meanwhile, the STS and fusiform gyrus are recognizing and decoding features of the face and communicating back and forth with the emotion areas to process the emotional expression in finer grain.

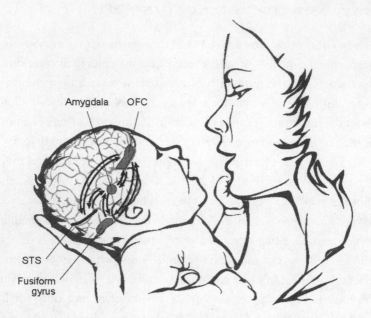

Brain circuits that develop to process emotional signals from faces include emotion-processing regions (amygdala and OFC) and face recognition areas (STS and fusiform gyrus).

As babies gain experience with faces, they begin to notice how expressions link with sounds and events—smiles are paired with cooing sounds and soothing sensations, fearful faces are paired with worried tones and perhaps followed by a painful fall. This tuning of the mind has a biological counterpart in the circuits of the brain. The wiring of the face recognition network is refined and strengthened while neurons and synapses that are superfluous are pruned back. In the end, we emerge from this sensitive period with a neural system that we will rely on for the rest of our lives to judge whether other people are angry or happy with us, whether they are threatening or welcoming. Over the course of our years, the personal adversities and fortunes we encounter will continue to calibrate this system through a process of experience-dependent learning. But as we'll see in the next section, adversity and trauma can distort the developing brain in ways that cast a long, dark shadow.

FACIAL PROFILING

IN THE LAST CHAPTER, WE SAW THAT EACH OF US HAS A SET OF genetic variations that make us more or less prone to perceive and feel certain emotions. These genes help create subtle biases in how we sense and respond to stress, unfamiliarity, and reward. But experience itself plays on those biases in crucial ways.

The most striking demonstration of this comes from research on the effects of childhood adversity and maltreatment. Child abuse and neglect are known to have long-lasting effects on behavior and emotional reactivity, and each year in the United States, nearly 2 percent of children under age three are victims of maltreatment. Maltreated children are more likely to have insecure attachments to other people; they have difficulty understanding the feelings and thoughts of others and form fewer close friendships; and they are prone to anxiety, depression, aggression, academic failure, and antisocial behavior later in life.[41]

How does early adversity cast such a long shadow over its victims? The answer emerging from biological studies is that the environment actually biases the brain's emotional circuitry in ways that refocus the lens through which we experience the world. When bad things happen, they can literally change the way we see other people.

Developmental psychologist Seth Pollak and his colleagues at the University of Wisconsin have shown that abused children have a perceptual sensitivity for anger.[42] In one study, they showed children pictures of emotional faces (fearful, angry, sad, and happy faces) that morphed from fuzzy to focused over the course of the experiment. The clarity of each picture was gradually increased at three-second intervals and after each interval, the children were asked to say what emotion, if any, they saw. Children who had been abused were able to pick out an angry face much sooner than children who hadn't been maltreated.

In other studies, Pollak's group has shown that the effect was specific to angry faces.[43, 44] They showed abused and nonabused children (the controls) faces that gradually morphed between two expressions: happy ➡ fearful, happy ➡ sad, angry ➡ fearful or angry ➡ sad. The researchers wanted to see whether abused children and controls perceive the transition from one category of emotion to the other at the same point in the morphing continuum. In fact, the two groups were identical in their ability to categorize the transition from happy to sad or fearful faces. However, there was a dramatic difference in how the two groups perceived angry faces. The controls stopped seeing the angry face when only about 30 percent of the expression was sad or fearful (and 70 percent was angry), but the results were opposite for abused children. They kept seeing anger until about 70 percent of the facial expression was sad or fearful (and only 30 percent angry). Sadly, their childhoods had made them experts in anger. They needed much less information to see an angry face than other kids. Their brains had adapted to their environments to give them an edge—a little extra time to see danger coming. That brief head start might mean the difference between getting hit and avoiding abuse by appeasing a parent, shielding their bodies, or running away.

Researchers who study the effect of adversity on brain development face a problem. If children raised in emotionally or socially impoverished environments see the world differently from other children, it is still possible that the difference depends more on their genes than on their experience. After all, parents are not only the most powerful environmental influence on their infants, they also pass on their genes. Maybe the same genes that predispose a parent to be less nurturing also predispose their children to have emotional and behavioral problems. If you really wanted to test the effect of the environment, you'd need to do some kind of controlled experiment where you randomize infants to grow up in either an environment that is nurturing or one that is not. But surely that would be impossible.

FOSTER CARE FOR THE MIND

AS IT TURNED OUT, A TRAGIC SERIES OF EVENTS COMBINED WITH the passion of a few scientists to set the stage for just such an experiment. In 1966 Nicolae Ceausescu, head of Romania's Communist Party, issued a decree banning abortion for women under forty-five. His motivation was not religious but political. Ceausescu was determined to expand Romanian communism by expanding the number of Romanian Communist workers. Government agents, nicknamed "the Menstrual Police," rounded up women under forty-five years of age for compulsory pregnancy examinations.[45] Restrictions were also placed on contraception and divorce.[46] As Ceausescu told his countrymen, "the fetus is the socialist property of the whole society."[47] He issued a law requiring women under forty to have five children. Financial incentives were provided for those who complied and severe penalties for those who did not, including tax penalties of up to 20 percent, a so-called çelibacy tax.[46, 48]

Ceausescu's economic policies also decimated the Romanian economy, and the consequences were tragic. The prohibition on abortion spurred a huge increase in illegal abortions that were often primitive and lethal. By the time Ceausescu was overthrown in 1989, Romania had achieved the distinction of having Europe's highest maternal and infant mortality rates.[49] Because of widespread economic hardship, those who did not obtain abortions were often unable to support additional children and had to abandon them to state-run institutions. At the end of Ceausescu's reign, more than 150,000 children were housed in orphanages that were notorious for their appalling conditions.[50, 51]

In the late 1990s the Romanian minister for child protection was seeking alternatives to institutionalization for the thousands of children who remained abandoned in Romanian orphanages. At the time, few alternatives existed; there was almost no government-sponsored foster care. Around the same time, the developmental neuroscientist Charles Nelson and his colleagues Charles Zeanah and Nathan Fox were trying to study how experiences in infancy affect the development of the brain and the consequences for cognitive, emotional, and social development. The Romanian government invited them to visit Bucharest to discuss possibilities for collaborative research. Soon thereafter, the Bucharest Early Intervention Project (BEIP) was launched.

The research conducted through the BEIP stands as one of the boldest and most important studies of the impact of early environment on the development of the mind. While previous studies had shown that children reared in institutions had developmental delays and altered activity in specific brain regions when compared to children who were adopted into families, they all suffered from a possible selection bias. It's likely that children who are adopted away from institutions are psychologically and physically healthier than those who remain behind. As a result, any differences between these two groups could be due to differences in the children, not the effect of their caregiving environment.

The BEIP avoided this problem by constructing a unique experiment. After extensive ethical review and collaboration with governmental and nongovernmental organizations, the BEIP researchers randomized 136 babies (age six to thirty-one months) who had been abandoned to an orphanage at birth to one of two groups. Half of the children were randomized to remain in institutional care while the other half were assigned to foster care at an average age of twenty-one months.* For comparison, they also included a third

---

* Because there was very little foster care available in Romania at the time, the investigators actually created their own foster care program. They established a child care network, including fifty-six foster families, and provided financial support, training, and close supervision in collaboration with trained Romanian social workers.

group of eighty children who were born in the same hospitals as the institutionalized children but who were living with their biological parents and had never been institutionalized.

In 2007 the BEIP group published the first major results of their study in *Science* magazine.[52] They compared cognitive development among the three groups before and after the intervention and tracked them up to age four and a half. The findings were unequivocal. The institutionalized group was significantly worse off than both the foster care group and the never-institutionalized control group on a whole range of developmental tests: IQ, sensorimotor abilities, and language development. Indeed, the institutionalized group's test scores placed them in the range of borderline mental retardation while the foster care group had caught up to the controls by the time they were three and a half years old.

The researchers also found that the earlier a child was placed in foster care, the greater the gain in cognitive abilities. Those placed before they were two years old performed as well as the never-institutionalized group. When the researchers reexamined the children at age eight, the IQ-boosting effect of foster care placement was still detectable and those who entered foster care before age two seemed to do best.[53] The results supported the idea that there is a sensitive period (before the age of two) when a change in nurture can have dramatic effects on the brain.

In other analyses, the foster care group was found to express more positive emotions and had better attention than the institutionalized group.[54] These differences emerged quickly after children in the foster care group were removed from the orphanages. It was as though the children were stuck in an experience-expectant phase with their brains on hold, waiting for social interaction. Once they were provided with social stimulation, they rapidly responded, unleashing their capacity for joy.

The effects of the foster care intervention could even be seen in how the children's brains functioned. Before and after the intervention, the children were shown emotional faces while their brain activity was measured with EEG electrodes. Compared to the children reared in their own families, children who remained institutionalized had diminished brain responses to seeing the faces and these persisted through the last measurement when they were four and a half years old. On the other hand, the children who were placed in foster care had normalization of their brain responses, although by four and a half years, they had still not caught up to the never-institutionalized group.

The deprivation that came with being raised in an institution had significant effects on the children's risk for psychiatric disorders. By age four and a half, they were more likely to have both "internalizing disorders" (anxiety and depression) and "externalizing disorders" (behavioral and impulsive disorders like ADHD and conduct disorder). Those who were placed in foster care by age two had lower levels of internalizing symptoms, but their risk for ADHD and other behavioral disorders was unchanged. Sadly, the window for reversing these symptoms had apparently closed before they left the orphanages.[55]

The results of the BEIP project were so compelling that they achieved something that few experiments do: they changed national policy. Several years after the study began, the Romanian government passed a law prohibiting the institutionalization of children under two years of age unless they were severely handicapped.

STUMBLING ON SADNESS

THE BEIP PROVIDES DRAMATIC AND CONVINCING PROOF THAT EXtreme deprivation in early childhood can have lasting effects on the development of intelligence and mental health. By the time they were four and a half years old, the children who remained institutionalized group were nearly three times more likely to have depression or anxiety disorders compared to the foster care group.[56] Adversity changes the brain and in the process bends the trajectory of human development. The enduring impact of childhood adversity is well known to every psychiatrist.

Deidre Ward came for help in a moment of crisis. Two months earlier, her boyfriend of three years had ended their relationship, saying she was too "needy and stressed-out all the time." At thirty-five, she had found herself alone again and sank into a bout of depression, much like those she had struggled with since she was a teenager. She was spending most of her days lying awake in her bed and had been unable to go to her job as a paralegal for three weeks. Crying jags, panic attacks, and thoughts of death had come to dominate her life. She was increasingly convinced that her worst fear was coming true: she would always be alone and would never have the chance to have her own family. Deidre said she'd always suffered from low self-esteem and an abiding sense of insecurity about her appearance, her intelligence, and her ability to sustain an intimate relationship. Her boyfriend was right about her, she said with a sadness that was heartbreaking. Whenever she allowed herself to become close to anyone, her anxiety would over-take her. Her desperation to hold on to a relationship would ultimately sabotage it and drive her boyfriend away. The reality was that Deidre was an attractive and accomplished young woman who had endured substantial adversity in her life.

Deidre was born in a lower-middle-class neighborhood just outside Baltimore. When she was three, her father abandoned the family, leaving her mother to care for Deidre, her six-year-old sister, and eleven-year-old brother. Her mother was overwhelmed, both emotionally and financially. She tried to find work, but with three children to raise and limited job experience, it proved too difficult. Over the next several years, Deidre's mother seemed to be increas-ingly stressed and erratic. It was never clear what would set her off, but when her mother got upset, she would blame and berate the kids for her troubles. When Deidre was a little older, her mother would often go out, leaving Deidre's brother in charge of the girls, or some-times leaving them with a neighbor, but they were never sure where she was or how long she'd be gone.

When Deidre was about ten, her mother remarried a man who was gruff, irritable, and seemingly resentful of the kids. When Deidre was distressed, her stepfather would tell her mother not to "baby her" and to "let her cry—she needs to learn to be tough." Her stepfather had a nasty temper and though he never hit the kids, Deidre often lived in fear that he would. In high school, she had few friends and was teased for being socially awkward, but she yearned for some kind of connection. She turned to books for escape and ended up doing well in high school, but she always carried with her a vague sense of dread that periodically bloomed into a full-blown depression. Minor setbacks often seemed catastrophic, and she found herself constantly worried about her school performance, her weight, and her social life. She began dating in college, but her relationships were brief and she felt tense in romantic situations. She seemed to be always on guard for any sign that her boyfriend was angry with her or wanted out. This latest relationship was the longest she'd had and she was beginning to feel hopeful that it would last. Now, left alone again, she felt ashamed for believing they had a connection—like the daydreaming child in a store who tugs at her mother's dress only to find out that it isn't her mother.

Deidre couldn't recall a time when she'd been carefree or really happy. Looking back at her childhood, she felt as though she'd never quite gotten a solid footing and had been stumbling ever since. "I feel haunted," she said.

Deirdre's childhood seemed to have set her on a collision course with suffering. As she passed through sensitive periods of emotional development, her brain tuned into a world that was threatening, chaotic, and unreliable. The results seemed to reverberate in her troubled relationships and bruised self-image ever after. Not surprisingly, researchers have found that childhood adversity and trauma are among the strongest risk factors for depression.

EXPRESS YOURSELF

WHEN WE SAY THE ENVIRONMENT AND EXPERIENCE AFFECT BRAIN development what does this mean exactly? How does the environment get into the brain?

The answer emerging from recent research is that experience actually changes how our genes behave.

To explain the nuts and bolts of all of this, we have to go deep down to the molecular level and into the world of *gene expression*— the process of turning a gene's instructions into a usable product (RNA and proteins). Our genes carry the set of instructions for making the proteins that run the cells of our bodies. But how those instructions play out depends on the details and timing of gene expression. In case you've ever wondered, that explains why your brain doesn't have teeth and your kidneys don't salivate. Every cell contains the same genome* and yet some cells become neurons while others end up making enzymes in the pancreas or make the heart contract. This process of specialization occurs because only certain genes are actively expressed at specific times and in specific places. And that's how experience plays its hand: it can shape how we develop, including the wiring and rewiring of the brain, by affecting where and when specific genes are expressed.

Several factors control how and when genes are expressed. For example, DNA sequences called *promoters*, typically located on the front end of a gene, are docking stations for proteins called *transcription factors* that are made by other genes. When transcription factors bind to the gene promoter, they can turn on the gene. You and I may have different promoter sequences on a given gene that make it easier or harder for transcription factors to turn it on. So, your gene might be more active—more likely to be expressed—than mine. We saw an example of this in Chapter 2: some people have the "long" version of the serotonin transporter gene promoter while

---

* Actually, there are exceptions: mature red blood cells and platelets have no nucleus and thus lack a genome.

others have the "short" version. The "short" version makes the serotonin transporter gene less likely to be expressed, so people carrying
that version make less of the serotonin transporter protein. And, as I
discussed in Chapter 2, that difference may contribute to anxiety by
making the amygdala more sensitive to threats in the environment.

But that difference is fixed—you either have the short version
or you don't, and the environment isn't going to change that. So,
if we're talking about experience changing gene activity, we need
a mechanism that allows the circumstances of our lives to activate
or deactivate genes. One solution nature has arrived at has to do
with what scientists call *epigenetics*. Epigenetics refers to the study
of gene expression changes that are not due to variation in the DNA
sequence itself.[57]

Some epigenetic effects involve chemical modifications of the
chromosomes—you can think of them as chemical "dimmer"
switches that get attached or removed from our chromosomes.
These modifications make it more or less difficult for transcription
factors to either turn on or silence genes. And this is one key way, at
a molecular level, by which the environment directly regulates how
our genes function. In essence, the environment can "mark up" the
genome, annotating the basic text with instructions on where and
when it can be read.

Two of the best-understood epigenetic mechanisms are DNA
methylation and chromatin remodeling. DNA methylation involves
the addition of a methyl group—a simple molecule that consists of
one carbon atom bound to three hydrogen atoms—to a gene. When
methyl groups attach to specific DNA sequences, they act like a lock
or an off switch, preventing transcription factors from binding to
the gene and turning on expression. With chromatin remodeling,
chemical groups (including methyl or acetyl groups) are added to
or removed from proteins that our DNA is wrapped around. These
changes affect how DNA interacts with the cell's gene expression
machinery so that specific genes are turned on or off. In order to

understand how chromatin remodeling affects gene expression, we need to understand how chromosomes are packaged.

Our DNA doesn't just lie naked in the nucleus of our cells. Rather, the long strands of DNA that make up our chromosomes are tightly spooled around proteins called histones. That packaging is essential because if the chromosomes weren't tightly wrapped, they simply wouldn't fit. The nucleus of a cell is about six millionths of a meter across—about four thousand times smaller than the size of a single uncoiled chromosome. And of course each nucleus has to accommodate the twenty-three pairs of chromosomes that make up our genome.

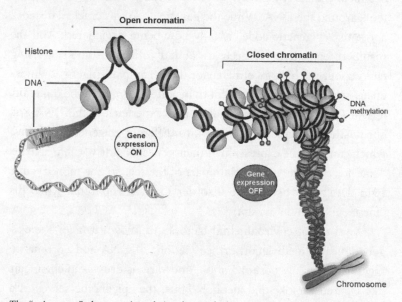

The "epigenome" plays a major role in when and where our genes are expressed.

So how do you fit twenty-three pairs of chromosomes into a space 1/4000 their size? You pack them really, really tight. The chromosomal DNA wrapped around histones and related proteins form what scientists call *chromatin*. The activity of genes depends in part on how tightly wound, or condensed, the chromatin is. Transcription factors and other components of the cell's transcription machinery have a harder time getting through regions of condensed chromatin to turn on the genes that lie underneath. Conversely, genes in regions where the chromatin is relaxed are open to transcription factors and more likely to be expressed. These shifts in chromatin can occur when the histone proteins are chemically marked up (by methyl, acetyl, or other chemical groups), with the result that gene expression gets dialed up or down.

So the environment and our life experiences can fine-tune the expression of genes in our brains by at least two routes: by triggering chemical changes that mark up the DNA itself or by modifying proteins around the DNA. While the genetic code was cracked decades ago, this "epigenetic code" is only now being deciphered. And the results of epigenetic research are already offering some surprising clues about the biology of behavior. For example, epigenetic differences help explain why identical twins often turn out to be quite different. Over time—due to chance and experience—the DNA and chromatin of identical twins acquire different epigenetic switches, which can alter the expression of their genes, and drive them apart.[58] That, for example, may explain part of the reason one monozygotic twin develops a psychiatric disorder like schizophrenia while his "identical" co-twin doesn't.[59]

A variety of environmental factors are known to alter the epigenetic state (methylation or acetylation) of DNA and chromatin, including diet, low-dose radiation, and various drugs and chemicals like cigarette smoke and alcohol. Thus, the epigenome serves as a gateway by which the world around us can change how our genes express themselves. Recent research is showing that early life experiences can also affect the brain's epigenome. And as we'll see in the next section, that may hold a key to how nurture shapes the workings of the mind.

"I'M BECOMING MY MOTHER"

MICHAEL MEANEY AND HIS COLLEAGUES AT McGILL UNIVERSITY pioneered this research by studying the effect of maternal care on rat mothers and their pups. Some rat moms are very nurturing: they lick and groom their newborn pups a lot and they arch their backs and crouch over the pups when they nurse, making it easier for their babies to feed.[60] Other rat moms are more cold and distant: they don't lick and groom much or make it easy for the pups to nurse. Meaney and colleagues found that this difference in maternal care had a profound and lasting effect. What's more, the dye is cast within the first week of life—a critical period that corresponds roughly to human infancy. Nurturing mothers set the development of their pups' brains and stress hormone systems on a lifelong path that helps them cope well with stress. However, offspring of mothers who are absent or less nurturing grow up with hyperreactive stress systems and a lifelong tendency to be fearful.

All of this is due to how the offspring are raised and not the genes they inherit: when rat pups of low-licking-grooming (low-LG) mothers are taken at birth and raised by high-LG moms, their behavior and biology matches that of the biological offspring of high-LG moms (and vice versa).

Meaney's group discovered that maternal care programs the infant's stress systems by changing the chemistry of its chromosomes. Recall that methylated DNA is closed off to transcription factors that regulate gene expression. On the first day of life, part of a key stress response gene in the newborn's brain, the glucocorticoid receptor (Nr3c1) gene, is locked up by methyl groups. This gene, which makes the receptor for the stress hormone cortisol, will go on to play a crucial role in the development and regulation of the stress response system by determining how quickly and effectively the brain copes with adversity.

By the end of that first week, the Nr3c1 gene will either stay locked or it will be unlocked, and the key is a mother's touch (or, more specifically, how much the mother licks and grooms her pups and how she nurses them). Meaney's group found that animals born to nurturing mothers produce higher levels of an enzyme that unlocks (demethylates) the gene, setting the course for the development of healthy stress responses. But those born to distant mothers end the week with the lock intact, beginning their lives with a brain less equipped to manage stress and set on a lifelong course of fearfulness and hormonal dysregulation.

This mothering effect reaches across the generations: by dampening the expression of stress response genes in their daughters, less-nurturing mothers produce offspring who have emotional and behavioral problems. As a result, the daughters themselves become less-nurturing mothers and go on to raise fearful offspring who go on to be less-nurturing mothers and the cycle continues. These behavior patterns are transmitted without any change in DNA sequence. Going one step beyond Freud, the implication is that a mother's behavior can influence the emotional development not only of her child but her grandchildren.

There is emerging evidence that early maternal care can shape the developing brain of human infants as well. In one study,[61] a team from the University of British Columbia found evidence at a molecular level that the same kind of epigenetic effects found in rat studies of maternal care can be seen in human infants. They compared infants born to depressed mothers to infants whose mothers weren't depressed. Babies whose mothers were depressed during the third trimester of pregnancy had increased DNA methylation of the human *NR3C1* gene at birth, the same gene that was methylated in the offspring of low-LG rat mothers. What's more, these babies had exaggerated stress hormone responses when they were tested at three months old.

The long arm of epigenetics was strikingly demonstrated in another study from Meaney's group.[62] Child maltreatment is a potent risk factor for depression and suicide. It's also known that people with severe depression, like the offspring of low-licking/grooming rats, tend to have high levels of the stress hormone cortisol and lower than normal brain expression of the glucocorticoid receptor (*NR3C1*) gene. Could the same epigenetic switch that derails the rat stress hormone system be the link between child abuse and suicide?

To answer this question, Meaney and his colleagues[62] looked at the human version of the glucocorticoid receptor gene (*NR3C1*). They obtained brain tissue from adult suicide victims who either had or had not suffered child abuse as well as from controls who had not died by suicide. When they looked at the *NR3C1* gene, they found no differences between the groups in terms of DNA sequence. But at the level of epigenetics, the results closely matched what they had found in rats: the promoter of the gene was much more highly methylated in those with a history of abuse. And, as in rats, this methylation blocked expression of the gene by making it less responsive to transcription factors that normally activate it. Basically, the gene was switched off.

The story of how early adversity influences the epigenome is more complicated than the *NR3C1* gene story implies. For one thing, it's becoming clear that early stress and deprivation can cause epigenetic changes across a much broader range of brain genes and the effects on behavior and stress responses likely depend on much more than the *NR3C1* gene.[63–65] But the important point is that researchers are beginning to unravel, at a molecular level, how it is that life experiences shape the trajectories of behavior and stress responses. During a sensitive period early in life, subtle and not-so-subtle differences in how parents treat their infants can change the chemistry of the chromosomes in ways that alter how stress response genes are expressed. This sets off a cascade of cellular events that may govern how a child's brain and stress hormone system responds to challenges and threats for the rest of her life. We saw that brain development involves a set of neural commitments—the selection of one path or another—that shape how an animal (or person) approaches life. And here, at a fundamental biochemical level, is one way that the brain makes a "commitment" to a particular life trajectory. Early experience programs the stress response system, predisposing the child to a temperamental or personality style that may last a lifetime.

This is perhaps the clearest demonstration of why the age-old dichotomy between nature and nurture is a false one. It's hard to imagine anything more fundamental to nurture than how parents treat their children. And it's hard to get closer to nature than the molecular biology of gene expression. But now we see that these two pieces of the puzzle are inextricable. Parental care (nurture) can affect child development by regulating gene expression (nature) and altering how the brain and stress response system function. Somehow I suspect even Freud would find this satisfying.

There's an important point to be made about normal here. It's tempting to see this research in simple terms: a nurturing environment promotes normal development whereas adversity and deprivation produce children with abnormal or broken stress response systems. But we have to remember that development is a process of adaptation to the world. What's normal depends on context. The developing brain makes an educated guess about what life will be like based on what it's been like so far: Is the world likely to be nurturing and predictable or threatening and chaotic or somewhere in between? During sensitive periods of development, epigenetic changes and patterns of gene expression start calibrating the brain and the mind to the expectable world. If you're born into a world where your caregivers are stressed, absent, or unpredictable, being vigilant and having hair-trigger stress responses may be the best way to go. In this sense, the fearfulness of Meaney's rats and the anger-sensitivity of Seth Pollak's maltreated children may be "normal" adaptations. But as Deidre Ward found, these adaptations can come at a cost: a predisposition to distress, anxiety, and even depression later in life.

THE PARENT TRAP

THE FINDINGS ON THE BIOLOGICAL IMPACT OF ADVERSITY WOULD seem to encourage the obsessive worry that many parents already have about making sure that every experience in their baby's life is optimal. After her son got into trouble at school, a colleague of mine joked that his behavior problems now made sense: "I didn't lick and groom him enough when he was a baby!"

But her comment was only partly facetious. Over the past decade, everyone from child advocates to marketing executives have used the results of developmental neuroscience to warn parents that their children may be permanently damaged if they don't provide the perfect environment for brain development.

On a spring weekend, my wife and I were surfing the web, looking at strollers for the baby we were expecting in several months. We came upon the website for "Orbit Baby," "the world's first rotating stroller."[66] Most strollers have the baby facing either outward or toward Mom (or Dad). The Orbit Baby allows you to rotate the baby's seat while she is still in the stroller. The product's website claimed that "parent-facing strollers are better for child development" and cited research to argue for the importance of their product. The research in question was a report entitled "What Is Life in a Baby Buggy Like?" and its author, Dr. M. Suzanne Zeedyk of the University of Dundee, noted that no published studies have examined the impact of stroller design on parent-child interactions and infant stress. She explained the motivation for this research by alluding to the science of early emotional development:

> Infants are born with brains that are already tuned into, and dependent upon, social responses from other people. Thus, on every occasion that a baby has a need for a communicative response from his or her parent, but is unable to obtain it, this creates a low-level stress response in the infant. When such instances of stress occur repeatedly

and frequently, they become damaging to infants' neural,
physiological, and psychological development. The pres-
ent research project arises out of recent suggestions that
baby buggies may inadvertently be generating such stress-
ful circumstances for infants. (p. 4)

That sounds alarming—could such "low-level stress" really damage neural development? To support the claim, Zeedyk reports a study in which she had volunteers observe the interaction between adults and their children as they pushed them in strollers of various designs on the streets of fifty-four UK cities. She found that parents and children spoke less when the children were facing away. Children facing their parents were also more likely to be sleeping, which Zeedyk took to be an indication that they were more relaxed and less stressed. The study also found that

there were a small minority of children who were attempting
to get their parents' attention, but failing to do so. These chil-
dren will have even higher stress levels, as they seek out par-
ents either through crying or through turning around, yet fail
to obtain a response. For these children, a buggy ride may
go from being stressful to being traumatic. This is not too
strong a statement, because young children's coping systems
are immature. To be left on their own, coping unassisted
with discomfort for too long, constitutes trauma for a young
child. If parents cannot easily see their infants' faces, they
may not realize in just how much distress their children are.

To examine the question in more depth, Zeedyk's team conducted a second small study in which she assigned twenty mothers to walk with their infants (age nine to twenty-four months) in either toward-facing or away-facing strollers.

Zeedyk reported that the toward-facing strollers won out: infant heart rates were lower (perhaps because they were less stressed), mothers and infants laughed and talked more, and mothers preferred the experience. While noting that definitive claims would be premature, she concluded that "infant development may be negatively affected by buggy design" and that life in a baby buggy "may be more emotionally impoverished than is good for children's development . . . If there is even the possibility that baby buggy design is aggravating children's stress levels, then this is a cause for concern" (pp. 26, 27).

We may never know how many children have been damaged by buggy-induced stress, but the scourge of strollers seems more like a bugaboo. There's a larger point here, and one that should reassure parents who fear that they must cleanse their children's lives of any distress. Research has certainly established that major adversity early in life—trauma, abuse, neglect, and the major deprivations that come with poverty—can have enduring and problematic effects on brain development. But those insults are a far cry from the minor vicissitudes that are inevitable in any child's life.

The notion of a distress-free development is not only an unattainable ideal, there's reason to believe it's not a worthy one. While extreme or prolonged adversity is clearly not good for development, a considerable body of research suggests that moderate stress can promote resilience and that being sheltered from all adversity may not be such a good thing. Psychologist Mark Seery and his colleagues suggest that "without adversity, individuals are not challenged to manage stress, so that the toughness and mastery they might otherwise generate remains undeveloped" (p. 1096).[67] Friedrich Nietzsche may have been right when he said, "That which does not kill me makes me stronger."

"NEVER SAY THIS IS YOUR LAST JOURNEY"

THERE ARE STILL MANY MYSTERIES THAT NEUROSCIENCE AND PSYchology have not fully explained. Why are some people sensitized by adversity while others are immunized?

Many studies of abused and neglected children have documented how adversity causes their brains to become sensitized to fear and anger. Where others see a challenge with adversity, they see a threat. But there are others for whom hardship doesn't just make them sensitized to fear and anger. Surviving adversity has given them perspective and fostered a kind of resilience that allows them to not sweat the small stuff.

In 1943 Mike Bornstein was three years old, living with his parents, older brother, and grandmother in the town of Zarki, Poland. One day, his family was rounded up and sent to a labor camp, where his older family members were forced to work in a munitions factory. After several months, they were shipped in cattle cars across the Polish countryside to a cold and frightening place whose name has become a symbol of brutal inhumanity and mass murder: Auschwitz. His father and brother were murdered shortly after they arrived. Mike, his mother, and his grandmother passed day after day in a state of hopelessness, surrounded by filth and starvation. They lived in barracks that were converted stables, with nothing in them but rows of cramped wooden bunks, stacked like shelves—each three-meter bunk packed with four people. Yards away from their cells, continuous plumes of smoke rose from the crematoria where the bodies of thousands of men, women, and children were incinerated.

During Mike's stay at Auschwitz/Birkenau, more than ten thousand were killed each and every day. The moans of the sick and dying were unrelenting. The latrine, an open bunker with a row of holes in a bench, was thick with the stench of urine, feces, and the diarrhea that came with epidemic typhus. The inmates were lucky to receive 100 grams of bread per day, and Mike became emaciated from starvation. Whenever she could, his mother would find him and give him some of her bread, but many times she was discovered by the guards and beaten for doing it. One day she disappeared. He later learned she had been shipped to a work camp in Austria.

As she watched men, women, and children being slaughtered

around her, Mike's grandmother feared he would be next. One day, she sneaked Mike into her bunk and hid him in a mattress, where he stayed hidden and survived on the rations she shared with him.

When the camp was finally liberated in 1945, Mike and his grandmother walked out, barely alive, and made their way to Czestochowa, a town near Zarki. Of the 230,000 children deported to Auschwitz, Mike was one of only 700 who were liberated. His grandmother, with little education, struggled to find work so they would have food to eat. Having nowhere for him to stay during the daytime, she would leave Mike in a chicken coop by himself. After several months Mike's mother, Sophie, who had been liberated from the Austrian work camp, made her way back to Poland and searched for him. She finally arrived in Czestochowa, and they were reunited. But the horror was far from over.

"I was very sick, and my mother took me to Germany, where there was better medical help." Mike told me. "She didn't have any vocation, so we lived in Munich in one room that my mother rented." They had no kitchen privileges. They lived for six years in that one room, wary of the landlady, who wore a swastika charm around her neck. His mother made a little bit of money teaching Hebrew, but it wasn't enough, so she smuggled food and sold it on the black market in Munich. "She would buy flour and chocolate and nylons from American soldiers and sell it on the black market. It was a very scary time for me." Sometimes after school Mike helped his mother smuggle food. He lived in constant fear that they would be arrested.

Life in Munich was lonely and frightening for Mike, as it was for many of the survivors who had relocated there. Several of his friends at school committed suicide, jumping to their deaths. He felt like an eternal outsider. Perhaps frightened by the emaciated state she had found him in, his mother overfed him, and he was now severely obese. "I looked different from other kids; I acted different. It was easy for them to make fun of me." During the day, he would hitch a ride between home and school. One day, a man picked him up in truck and tried to sexually molest him, but he escaped.

When Mike was eleven, he and his mother applied to emigrate to the United States. They arrived penniless and spoke no English. Through charitable services, they were put up in temporary housing in the Bowery, in Lower Manhattan. They eventually found a one-room apartment, and Mike took odd jobs to help support himself and his mother, who made $30 a week as a seamstress. At some point, Mike took a job in a pharmacy on the Upper East Side, delivering medicines, sweeping the floor, and doing whatever was asked of him. The pharmacist, a severe taskmaster, berated him for any mistakes he made but also trained him to be meticulous about his work.

The pharmacy was an exciting place for Mike, and his experience there would prove pivotal to his future in the United States. He learned English and managed to do well in high school, thanks in large part to his mother's dedication. "My mother would sit up with me late into the night—we only had one room—but she'd make sure I did my homework. She didn't know how to check it, but at least she'd make sure I'd stay up till ten o'clock after getting home from the pharmacy. She basically did everything in my interest. She didn't have much, but whatever she had, she gave to me."

When he was eighteen, Mike was accepted to Fordham University. He studied pharmacy there and then received a PhD from the University of Iowa in pharmaceutics and analytical chemistry. One day in Iowa, while at the local synagogue, he ran into a girl named Judy, whom he knew through a mutual friend. Within two years, they married. Mike had a number of jobs working as a scientist for large chemical and pharmaceutical companies in the Midwest before becoming an executive at Johnson and Johnson in New Jersey. Mike and Judy had four children—all of whom became successful professionals—and nine grandchildren. If you were to look at them, you would see the prototypical Midwestern family: close-knit, happy, and successful.

That's not to say that Mike is in denial. He is able to talk about his experiences, has shared them with his children and even lectured

about them in the Indianapolis schools. And there is one constant
reminder of them that he sees every day: a number engraved on his
forearm.

One day in 1981 Mike and Judy went to the movies to see *The
Chosen*, a film adapted from the book by Chaim Potok about two
boys, one Hasidic, the other a Reform Jew, who forge a friendship
in 1940s Brooklyn. In one scene, a newsreel of the liberation of
Auschwitz is shown—there are heaps of dead bodies and piles of
shoes and eyeglasses. Finally, the camera shows a group of children
crowding together with blank faces. One little boy rolls up his sleeve
to bear the number on his forearm. Sitting in the theater, Mike was
startled—the number was his. The child was Mike.

Mike and Judy Bornstein with one of their daughters. The background photo, taken in
1945 shows Mike (*right*), age five, at the moment of his liberation from Auschwitz.

I asked Mike whether he thought of himself as resilient. He
thought for a moment, as though he'd never considered the question, and said that he supposed he was. He wasn't sure how he had
managed not only to survive but to thrive after such a traumatic
childhood. "There are two things that I keep in mind when things

don't go the way I want them to go. One of them is a watch that my mother gave me when I was eighteen. She brought a couple of things from Germany and one was an 18-karat gold Schaffhausen watch that she gave me when I turned eighteen. On the back of the watch, she had inscribed the Hebrew letters *gimmel* and *zayin* which stands for *gam zeh ya'avor* meaning 'this too shall pass,' and I try to remember that if things go real bad. And the other thing—if things go really bad—I like to sing a song. In Yiddish it's called 'Zog nit keyn mol az du geyst dem letzten veg,' which means 'never say this is your last journey.' And I like to sing that and it helps me overcome things that look pretty bleak. I just let things fly off me and start anew."

RISING TO THE CHALLENGE

HOW IS IT THAT A CHILD WHO SPENT HIS EARLY YEARS IN A NAZI death camp and then endured poverty and social isolation ends up happy and well-adjusted?

The question of why some people are particularly resilient in the face of stress is just as important as understanding why some people are particularly vulnerable. We need to know what genetic and experiential factors are protective rather than simply identifying risk factors. And yet we know much less about resilience than about vulnerability.

One clear resilience factor is the support and nurturing we get from other people. In the rodent studies of Michael Meaney and others and similar studies in monkeys,[68] the buffering effects of maternal care are clear. Maternal nurturing seems to help program stress response systems that are flexible and efficient—turning on when they're needed, and, importantly, turning off when they're not. And in humans, close relationships with parents and social support can buffer the effects of adversity even among children who have genetic risk factors for depression.[69, 70] Secure attachments to our caregivers can sustain us when the going gets tough, and have lasting effects on our development (about which I'll say more in Chapter 5). One of the ingredients in Mike Bornstein's story of resilience was the bond he had with his mother that had been developing even before he arrived at Auschwitz as a three-year-old—a mother who endured beatings to bring her son a crust of bread.

With the tools of molecular biology, genetics, and neuroimaging, researchers are just beginning to dissect the biological origins of resilience.[71] And, again, the evidence points to remarkably subtle effects on how the brain responds to experience.

At one level, resilience may be related to the brain's ability to renew itself. Around the time you're born, the process of generating neurons (called *neurogenesis*) that build the brain is largely over. While synapses between neurons continue to be remodeled throughout life, the neurons themselves never regenerate. Or so scientists thought until recently. It's now known that neurogenesis continues throughout adult life from neural stem cells in just two locations. The first is in the walls of the brain's lateral ventricles— part of the system through which cerebrospinal fluid flows between the brain and spinal cord. New neurons born here migrate to the olfactory bulb, where our sense of smell is processed. The second is located in a part of the hippocampus called the *dentate gyrus*. The hippocampus is well known to be crucial for learning, memory, and regulating stress responses, and neurogenesis here is part of the brain's response to new experiences.

Shortly after these new neurons are born, they are especially plastic—that is, responsive to stimulation and able to form new synapses with other neurons.[72] As they integrate into brain circuits in the hippocampus, that extra plasticity may help them build connections that allow us to adapt to new and stressful situations. Animal studies have found that neurogenesis is crucial for the normal ability of the hippocampus to buffer the effects of stress by keeping stress hormone levels from going out of control. When neurogenesis is blocked, levels of the stress hormone cortisol stay abnormally high and animals exhibit behavioral signs of depression.[73] In other words, resilience may depend in part on the brain's ability to generate new neurons in the hippocampus.

At the same time, animal studies have shown that stress and early life adversity themselves inhibit neurogenesis. That means that stress itself can overwhelm the brain's own resilience and coping mechanisms, taking a bad situation and making it worse. But when these normal coping mechanisms fail, there may be ways to restore them. For example, SSRI antidepressants like fluoxetine (Prozac) work in part by stimulating neurogenesis in the hippocampus.[74] And even physical exercise, which also has antidepressant effects, has been shown to promote neurogenesis in animal studies.[75]

Neuroscientist Eric Nestler and his colleagues have discovered other key elements of resilience pathways using mice to create a model of chronic stress. In their "social defeat" model, mice are exposed repeatedly to the stress of an aggressive intruder mouse. Most mice end up falling into the mouse equivalent of despair—avoiding social contact, losing weight, losing interest in reinforcing stimuli, and showing more anxiety-related behavior. But some mice seem to be immune to the stress. Through a series of detailed experiments, Nestler's team was able to identify molecular signatures of resilience.[76-78]

Nestler and his team found that, in the face of stress, resilient mice were able to turn on genes in a key reward circuit, blocking a cascade of chemical events that produced anxiety- and depression-like behaviors in vulnerable mice. One of the genes encodes a transcription factor known as ΔFosB that, in turn, sets in motion synaptic changes that appear to protect brain circuits from encoding adversity. It turns out that people with depression have lower levels of ΔFosB in these reward centers, further suggesting that these chemical cascades play a key role in shifting the brain between vulnerability and resilience. Intriguingly, treatment with the antidepressant fluoxetine (Prozac) was able to turn vulnerable mice into resilient mice by switching on ΔFosB.

These and other studies have helped us develop an understanding of how genetic and epigenetic variations can have either "buffeting" or "buffering" effects on how we cope with adversity. The accumulation and interaction of these effects calibrate set points for brain circuits that influence how we appraise the challenges we face.[71] For those fortunate enough to have more buffering than buffeting, the world becomes less threatening and more manageable.

Each of us lives out a unique configuration of human possibility. And it's true that some important influences on our individual trajectories can be subtle, especially early in life. They can have a "butterfly effect" in which small perturbations can have cascading effects that amplify over time. How much any given perturbation or experience will matter for an individual can be hard to predict. Clearly, beginning life in a Romanian orphanage is not the same thing as riding around in a suboptimal stroller. Regardless, Mike Bornstein's story underscores the broad sweep of normal, encompassing the vast scope of human vulnerability and resilience. One of the useful implications of exploring the biology of normal is that it broadens our perspective beyond a focus on pathology, even if, as it stands, science has yet to fully account for the remarkable resilience of the human spirit.

AN EMBARRASSMENT OF RICHES

LET'S RETURN TO WHERE WE STARTED. WE'VE SEEN THAT EARLY deprivation and adversity can have lasting negative effects on brain development. But what about children who aren't deprived or maltreated? Could we boost their cognitive and emotional development and even give them an edge by providing their brains with the right kind of stimulation and experiences? This is the logic that fueled the appeal of Baby Einstein and Governor Miller's push to send newborns home with classical music CDs. But there's a problem with this logic. The research on sensitive periods suggests that more of a good thing won't necessarily get you a better outcome. Let me explain with an analogy.

In the 1990s Americans were introduced to a new kind of bar scene that had its roots in Japan. Instead of serving alcohol, these bars offered something for the health-conscious set. They served oxygen. Patrons who bellied up to the bar were given oxygen through a nasal cannula (prongs that fit in the nose) for which they paid about $1 per minute. Proponents claimed that inhaling extra oxygen has a wide range of benefits, from detoxifying the body and reducing stress to boosting the immune system and improving mental abilities. In 1997 actor Woody Harrelson and a business partner opened "O2" on Sunset Boulevard, charging patrons $13 for twenty minutes of oxygenated air. For an extra couple of bucks, you could get your oxygen spiked with aromas like the rose-scented "Joy" or the eucalyptus inspired "Clarity."[79] By the decade's end, the oxygen bar trend was in full swing, with outlets spanning the country.

As the *New York Times* reported, oxygen bars traded on a simple idea: "if oxygen is good for life, more oxygen must be better."[78] After all, we know that when people are deprived of oxygen, their bodies and minds suffer and, if the deprivation is severe enough, they die. If you are deprived of oxygen or have a lung disease or heart disease that interferes with your body's ability to get oxygen, extra oxygen could make a big difference.

But any physician could tell you that as long as you don't have respiratory problems, you get plenty of oxygen from simply breathing the air around you. Oxygen is carried throughout the body by hemoglobin, and hemoglobin has a certain capacity or limit to the amount of oxygen it can carry. Under normal circumstances, that capacity is nearly saturated. If I measured the oxygen-saturation level of your blood right now, it would likely be somewhere in the range of 97 to 99 percent. And that's all you need to deliver adequate oxygen to your brain and other tissues. Inhaling extra oxygen

might push you from 97 percent to 100 percent, but that difference wouldn't matter. The human body has evolved to be efficient
at extracting oxygen from ambient air and giving it supernormal
shots of oxygen won't make your body supernormal. In fact, too
much oxygen can be harmful. The fallacious "more is better" premise of the oxygen bar resembles the claims made by those promoting supersized stimulation for young children. Some advocates of
early enrichment claim that providing extra cognitive, emotional,
or social stimulation during sensitive periods can give children the
edge they need to outpace their peers later in life.

But remember, the sensitive periods for cognitive and social
development occur because children are passing through an
experience-expectant phase of development. Evolution has prepared their brains to be open to an expectable environment. If the
basic elements of that expectable environment are present, the brain
gets what it needs. There is little evidence that going beyond the
expectable environment (and exactly what that would entail is not
clear) will make the brain excel. The Romanian orphanage studies
have shown that social deprivation, like oxygen deprivation, can be
harmful and that providing a normal social environment can result
in dramatic benefits. But that doesn't necessarily mean that starting with a normal environment and trying to make it supernormal
would have any detectable effect. The impact of enrichment depends
on where you're starting from.

Advocates of cognitive enrichment for everyone—including
marketers trying to sell baby products designed to make your baby
smarter or happier—often point to animal studies that have suggested that increasing the complexity of the environment can promote brain connections and enhance cognitive performance. I'm
not disputing that, but let's dig a little deeper than the headlines.
Part of the anxiety parents have about optimizing their children's
first few years of life comes from a misunderstanding about brain
development. As John T. Bruer describes in his book *The Myth of*

the First Three Years,[81] journalists, policymakers, and activists have abetted these problems by reading too much into the science.

First, there is the misconception that children have a fixed window of opportunity (the first three to five years of life) during which they must develop key cognitive and social skills. In a 1997 address to the National Governor's Association, Rob Reiner, the actor/director who became a child development activist, claimed that "By age ten, your brain is cooked."[82] But that isn't true. We know that brain development is not restricted to an early critical period. Through the mechanisms of experience-dependent learning, we continue to learn and adapt to our environment throughout our lives.

It's true that many studies of rats have reported that environmental enrichment boosts their cognitive skills and that this is accompanied by synaptic plasticity and changes in the wiring of their brains.[83] It may be true that these rats' experiences have been enriched, but the question is, enriched compared to what?

Typically, enrichment in these studies refers to adding complexity to a rat's environment beyond what they get from standard caged housing. Even William Greenough and his colleagues, who reported some of the influential studies of this phenomenon, acknowledged that "these conditions represent an incomplete attempt to mimic some aspects of the wild environment and should be considered 'enriched' only in comparison to the humdrum life of the typical laboratory animal" (p. 546).[32] In this sense, going from a standard cage to an enriched one might be more analogous to taking an institutionalized infant and placing her in foster care. That is, it's really more like going from a deprived environment to a normal, expectable environment. We already know that alleviating deprivation is good for the brain. But that's a far cry from saying we can make brain development better than normal by manipulating the environment. So the lesson isn't that intervening during sensitive periods can't enhance brain function. But, again, it matters where you're starting from. There is clear evidence that early educational interventions can have lasting cognitive and behavioral benefits for economically disadvantaged children.[84] Enrichment can certainly be powerful when the environment is not good enough; but trying to go beyond that may have diminishing returns.

On the other hand, for children who are disadvantaged or raised in highly stressful environments, the biology of sensitive periods offers an important policy insight for optimizing intervention programs. Educational and social programs that aim to give these children a head start should be informed by what we are learning about the timing of brain development. Rather than simply enriching the environment as early as possible, these programs could tailor interventions to specific sensitive periods when they are likely to have the most potent effects. Key mental functions—language, attachment, social cognition, executive functions, and so on—have their own developmental periods during which the brain is exquisitely sensitive to the environment. It stands to reason that efforts to foster and protect these functions will have the biggest bang for the buck if we work with the brain's own timetable of plasticity.

A GPS IN THE BRAIN

AND, FINALLY, IT'S IMPORTANT TO REMEMBER THAT PLASTICITY doesn't end when sensitive periods pass—the brain is not "cooked" by the time we bid childhood good-bye. In fact, any episode of learning from experience involves changes in the brain. That's what the idea of experience-dependent plasticity is all about. Even if our past experiences deprived us of a "good enough" environment, we can change and learn new ways of coping. In essence, that's the premise and promise of psychotherapy.

Some of the most dramatic demonstrations of how learning can change the brain involve the development of specific skills and expertise. Take the case of spatial memory—that is, our ability to remember where things are. Since the mid-nineteenth century, a select group of people in London has acquired what may be the greatest talent for spatial memory ever achieved. These people are not supergeeks with outsized IQs. They are cabbies. And the body of knowledge they have mastered is known, appropriately, as "The Knowledge."

If you want to get a license to operate a black cab in London, you have to memorize the layout of twenty-five thousand city streets and thousands of destinations.[85] In other words, aspiring cabbies have to develop a GPS system in their brains for every route in the London area. The daunting process of acquiring "The Knowledge" typically takes two to four years of training. Those who get through it and pass written and oral exams are granted a license by the Public Carriage Office. By that point, the cabbies have an encyclopedic knowledge of London's streets. But they also have something even more remarkable: remodeled brains. Brain-imaging studies have shown that cabbies with the Knowledge have thicker gray matter in the posterior hippocampus, a brain region involved in processing spatial memory, compared to controls. The longer they had been cabbies, the thicker their gray matter was.

They have accommodated the demands of storing the details of London's roadways by devoting more brain territory to it. That visible change in brain structure seems to be a result of the vast body of spatial information they have to acquire. In contrast, bus drivers—who have to memorize a few specific routes—don't show such changes, nor do people who memorize nonspatial information like doctors or even contestants in the World Memory Championships.[85]

Similar findings have been reported for a wide range of skills and talents. Structural and functional changes are visible in brain scans of ballet dancers, golfers, basketball players, and people who learn new languages or musical instruments.[86-90]

WRITTEN IN PLASTIC NOT STONE

IN THIS CHAPTER, WE'VE SEEN THE POWER OF EXPERIENCE TO shape the function of the brain and the architecture of the mind. Ultimately, our sensitivity to experience is an evolutionary solution to two daunting problems: a finite genome and a changing world. Our genetic endowment provides a general plan for wiring brain circuits. But the genome can't prepare us for the full range of contingencies we might encounter. If the wiring of the nervous system were fixed by a genetic program, we'd be up a creek—totally unprepared for the challenges the environment can throw at us. And so natural selection has solved this dilemma in an ingenious way. Instead of trying to cover all the possible "if-then" instructions we might need, the genome encodes the ingredients for plasticity.

Plasticity has a sequence of its own. For certain key capacities—learning to see, learning a language, forming emotional attachments, understanding social cues—the brain uses the environment to wire itself. Because these functions are so important, little can be left to chance. And so, early in life, we have sensitive periods when the brain is exquisitely attuned to the environmental inputs it needs to set up these capacities. Like the most sophisticated computer imaginable, the brain mines the data that it receives to build representations of the world around it. It's a remarkably efficient plan for getting a lot done quickly while adjusting to the facts on the ground. Of course, it also creates windows of vulnerability.

If the inputs are corrupted or the wiring goes awry, trajectories can be distorted in ways that have cascading effects. Caught early on, while the brain is still capable of large-scale adaptation, these distortions may be remediable—as the Romanian orphan studies have shown.

But ongoing experience-dependent plasticity allows us to fine-tune the wiring of circuits as we adapt to the circumstances of our lives. And so, as I suggested at the beginning of this chapter, both

Freud and Watson had a point. As Freud would have it, infancy and childhood are privileged periods. The existence of sensitive periods means that fundamental aspects of how our lives turn out depends on what the world was like as we passed through these windows. But as Watson emphasized, we can learn and change in important ways well beyond those early years.

In this chapter and the last, I've introduced the major players that drive the biology of normal: natural selection, genetic variation, and experience. In the next several chapters, we'll explore what happens as we bring these influences to bear on some of the universal challenges we face, how they shape the trajectory of our lives, and what happens when that trajectory goes awry.

CHAPTER FOUR

DOGS, POKER, AND AUTISM: THE BIOLOGY OF MIND READING

O N AUGUST 11, 2006, A *NEW YORK TIMES* ARTICLE about the finals of the World Series of Poker began:

> *When Jamie Gold bluffed, his opponents folded.*
> *When he had the best hand, they threw in all their chips.*
> *With a run of cards, a huge chip stack, and an uncanny*
> *knack for reading other players, Gold, a talkative former*
> *Hollywood talent agent, cajoled his way to victory Friday*
> *at the World Series of Poker for the $12 million grand prize.*

Between 1903 and 1910, Cassius Marcellus Coolidge painted a series of pictures that are still among the most widely reproduced and copied American oil paintings. The image of dogs playing poker may be a symbol of kitsch for most, but it could also serve as an emblem for the fascinating science of social cognition. It turns out that dogs and champion poker players have something in common: they're both skilled at reading people's minds. Far from being a dubious power of those who claim to have ESP, mind reading—that is, deciphering the thoughts and feelings in other people's minds—is in fact a universal, even essential, skill of the human brain. Think about it—how could you function if you were unaware that your spouse had her own thoughts and feelings? (When I posed that question to my wife, she said, "You seem to function just fine.")

At some point in your childhood, you came to understand that others have their own thoughts, intentions, and beliefs. That understanding, which psychologists refer to as a *theory of mind*, is essential to almost every social encounter we have from early childhood onward—from appeasing a bully on the playground, to getting a date in high school, to negotiating with the boss at work.

As we'll see, the human brain has been shaped by natural selection to be able to monitor and respond to the activity of other human brains. Evolution has given us a social sense, specialized for navigating the world of human interaction. Our theory of mind and capacity for empathy are mental tools that allow us to compete and cooperate in a social world. They are so essential to how we operate that they have to be effortless; they come standard as part of the brain's basic package. This chapter is about how this social sense develops normally and what happens when it doesn't.

"FACE" FACTS

OUR JOURNEY INTO OTHER PEOPLE'S HEADS BEGINS WITH THEIR faces. Facial expressions are the outer windows into other people's minds. We are able to form emotional first impressions based on viewing a face for only thirty-nine thousandths of a second.[1] But faces also give us vital information about the social environment at any moment. The ability to recognize faces quickly allows us to instantly judge whether someone is kin, friend, or stranger. We look at their eyes to figure out what they are attending to: Is she watching me? Is he looking at something that I need to know about—an approaching threat? A source of food? Not surprisingly, we are expert at recognizing and decoding faces. We have to be. But how do we get there?

One view is that we're particularly adept at recognizing and reading faces because we have to do it all the time. In other words, the brain has the capacity to process all kinds of things, but it *acquires* face expertise because it's called on to process faces every day. If we lived in a world where we had to recognize luggage every day, we'd become equally expert at that (if you think back to the last time you tried retrieving your black suitcase at an airport baggage claim, you'll realize we lack this skill). The other view is that our expertise for faces is an innate skill that develops early in life.

Either way, a large body of research now shows that our brains have a biological system for processing faces. For one thing, brain abnormalities can selectively knock out face-recognition skills. People who have this condition, known as prosopagnosia, have problems recognizing faces even though they can recognize other objects.[2] The problem can be acquired—for example as a result of brain injury or stroke—but the most common form, developmental prosopagnosia, is present from birth. People with developmental prosopagnosia have no apparent brain damage—they grow up with this face-blindness and may not even realize they have it until their deficit collides with social norms. Bradley Duchaine, an expert in developmental prosopagnosia, has heard plenty of these stories— many of them offered by visitors to his website. As one woman put it, "This week I went to the wrong baby at my son's daycare and only realized he was not my son when the daycare staff looked at me in horrified disbelief"(p. 166).[3]

After Duchaine and his colleagues' work received media coverage, a strange thing happened. They began to hear from people who said they had the opposite of prosopagnosia. Instead of not being able to recognize faces, these people claimed to have supernormal powers of face recognition. Intrigued, Duchaine, along with Harvard colleagues Richard Russell and Ken Nakayama, decided to test the supernatural claims. They brought four of these people into the lab to test their face-recognition skills. The subjects told stories of how their superskills were a decidedly mixed blessing. As one said:

"I've learned to stop surprising people with bizarre comments like, 'Hey, weren't you at that so-and-so concert last fall . . . I recognize you.' Before that, I'd occasionally make people uncomfortable with my recognition." Another said, "I do have to pretend that I don't remember [people], however, because it seems like I stalk them, or that they mean more to me than they do when I recall that we saw each other once walking on campus four years ago in front of the quad!"[4]

To see if these people were really better than normal at recognizing faces, the researchers had to develop special tests. One of the tests was straight out of the pages of *People* magazine. They showed subjects pictures of famous people "before they were famous" and asked them to identify each celebrity. Some of the pictures were photos from childhood and were cropped to make them extra hard to identify (the figure below shows four examples from the test set—see if you can identify them).

Examples from the "Before They Were Famous" test. See end of chapter for answers.*

The tests showed that the subjects were extraordinarily good at face recognition. The researchers dubbed them "super-recognizers." They far outperformed normal control subjects. And, in fact, they seemed to be as far from normal on the superior side as prosopagnosics were on the impaired side.

The existence of prosopagnosics and super-recognizers may be more than just a biological curio. Perhaps these individuals define the extremes of a basic mental function that we all use to establish social connections. In fact, there does seem to be a spectrum of face-recognition ability—some of us are better at it than others, and a study of twins found that where people lie on this spectrum is almost entirely due to genetic differences.[5]

In the late 1990s MIT neuroscientist Nancy Kanwisher began using fMRI to search for the brain's face recognition center. She found that an area known as the fusiform gyrus in the temporal lobe responded specifically to pictures of faces.[6] This fusiform face area (FFA) is the hub of a cortical brain network that activates when we look at faces.[7] These areas communicate with subcortical regions, including the amygdala, that provide a fast read of the structure and emotional salience of faces.

As it turns out, learning how to process faces seems to involve both innate, face-specific brain mechanisms and learned expertise—both nature and nurture. In children, the development of what has been called "the social brain" follows a path from simple attention to faces and emotion perception to sophisticated mind reading and empathy within a span of just a few years.

Within minutes of being born, neonates are drawn to face patterns. We're born with the basic circuitry to process social information, but experience tunes it to the social world around us. In fact, the face processing network continues to develop after early childhood and doesn't become fully specialized for reading faces until we're about ten years old.[8]

As we encounter the social world, our neural networks sharpen their responses and become highly efficient and specialized. In essence, the social environment trains the brain using an innate network that is loosely wired from birth. As we saw in Chapter 3, for key functions of the mind and brain like reading faces or learning a language, we begin with a slate that is not blank but broadly tuned—a brain that is biased, as a result of natural selection, to attend to certain expectable cues (like faces or speech) from the environment. During experience-expectant phases, these environmental cues help strengthen some synaptic connections and let others get pruned away. Experience guides brain circuits to make commitments by focusing on some information at the expense of others.

Around six months of age, with some face time under their belts, babies typically begin to show several milestones of social cognition. They are able to recognize the face they've had the most experience with—Mom's—and they begin to distinguish positive and negative emotional expressions on other people's faces.[9] They also develop an ability to read eye gaze—that is, to follow the gaze of an adult who has made eye contact with them.[10]

Soon after gaining these rudimentary abilities, infants begin to acquire a set of cognitive skills that are characteristically (and, perhaps, uniquely) human. One of these is the capacity for joint attention—a mental breakthrough that transports the child beyond the world of just "you" and "me" to the realm of "you," "me," and "that." "That" is something we are both looking at or paying attention to. By twelve to fifteen months, typically developing infants understand that adults are not only looking or pointing somewhere, but directing their attention at something interesting.[11] Not coincidentally, it's at this age that infants around the world acquire one of their own tools for joint attention—they begin to point.[12] Joint, or shared, attention is a deceptively simple concept that represents some pretty sophisticated mental abilities. It implies that you and I are distinct beings and that there is a world outside us. Joint attention also requires tracking your attention relative to mine, recognizing the significance of your eye movements or pointing, and coordinating our attention to focus on something else.[13] It also precedes and perhaps enables a whole suite of activities that only humans do. Without an ability to share attention and information, human societies as we know them would never have happened.

There are a few features of human life that are qualitatively different from the rest of the animal kingdom, and the creation and persistence of human cultures is perhaps the most dramatic and far-reaching of these. Other groups of mammals develop sporadic local traditions and wild chimpanzees even exhibit complex customs that spread throughout a community—particular styles of tool use, foraging strategies, and even social customs like grooming behaviors. Some even call these traditions "cultures."[14, 15]

But the richness and variety of human activities, the breadth of their reach throughout human populations, and their propagation across the centuries are unparalleled. Human societies have a vast array of social and behavioral customs and values—eating habits, bathroom habits, food preferences, religious beliefs, aesthetic ideals, dating and mating customs, child-rearing practices, and moral proscriptions. Some of these cut across geography (the use of utensils for eating or the use of symbols for communication) and others may be unique to a particular society (how to hold a fork "properly" or how to address an elder). Our success or failure as members of a culture and even our survival may depend on mastering these things.

But life is short, and it would be simply impossible to assimilate everything we need to by trial and error or even simple observation and imitation. How do we do it? How does a child climb this impossibly steep learning curve and still have time to sleep? The answer, in large part, is that our brains are structured to make use of a uniquely human short cut: pedagogy.[16]

SEE ONE, DO ONE, TEACH ONE

ALL MAMMALS LEARN, BUT ONLY HUMANS TEACH. ONE CHIMP CAN learn to use a stick to pull ants out of an anthill by imitating another chimp. But chimps don't have a way of communicating, "Here's what you need to know . . ." or "Show me that again" or even "Watch this!" And it's not just because they lack the words.

Only humans seem to have the conscious and deliberate motivation to share information.[11, 12, 17] And a fundamental building block for this uniquely human brain adaptation is joint attention—it allows you and me to exchange information about the rest of the world. Coupled with another uniquely human capacity—spoken language, which, like joint attention, begins to develop at twelve to fifteen months—we can share our knowledge, transmit it across generations, and build the complex structures of human culture. Babies expect this kind of information because their brains are tuned to recognize teachable moments.

Sharing attention is also the germ for sharing intention, the basis of cooperation and collaboration. Humans share goals and make plans in ways that even our closest primate cousins—chimpanzees and bonobos—don't seem to. Despite the fact that their genomes are more than 99 percent the same as ours, chimps lack the key mental capacities that we use effortlessly to collaborate: they don't speak, point, or even smile—behaviors that human infants universally exhibit by about fourteen months of age.

So, from very early on, our brains are wired to process faces, to engage with others, and to communicate with one another about the world. But developing a theory of mind—an understanding of the mental states of others—requires something more than that. We need to think about thinking.

TALK TO THE BANANA

IMAGINE THE FOLLOWING SCENE: YOU'RE SITTING IN A RESTAURANT, about to have brunch with your mother and one of her friends. Things are going well until her friend picks up a banana and presses it against the side of her head. Next thing you know, she's staring at you with an exaggerated grin and talking into one end of the banana. Instead of being alarmed, your mother picks up another banana, puts it to her ear, and seems to be getting messages from the banana. Then she presses it to your ear and says, "It's for you."

What the hell is going on? Are these people psychotic? As a psychiatrist, if I were passing by the table and witnessed this, the thought might cross my mind. Except for one thing: Did I mention that you are two years old? That little detail transforms the scene from being rather bizarre to utterly unremarkable. Chances are a scene like this happened thousands of times in your own childhood. It's called pretending and it's a universal part of childhood. There's nothing strange or puzzling about it. Or is there?

If you think for a moment, pretending could be a disaster for children. As a two-year-old, you're immersed in a crash course on reality. Your job is to learn how the world works, understand what things mean and what they do, and learn how to predict the behavior of those around you. If half the time your parents, siblings, and peers are talking into bananas, staging faux tea parties, and mooing like cows, how in God's name are you supposed to make sense of the world?

The truth is, of course, that children don't get cognitively derailed by pretense. In fact, pretending may be an essential step along the way to developing a social brain. Babies around the world begin pretending by eighteen to twenty-four months old, whether or not their parents encourage it.[18] Alan Leslie, a psychologist and expert on the development of pretend play, has pointed out that pretending is an inherently social activity that virtually every child begins to engage in around the same age.[19]

Leslie suggests that when a child is engaged in pretending, she mentally puts quotes around some behavior or thing, a capacity that involves creating a "metarepresentation."[20] Our minds must conceive of a mental world that can be different from the physical world. When we perceive things or people, we form a mental representation of them ("that's a banana"). This kind of direct representation allows us to learn about the world. But when we think about the contents of other people's minds, we need another layer of representation—we need to put quotes around something: "she is pretending that the 'banana is a phone.'"

This capacity to form metarepresentations is what links pretending to the more general human capacity for theory of mind. There is only a small step between "she is pretending that" and "she believes that." That's why they call it *make believe*. In Alan

Leslie's account, pretending is the playground where we learn to think about thinking.

THAT'S WHAT YOU THINK!

HOW DO YOU KNOW IF A CHILD UNDERSTANDS THAT OTHER PEOPLE have their own minds? To many researchers, the strongest test is whether the child understands that someone else can have a *false belief*. For a child to have this capacity, he must recognize that someone else can see the world differently, and he must also imagine what's in someone else's mind (a metarepresentation). The classic demonstration of this developmental milestone involves a kind of pretend game called the "Sally-Anne task." In this game, an adult shows a child two dolls—Sally and Anne—and has the Sally doll place a marble in a basket. The adult then takes Sally away and has Anne move the marble from the basket to a box. Then the adult asks the child, "When Sally returns to the room, where will she look for the marble?" A typical three-year-old will say "In the box," which is where the marble really is. But by age four or five, most children understand that Sally will have the false belief that the marble is where she left it—in the basket. The child understands that Sally's mind contains beliefs that can be different from reality (and different from his own).

More recent research indicates that children can attribute false beliefs to others quite a bit earlier than age four or five. By designing experiments that don't require the child to manage lots of information at once, studies have now shown that false belief detection can be present as early as thirteen months of age, although theory of mind skills clearly become more sophisticated as the child develops.[21]

That simple transition to understanding that other people have their own thoughts and beliefs opens the door to the whole world of social relationships. The operation of our theory of mind has been called *mentalizing* or *mind reading* to emphasize that it's about getting inside someone else's head, reading or inferring their mental states. A theory of mind allows us to cooperate and compete, to recognize other people's motivations and beliefs, to predict how they will behave, to empathize and trust, to deceive and avoid being exploited.

Without the capacity to infer motivations, beliefs, feelings, and other mental states, we wouldn't be able to create or appreciate literature, theater, or art. In fact, a theory of mind is so fundamental to how we function that it was only identified as a subject for research about thirty years ago. Like the purloined letter of Poe's story that I mentioned in the prologue, it was so self-evident that it was almost invisible. We do it effortlessly. We can't help but mentalize—that is, infer mental states—when we see behavior. In 1944 psychologists Fritz Heider and Marianne Simmel provided the classic demonstration of this phenomenon when they showed people a two-minute film of two triangles and a circle moving around a rectangle and asked them to describe what they saw.[22] Almost all of them described the action in terms of animate beings that had feelings and intentions. If you want to see how automatic the impulse to ascribe mental states is, put the words *Heider* and *Simmel* into YouTube and watch the film. Even if you try not to, I'll bet you can't resist seeing the large triangle as menacing.

THE EVOLUTION OF MIND READING

WHERE DOES OUR THEORY OF MIND COME FROM? AT LEAST PART of the answer seems to be that natural selection created the genetic blueprint for brains that can peer into other brains.

The phrase "theory of mind" first appeared in a 1978 paper entitled "Does the Chimpanzee Have a Theory of Mind?" by David

Premack and Guy Woodruff at the University of Pennsylvania.[23] They called the capacity to impute mental states a theory of mind "first, because such states are not directly observable, and second, because such systems can be used to make predictions, specifically about the behavior of other organisms" (p. 515).[23]

In their original study Premack and Woodruff tested a laboratory chimp named Sarah by showing her videos of a man in a problematic situation—for example, standing under a bunch of bananas that were suspended from the ceiling, out of his reach. Then Sarah was shown two sets of pictures, one which showed the man solving the problem (for example, by standing on a box), and the other which did not. Her task was to choose the "correct" picture. Admittedly, Sarah was no ordinary chimp—for one thing, we're told, "she had had extensive prior experience with commercial television." Sarah chose the correct solution almost every time, implying that she understood that the human wanted the food and was trying to solve the problem. Their study certainly didn't settle the issue of whether chimps have a theory of mind, but it launched a whole field of research.

Summarizing the thirty years of studies that followed,[24] psychologists Josep Call and Michael Tomasello concluded that the answer to the question of whether chimpanzees have a theory of mind is "yes and no." Chimps seem to be able to tell when a human is *choosing* to do something as opposed to doing it unintentionally and they seem to be able to distinguish positive and negative emotional expressions.[25, 26] Like human infants, they are also capable of appreciating someone else's perspective. For example, given a choice between reaching for food that a rival can see and food that is hidden from a rival's view, chimps will reach for the hidden food. In other words, they can track what someone else sees, hears, or knows and use that information to avoid a fight.[24] Some chimps will even use deception by going out of their way to hide their efforts to get at the food if a rival is watching.[27]

But can chimps pass the classic test for a full-blown theory of mind? Can they understand the concept of false beliefs? Here, at last, may be the Rubicon separating our minds from our hairy cousins'. When chimpanzees are given tests analogous to the Sally-Anne task, they don't get it—they can't conceive that someone else believes something that isn't true.[24] So the evidence supports David Premack's rule of thumb: "Concepts acquired by children after three years of age are never acquired by chimpanzees."[28]

Why do we care if apes are able to think about thinking? Well, for one thing, it tells us something about the evolutionary history of our own minds. The chimp research suggests that theory of mind is a relatively recent evolutionary development. Tomasello and his colleagues speculate that about 150,000 years ago, when humans lived in small groups, the fitness advantages of cooperation created a selection pressure to collaborate.[11] Groups that hunted and gathered together beat out those that lived by the creed of "every man for himself." This shift toward cooperation required not simply predicting what another member of the group will do, but understanding their goals and intentions and aligning yours with theirs. And with that, the modern human mind was born—a mind that could read other minds and that was motivated to share information (the building blocks of human culture). But there is also fascinating evidence that some elements of mentalizing developed independently—by a process biologists call convergent evolution—in animals that are far more distant from us than the apes. For example, the Western scrub-jay, a bird in the crow family, is able to hide and guard its food by keeping track of what rival birds have seen and know about where the food's been stashed.[29] But there's another species whose social cognition skills may surpass even those of chimpanzees and chances are, at some point, you've had one of these creatures in your home.

"TIMMY'S IN THE WELL?"

JON PROVOST, WHO PLAYED TIMMY ON TV'S *LASSIE*, TITLED HIS autobiography *Timmy's in the Well* as a nod to the iconic scene in which Lassie saves Timmy by getting help:

> LASSIE: Bark! Bark-Bark! Bark!
> ADULT: What, girl? Timmy's in the well? Go get a rope!
> LASSIE (returning with rope in mouth): Bark-bark!

The irony is that, of the many scrapes Timmy got into, falling in a well was not one of them. But Lassie's uncanny ability to read people's thoughts, empathize, and engage in interspecies communication was the essence of the show. In a sense, *Lassie* was a TV show about doggie social cognition. Lassie had it all: joint attention, shared intentionality, and a sophisticated theory of mind. And recent scientific evidence suggests that there was a kernel of truth to Lassie's mental sophistication.

It turns out that the domestic dog has some humanlike social skills that even apes can't match. The clearest demonstration of this involves an experiment called "the object choice task." The experimenter places two opaque containers on the floor and puts a piece of food under one of them. The test subject, say a chimpanzee, is brought into the room and the experimenter cues him about the food's location by looking at the correct container or pointing to it. Despite their mental talents, chimpanzees simply don't get it—they can't make use of human communication signals. But most dogs have no problem picking the right container.[30, 31]

"But wait a minute," you might say. "Of course dogs do better than chimps—most dogs spend lots of time around humans, so they learn how to read human signals." Makes sense, but that doesn't seem to explain their social skills. In 2002 Brian Hare, along with Michael Tomasello and others, reported a series of experiments in the leading journal *Science* that tested whether dogs' ability to read human social cues is unique and innate.[30] First, they showed that on the object choice task, dogs outperformed both chimps and even wolves, their closest evolutionary ancestors. But are dogs just better than wolves or chimps at reading human social signals because they have more exposure to humans?

To answer that, the researchers went one step further. If reading human minds were all a matter of experience and training, dogs with more experience should do better than dogs with little or none. To test this, they ran puppies through the object choice task. Like adult dogs, the puppies were able to understand the human signals, and how much human contact they'd had made no difference. So there is something special about the social skills of dogs. They are better than chimps and even wolves at reading human behavior and, as the puppy experiments showed, their human-reading skills seem to be innate.

What accounts for this ability? Call it a kind of "unnatural selection." It's no accident that the domestic dog is "man's best friend": we made him that way. Modern genetic analyses suggest that dogs originated about fifteen thousand years ago when humans began domesticating wolves, their evolutionary ancestors.[32, 33]

At least two ingredients combined to provide the raw material for transforming wild wolves into domesticated dogs. First, dogs were exposed to human social groups, which collect and often discard food, providing a ripe opportunity for animals that were inclined to seek out and scavenge human leftovers. Migrating humans, in turn, could have used help in carrying, hunting, and guarding their resources. And so a niche was born.

The second ingredient—genetic variation—allowed some enterprising wolves to enter that niche. Presumably, some wolves carried genetic variants that allowed them to approach rather than avoid or attack the humans they encountered. They were rewarded with a replenishing supply of scavengeable food. The advantages bestowed on these "protodogs" helped them to flourish and favored the selection of those who could accommodate to the human environment. Once the process of domestication got under way, the theory goes, humans selected those dogs that were least aggressive and most cooperative. Somewhere between one thousand and five thousand years ago, the human-dog partnership took a leap forward when people began selectively breeding dogs based on their appearance, behavior, and ability to do work.* As the human-dog partnership strengthened, dogs developed brains with the specialized social cognition skills they needed to herd, work, and just know when we need a friend. In a sense, dog breeding became a kind of tool making in which the tool was another animal's mind.

Apart from providing a fascinating story, the mental skills of dogs are important because they support the conclusion that genetic selection is a key to understanding the biology behind mind reading. Domesticated dogs are better at reading social cues than their feral counterparts, suggesting that the genetic selection that occurred during domestication shaped the social brain. But, as compelling as it is, this conclusion is still largely based on circumstantial evidence, and recent work suggests that both dogs and wolves vary in their ability to read human cues depending on their own experience with people.[35] To really study whether genetic selection can shape social cognition, you'd want to measure social skills before and after a species undergoes domestication.

TWENTIETH-CENTURY FOX

ONE OF THE MOST DRAMATIC EXPERIMENTS IN THE EVOLUTION OF social behavior came from a most unlikely place—the fox farms of Estonia. In the 1950s a Russian geneticist named Dmitry K. Belyaev was rebuilding his career in the aftermath of a dark chapter for Soviet biology. Josef Stalin had placed Soviet science under the

---

* Most of the several hundred breeds recognized today were created by selective breeding that occurred within the last five hundred years.[34]

direction of Trofim Lysenko, an authoritarian anti-intellectual who rejected classical Mendelian genetics in favor of pseudoscientific theories about the inheritance of acquired characteristics. Challenging Lysenko's brand of genetics became a criminal offense and dissenters were imprisoned . . . or worse. Belyaev's research on classical genetics led to his removal as director of a fur-breeding laboratory in Moscow. He moved to Siberia, where he studied how to enhance fur breeding.[36, 37] Fortunately, his interests in genetics had a practical application. The silver fox was prized for its fur but the animals were aggressive and difficult to manage. Taming the silver fox would be a boon for breeders and farmers.

In 1959 Belyaev launched an experiment that continues to this day. He was intrigued by the fact that domesticated animals look quite different from their wild cousins. Belyaev hypothesized that the process of selection for tame behavior acts on genes that affect the development of both emotional and physical traits.

To test this theory, Belyaev obtained silver foxes from a fox farm and began breeding them for their behavior. At the start, most of the foxes were pretty nasty creatures—aggressive and fearful of humans.[38] From each generation, Belyaev culled and bred the most tame (least aggressive) foxes. The goal, he wrote, "was, by means of selection for tame behavior, to obtain animals similar in their behavior to the domestic dog" (p. 302).[38] After forty generations of selection, a remarkable thing happened: the silver foxes had become . . . dogs. They became playful, they cuddled and licked their human handlers, they wagged their tails to express pleasure. But even more startlingly, they took on the physical characteristics of dogs: their pointy, upright ears became floppy, their long bushy tails shortened and curled up like a dog's, they developed patches of light fur, wider faces, and shorter, doglike snouts.[37]

What's more, the domesticated foxes seemed to acquire the dog's ability to read human signals. In a head-to-head comparison on the object choice task, domesticated fox kits performed better than undomesticated fox kits and just as well as puppies at understanding human pointing gestures.[39] So the process of breeding for tameness—a form of temperament—seemed to have some dramatic side effects, including the emergence of social cognition related to theory of mind.

Putting the evidence together, Brian Hare and Michael Tomasello proposed that social cognition in dogs initially evolved as a by-product of selection pressure on temperament and its underlying emotional brain circuits (which I described in Chapter 2).[31] The main goal of domestication is to reduce emotional reactivity (aggression and fear behavior). But the side effect of this recalibration of emotional brain circuits and stress hormone systems was the development of a kind of social intelligence—the capacity to recognize and respond to the intentions and desires of other animals. If, as Hare and Tomasello claim, something like this also happened in humans, then the foundations of our theory of mind may have been a side effect of natural selection for anger management.*

As our primate and hominin ancestors faced the challenges of social group living, the ability to impute mental states to others would have provided a compelling, even transformative, fitness advantage. Animals able to mind read would be able to predict behavior, to cooperate, to deceive, and to teach. Once the rudiments of these skills took hold, there would be powerful selective pressure to enhance them into a full-blown theory of mind.

BRAIN ON BRAIN ACTION

WHERE IN OUR BRAINS DO WE THINK ABOUT WHAT'S GOING ON IN other people's brains? Rebecca Saxe, a neuroscientist at MIT, has been studying the neural basis of social cognition for most of her

---

* Not all experts agree with this "emotional reactivity" account, and there is vigorous debate about whether animals really have anything resembling a theory of mind.[40, 41] No one is claiming that canines, scrub-jays, or nonhuman primates have a theory of mind to match our own. But the evidence from comparative studies of animals does illustrate how the rudiments of mentalizing could have arisen from evolution's influence on brain function

career. As a graduate student, Saxe began searching for a region of the brain that is uniquely active when people think about the mental states of others. Because typical theory-of-mind tasks involve a whole host of features that stimulate a wide range of brain circuits—people behaving, responding to visual and social cues, conducting causal reasoning, and forming mental representations—her challenge was to separate out brain activity that reflects thinking about mental states per se. In an elegant study, Saxe and Nancy Kanwisher[42] presented subjects with a range of stories that systematically isolated each of these features and measured the subjects' brain function using fMRI. They discovered that a region of the brain at the intersection of the parietal and temporal cortex called the temporoparietal junction (TPJ) is specifically engaged when we think about another person's mind.

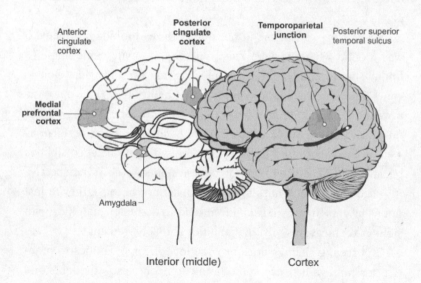

The key elements of the brain's social cognition network. The bolded areas are crucial for theory of mind.

Though the TPJ is essential to mentalizing, it is only one hub in a larger theory of mind network that includes the medial prefrontal cortex and the posterior cingulate cortex.[43, 44] And the more active these regions are in preschool children, the better they perform on theory-of-mind tests; by age four, this network has matured enough for most children to understand the possibility of false beliefs.[45]

NATURE, NURTURE, AND MENTALIZING

SO MIND READING APPEARS TO BE A MENTAL CAPACITY THAT IS SO important that our brains have circuitry dedicated to the task. It is a part of our universal human nature. But not all of us have the same level of skill when it comes to thinking about other people's minds.

Twin research has shown that as much as 67 percent of the variation in theory-of-mind abilities among three-to-four-year-olds is due to genetic differences, although after age five, life experiences play a larger role.[46, 47]

One of the key factors in our ability to read minds may be who was around when we were kids. One day, my wife came home to find me enjoying the last bites of an ice cream bar and said, "Ooh— can I have one?" I had to confess that I had only bought one. Annoyed, she huffed, "You're such an only child!" She may have been on to something. Studies show that only children don't perform as well on theory-of-mind tasks as children with age-matched siblings because they're not as good at taking someone else's perspective into account. And, I hasten to add, that's not because they're just not as bright. In fact, only children tend to do better than kids with siblings on measures of verbal abilities and achievement.[48]

But having siblings does provide lots of opportunities to practice (or rail against) accommodating someone else's thoughts and desires. For one thing, brothers and sisters engage in pretend play, which involves creating a shared mental representation that differs from reality.[49, 50] Sibling rivalry is filled with episodes of persuading, cajoling, and arguing—each of which requires an effort to work with someone else's thoughts and beliefs. Also, siblings have to learn to protect their "stuff" from an envious rival. When siblings fight, mom may try to settle the dispute by trying to get one child to understand what the other was trying to do or say. In doing so, she's likely to refer to their mental states—desires, goals, and feelings. Being exposed to another mind early on seems to help a child develop his mentalizing skills.

And, it seems, the more dissimilar that mind is, the better (up to a point). That was the conclusion of an intriguing study that compared theory-of-mind skills in three groups of four-year-olds: only children, twins, and children with siblings of different ages.[51] The groups were tested on a series of false belief tasks analogous to the Sally-Anne story. The sibling group did significantly better than the twin pairs who did about the same as only children. In other words, the results showed that it's not enough to have a sibling—you want to have a sibling whose mental perspective is substantially different. If you think about it, growing up with a twin is like growing up with someone whose mind is pretty similar to yours. Your brains are at the same developmental stage, you experience things at about the same time, and, if you're identical twins, you are genetically the same. This study and others[51] suggest that the best combination for developing a child's theory-of-mind skills is to have older or younger siblings of the opposite sex.

But as we'll see in the next section, for some people these subtle variations in mentalizing skills are painfully beside the point. They suffer from a form of mindblindness that can overwhelm the trajectory of their lives.

MIND BLIND?

THE SCIENTIFIC STUDY OF MIND READING PROVIDES ONE OF THE best examples of how understanding normal mental function has helped us make sense of dysfunction. Imagine what life would be like for someone whose theory of mind never fully developed. Without the ability to imagine that other people have their own thoughts and beliefs, the simplest social interactions would be bewildering. You walk into a store and on the way in, bump up against another customer. She frowns at you and stares expectantly and finally says, "Thanks a lot!" Without a theory of mind, you'd miss the sarcasm in her response. What would you say? "You're welcome"? You'd be liable to make all kinds of inadvertent faux pas. Your sister smiles and asks, "Do you think these pants make me look fat?" Well, they do . . .

Without a sense that people have their own agendas, you would be vulnerable to all kinds of exploitation and deceit. You'd also have a hard time sharing a laugh since most humor depends on things like irony, which in turn require recognizing the difference between what is said and what is meant.[52] And, without access to other people's thoughts or intentions, you'd have a hard time predicting their behavior.

In fact, there is a condition that involves an impairment in theory of mind: it's called autism.

Although autism was given its name by the child psychiatrist Leo Kanner in 1943,[*][53] it remained a mysterious and misunderstood condition for decades. Kanner considered the condition to be quite peculiar and rare—his original paper described eleven cases and noted that after completion of the paper, he had seen only two more. But from the start, he identified core features of the condition that remain central to its modern definition. The fundamental problem of autism, he wrote, "is the children's *inability to relate themselves* in the ordinary way to people and situations from the beginning of life. . . . There is from the start an *extreme aloneness*" (p. 242).

The current diagnostic criteria for autistic disorder involve three clusters of symptoms: (1) a substantial impairment in social interaction (e.g., eye contact, pointing, emotional connectedness, and failure to develop peer relationships); (2) impairments in communi-

---

* In the same year, Hans Asperger, another Austrian-born physician, independently described the same syndrome, which he called "autistic psychopathy." Although Kanner's is often recognized as the pioneering description, Asperger is probably more widely known for the syndrome of high functioning autism that now bears his name.

cation (e.g., speech, language, and pretend play); and (3) repetitive and stereotyped patterns of behavior. These problems have to begin before a child's third birthday.

In the past decade, autism has burst into public awareness, in large part because of highly publicized reports that the disorder has become dramatically more common in recent years. In the 1980s, autism was estimated to affect one child in 2,500. But estimates have been climbing ever since, and by 2006, estimates from the U.S. Centers for Disease Control (CDC) had reached 1 in 110 children.[54] And those high rates don't seem to be just the product of some American propensity to pathologize. A large 2011 study of Korean children[55] reported that more than 1 in 40 were affected with an autism spectrum disorder—nearly three times more than the latest U.S. estimates! While some have claimed that we are in the throes of an autism epidemic, the cause of the apparent increase remains controversial. A number of factors seem to be contributing to the increase, including increased awareness of the disorder, which in turn leads to increased surveillance and diagnosis.

One reason for the expansion of autism diagnoses has been the expansion of the syndrome itself. In fact, within the broader class of autism spectrum disorders, classic autism represents only about 20 to 30 percent of cases.[56, 57] Several conditions are now grouped under the rubric "autism spectrum disorders." In addition to classic autistic disorder, these include Asperger syndrome and pervasive developmental disorder, not otherwise specified (PDD-NOS). Asperger syndrome (sometimes called "high-functioning autism") is a milder form of autism in which there is no major delay in language or general intellectual development. PDD-NOS is diagnosed when there is an autismlike disorder that doesn't exactly fit the criteria for autistic disorder or Asperger syndrome.*

---

* In 2013 the new edition of the DSM (DSM-5) will appear. As of this writing, the proposed DSM-5 criteria include some major changes to the criteria. Most important, classic autism, Asperger syndrome, and PDD-NOS would all be collapsed into a single category called "autism spectrum disorder."

In 1985, Simon Baron-Cohen, along with coauthors Alan Leslie and Uta Frith, published an article whose title—"Does the Autistic Child Have a 'Theory of Mind'?"[58]—was a nod to Premack and Woodruff's original paper on chimps. The authors used the Sally-Anne task to test theory-of-mind skills in a group of children with autism, a group with Down syndrome, and a group of typically developing children. They found that 80 percent of the autistic children failed the false belief test, while 85 percent of both the Down syndrome and healthy control groups passed. The fact that the Down syndrome group performed as well as healthy children helped rule out the possibility that the autistic children's poor performance could simply be due to overall cognitive impairment. Baron-Cohen and his colleagues suggested that a theory-of-mind deficit is a core dysfunction in autism and could explain much of the social impairment that is a defining feature of the disorder.

Baron-Cohen coined the term *mindblindness* to describe the theory of mind deficit at the core of autism.[59] Children with autism also show impairments at each stage on the typical road to developing a theory of mind. As infants, they are less attentive to faces and facial cues and less likely to smile or follow another's eye gaze. As toddlers, they show less evidence of pointing and joint attention, and by the age of two, when other children are engaging in pretend play, children with autism spectrum disorders are much less likely to.

FROM REFRIGERATORS TO GENES AND SYNAPSES

WHAT CAUSES THE ABNORMALITIES IN SOCIAL COGNITION THAT WE see in autism? In his 1943 paper, Kanner conjectured about the causes of "infantile autism" based on what he had observed in eleven cases:

> One other fact stands out prominently. In the whole group, there are very few really warmhearted fathers and mothers.

For the most part, the parents, grandparents, and collaterals are persons strongly preoccupied with abstractions of a scientific, literary, or artistic nature, and limited in genuine interest in people. Even some of the happiest marriages are rather cold and formal affairs. Three of the marriages were dismal failures. The question arises whether or to what extent this fact has contributed to the condition of the children. The children's aloneness from the beginning of life makes it difficult to attribute the whole picture exclusively to the type of the early parental relations with our patients. We must, then assume that these children have come into the world with innate inability to form the usual, biologically provided affective contact with people, just as other children come into the world with innate physical or intellectual handicaps. (p. 250)[53]

Although Kanner recognized the likelihood that the disorder had "innate" biological roots, the suggestion that parents were at fault gave rise to the influential "refrigerator mother" account of autism. In the 1950s and 1960s, this notion that cold and distant parents were to blame found influential proponents. The child psychologist Bruno Bettelheim wrote in *The Empty Fortress*, "[T]he precipitating factor in infantile autism is the parent's wish that his child should not exist."

It's almost painful to read a statement like that, and from a twenty-first-century standpoint, it seems bizarre. It would also, by the way, require us to believe that parents were particularly averse to male children since 80 percent of children with autism are boys. Unfortunately, this kind of thinking was not a fringe view and caused untold anguish and guilt for the parents of autistic children.

And these ideas persisted until the work of geneticists and neuroscientists began to discredit them, using methods that only became available in the last twenty years. We now know that autism is a disorder of brain development and that genes play a major role.

Studies of twins have shown that the heritability of autism (the contribution of genetic variations to risk in the population) may be as high as 80 to 90 percent, making it one of the most highly heritable medical conditions. It was known for some time that a number of rare classic genetic disorders—like Fragile X syndrome and tuberous sclerosis—can cause autism in children who inherit the mutations that cause these diseases. But the genes contributing to the vast majority of autism cases remained unknown. In the past several years, new tools of genome biology have begun to change that. Now researchers can look at common and rare variations across the whole genome and ask: What variations are more likely to be found among people with autism?

Finding the genes responsible for autism has become one of the hottest areas of research in the field of genetics. Most of the gene variations that have been found are relatively rare mutations and involve genes that make proteins crucial to holding synapses together. But some of the most dramatic findings have pinpointed a kind of genetic difference between people that was relatively obscure before 2004. Known as copy number variations or CNVs, these are segments of DNA sequence spanning a few hundred to a few million bases that are either missing or duplicated. The deleted or duplicated stretches may contain several or even dozens of genes. In 2006 an international team of geneticists reported the first map of CNVs across the whole genome.[60] The results startled many in the genetics community because they suggested that more than 10 percent of the human genome contains these deletions and duplications. If our genomes are the book of life, that means that we differ not just in a letter here or there but in how many paragraphs we have. Many of these genetic differences are common and seem to be benign. But there are rare CNVs that can have profound effects on the brain.

It is now clear these rare duplications and deletions of DNA can be a cause of autism.* [61–64] As the evidence has accumulated, the emerging picture is one in which genes involved in the development of the brain are deleted, disrupted, or duplicated. These include genes involved in how neurons find their place in the brain, the formation of synapses, and the balance of excitatory and inhibitory connections[64]—all fundamental players in how brain circuits get wired up.

Given that, you might expect that disrupting these genes would have widespread and diverse effects on the brain. And that seems to be the case. One of the most striking findings to emerge from this genetic research is that many of the same CNVs that can cause autism can also cause other conditions where brain development has gone awry—including schizophrenia.[65, 66] That may also explain the finding that the risk of both autism and schizophrenia is higher in children born to older parents.[67, 68] It turns out that these CNVs are largely due to copying errors that occur when our cells replicate their genomes before dividing into new cells.[69] The older our germ cells are (the cells that give rise to sperm and eggs), the more likely they are to make copying mistakes that lead to these mutations. In fact, part of the recent increase in autism spectrum disorders may be due to people having children later in life.[70]

One consequence of disrupting genes needed for wiring the brain may be dysfunction in circuits needed for the development of theory of mind. Brain-imaging studies of individuals with autism have found altered function in theory-of-mind areas.[71] Michael Lombardo, Simon Baron-Cohen, and other scientists at the University of Cambridge compared brain activity in adults with Asperger syndrome to healthy controls while they had them make judgments about people's mental states or physical attributes and found a spe-

---

* So far, CNVs that have been associated with autism account for only a small proportion of cases—less than 10 percent. Other genetic and epigenetic variations have also been associated with autism, but in most cases, the causes and risk factors for autism are still unknown.

cific difference in the right TPJ—the same region that Rebecca Saxe and others have pinpointed as the crucial area for thinking about other people's minds. Whereas the controls had greater activation of the TPJ when thinking about others' mental states as opposed to physical traits, those with Asperger syndrome did not. In other words, the special mentalizing function of the TPJ was reduced in the Asperger subjects. And the less TPJ activity they had, the more social impairment they had.[72]

On the other hand, it's clear that alterations in theory-of-mind and social cognition networks don't capture all of the brain basis of autism. There is growing evidence of widespread disruption of connections in several brain systems. The emerging consensus is that autism involves a basic problem with the wiring of the brain and the formation of synapses in early development.

MIND READING AND THE SPECTRUM OF NORMAL

AUTISM MAY BE AN EXTREME VERSION OF AN IMPAIRMENT OF MIND reading, but the more we learn about theory of mind, the more we appreciate that there is a spectrum of individual differences with no bright lines between normal and abnormal. As we saw, the autism spectrum is now considered to be pretty broad, encompassing Asperger syndrome and even traits of social awkwardness that are seen in the "normal" population.[73] For example, relatives of individuals with autism spectrum disorders tend to score higher than other people on traits that make up what's called "the broad autism phenotype," which includes rigid and aloof personality traits, a lack of tact, and socially awkward speech.[74–77] And, when Simon Baron-Cohen and his colleagues gave a test of autism spectrum traits to students at Cambridge University, they found that science majors scored significantly higher than students majoring in the humanities or social sciences.[78] Math majors and winners of the UK Mathematics Olympiad scored highest.

Popular culture has recently embraced the stereotype of the somewhat odd person who has a narrow focus on numbers and technology; "nerds," "geeks," and "trekkies" used to be pejorative labels, but now they are worn with pride. The *American Heritage Dictionary* defines a "geek" as "a person who is single-minded or accomplished in scientific or technical pursuits but is felt to be socially inept." You can now find "geek chic" fashions, online geek-to-geek dating sites, and Geek Pride Day, which is held annually on May 25, to coincide with the anniversary of the first *Star Wars* movie premiere.

Of course, being a computer whiz is not evidence of a disorder. On the contrary, my point is that variation in the same brain systems that underlie normal social intelligence and information processing give us insights into how a syndrome like autism works. Whatever genes and environmental exposures contribute to the development of autism spectrum disorders likely act on the same brain systems that govern cognitive and social skills for all of us.[79] Some of the traits that are more common among people with autism spectrum conditions, especially high-functioning autism, can clearly be beneficial: an ability to think systematically, a facility with mathematical and technical concepts, an ability to recognize patterns, and an exquisite attention to detail.[80] These are the kind of skills that are essential to many of the most complex occupations in our modern world—computer technology, finance, and engineering, to name a few.

Temple Grandin, a professor at Colorado State University who was diagnosed with autism in early childhood, has become internationally known for her work in animal science and behavior. Named one of *Time* magazine's "100 Most Influential People" in 2010, she has attributed much of her success to being able to think in pictures and attend to details. And she has emphasized that "the world needs all kinds of minds,"[81] highlighting the importance of valuing and cultivating the skills that may be part of the broader autism spectrum.

It's clear that many people with autism are profoundly disabled.

But there is a broad spectrum of severity, and at the milder ends of the spectrum, there are no sharp borders between "normal" and "abnormal" social and cognitive functioning.*

Steve Parris seemed ill at ease and anxious during our first appointment. As I welcomed him into the office, he seemed unsure which chair to use, and when he chose one, he sat silently, glancing around the room. After a few moments, I suggested, "Maybe you could tell me what's brought you in today and how I might be of help." He spoke in a formal style, frequently interrupting his sentences to address me as "Doctor." He said he thought he might be depressed and that someone at work had suggested he "get it checked out."

He described a pattern of painful interactions with his coworkers at a publishing job he'd started eighteen months earlier. He felt excluded and suspected that other people were taking advantage of him. "I was supposed to go on a trip for work to New York, Doctor, but the day before the trip, my boss tells me that he's decided that another guy in my department is going. And he wouldn't tell me why." It was an experience he was familiar with. As a boy, he'd had few friends and was teased by his peers for being "weird."

But as we talked further, it became clear that the source of much of his pain was a relationship that ended four years earlier, when he was thirty-one. He had dated rarely in his twenties, but he always felt that he should be in a relationship. While living in California, he had a met a woman through work and they went out a few times over a period of months. "I finally had a real girlfriend, Doctor." But then, after about six months, she told him she was interested in somebody else and broke off the relationship. He was wounded and bewildered and for some months after that he continued to call her, visit her at work, and arrange things for them to do together, but she rebuffed him. She became increasingly blunt in her rejection and this only made him more persistent. He began to have anxious ruminations and difficulty

---

* Advocates for acceptance of "neurodiversity" have argued that the concept of "normal" or "neurotypical" is inherently stigmatizing and ignores the ubiquitous and natural diversity of mental functioning.[82, 83]

sleeping and concentrating at work and now he didn't know what to do. After meeting with him several times, I referred him for a neuro-psychological evaluation and the results were consistent with a diagnosis of Asperger syndrome. But that still left us with the challenge of how to help him tune in to the social cues that he was missing.

People with Asperger syndrome can learn to read social cues, but it requires effort. They need to learn the signals by studying people's outward behavior and decoding it. Reading social cues for someone with high-functioning autism is a bit like reading a book in a language that you're not fluent in. I studied French in high school, and I can understand it, but unlike a native speaker, I can't think in French. It's not intuitive. If I'm reading something in French, I need to translate the words into English in my mind. Learning the language of other people's minds is one of the elements of "social skills training" in which individuals with autism spectrum disorders are taught to decode social signals and respond effectively in social situations—how to start a conversation, make small talk, make eye contact, interpret facial expressions, imagine how other people might feel in a given situation, and so on. Studies of these approaches have had mixed results,[84-86] but it does appear that mind reading is a language that can be learned, even if it's not your native tongue.

BLUFFERS AND LIE DETECTORS: THE OTHER END OF THE SPECTRUM?

IF MIND READING IS REALLY A SPECIALIZED SKILL OF THE HUMAN mind and autism illustrates one end of that spectrum, what about the other end? Is there anything analogous to the super-recognizers whose face recognition skills are beyond normal? Are there people whose mind-reading skills are better than good? These would be people who have a superability to detect and reason about the mental states of others.

Trying to answer that question points up a fascinating asymmetry in what scientists have chosen to study about the human mind. We know a good deal about the mental and brain functioning of "healthy" volunteers (or "typically developing individuals," as they're sometimes called) and people suffering from disorders or impairments of those functions. But we know much less about whether the mind can do its job better than average.

In a sense, that's not surprising, because that asymmetry is in our DNA. One of the points I've made several times is that our brains are prepared by natural selection to acquire capacities that were tuned by the environment of our evolutionary past. But for most things that have really mattered to human evolution, our neural equipment is designed to make us good enough, plus or minus a little. On the other hand, not everyone is the same. There may not be many baby Einsteins, but there has been at least one. Science has given the study of talent short shrift.

So is there any evidence that some people are super–mind readers? Perhaps the closest researchers have come to asking this question can be found in studies of lie detection. Obviously, at some level, recognizing that someone is lying requires theory-of-mind skills. You have to understand that another person holds beliefs that are different from what they are communicating. And liars are essentially trying to create false beliefs in the minds of other people.

Typically developing children begin to lie at age three, around the time that they pass tests of false beliefs, but it's not until age seven or eight that they are able to sustain a lie with a consistent story.[87] In school-age children, the development of proficient lying is a sign that a child's theory-of-mind skills are developing normally. Not surprisingly, children with autism are not adept at telling or detecting lies, or even conceiving of the possibility that others are lying to them.

Actually, most of us are not very good at recognizing when someone is telling a lie. When people are shown videotapes of someone who is either lying or telling the truth and they're asked to catch the liars, they do no better than chance.[88] Paul Ekman, a psychologist who has made a life's work of studying the communication of emotion, wanted to see if some people are unusually good lie detectors. Would people who are trained to detect deceit—law enforcement agents, for example—do better than the rest of us? In one study, Ekman and his colleague Maureen O'Sullivan showed groups of people videotapes of ten young women and told them that about half of them would be telling the truth and the rest would be lying. Ekman and O'Sullivan told their subjects that all of the women would be describing their positive feelings about a nature film they were watching.[89] But in reality, only some of them were watching nature films. The rest were watching horribly gruesome films and lying about what they were feeling.

Some of the subjects in this lie detection test were average college students, but others were professionals whose jobs involved lie detection: members of the Forensic Services Division of the Secret Service, federal agents from the CIA and FBI, police investigators, a group of judges, and a group of psychiatrists. Only one group performed better than chance: the Secret Service agents. The best lie detectors were more likely to use nonverbal cues—like subtle features of facial expression—when they judged the truthfulness of the videotaped performances. In a subsequent study, Ekman used films of people lying or being truthful about their *beliefs* rather than their feelings and again found that expertise matters in being a good lie detector. The only groups that showed special skills were trained interrogators and forensic psychologists.[90]

While many studies have shown that ordinary people (mainly college students) don't do better at spotting liars than they would if they flipped a coin,[88] O'Sullivan and Ekman claim that there are rare individuals—mainly forensic professionals—who are "truth wizards."[91] The idea that some people are human polygraphs has an undeniable "cool" factor that hasn't escaped the entertainment industry. There was Jack Byrnes, the intimidating dad played by Robert De Niro in *Meet the Parents* who spends much of the film suspiciously sizing up his daughter's hapless suitor Greg Focker

(Ben Stiller). Turns out Jack is a former CIA profiler, and, as his daughter warns Greg, "a human lie detector." And more recently, there was the crime series *Lie to Me*, whose main character, Dr. Cal Lightman, is "the world's leading deception expert," and was actually based on Paul Ekman himself.

But there's another forum for mind-reading expertise that has lately become a national obsession: poker. In fact, poker is in many ways the apotheosis of theory of mind in action. Beyond knowing the hierarchy of winning hands and having a familiarity with probability and odds, the talent that separates great poker players from the rest of us is an ability to detect and manipulate mental states. And who better to teach you these techniques than a former FBI counterintelligence agent?

In his book *Read 'Em and Reap*, retired FBI special agent Joe Navarro tells us "that 70 percent of poker success comes from reading the people and 30 percent from reading the cards. . . ."[92] And in *The Theory of Poker*,[93] David Sklansky captures the complex theory of mind skills that a great poker player needs: "getting into your opponents' heads, analyzing how they think, figuring out what they think you think, and even determining what they think you think they think" (p. 236). To play at this level, you need to be attuned to the thoughts, emotions, and intentions of the other players despite their efforts to keep them hidden. You need to hide your emotions and suppress any tells about your own mental states. Mastering the poker face means overcoming a biological system for broadcasting our emotions that's been shaped by millions of years of primate evolution. And great poker players are also expert at creating and exploiting false beliefs in other people—the art of the bluff.*

Poker legend Jack Straus executed one of the most famous bluffs

---

* Here's a tip: the best poker face is not what you'd expect. In one study, subjects were brought into the lab to play Texas Hold 'Em poker.[94] The only information they had was their own cards and the face of their opponent. The subjects made more betting mistakes when their opponents looked trustworthy and approachable than when the opponents had a neutral poker face. So next time you're in a poker game and you want to annihilate your opponent, smile and try a little tenderness.

in poker history. Nicknamed "Treetop" because of his 6'6" frame, Straus was playing high-stakes Texas Hold 'Em poker and he was on a winning streak. In Texas Hold 'Em, players make their first bet after being dealt two cards facedown (the "hole cards"). The next bet comes after three more cards are laid out face up ("the flop"), and then another bet after the fourth card ("the turn card") and a final bet after a fifth face-up card ("the river") is dealt. The winner is the one with the best five-card hand or the last one standing.

An aggressive gambler, Straus decided he'd bet big no matter what two hole cards he was dealt. When he picked up his hand, he saw a 7 card and a 2 card of different suits—known to poker players as the worst possible hand. But he followed through with a big bet. Only one player at the table called his bet. The "flop" cards were a 7, a 3, and another 3. Straus bet again, and his opponent made a huge raise suggesting that he had at least a high pair. Things were not looking good for Straus. The turn card was dealt: it was a 2. So Straus had two low pairs (7–7 and 3–3), but his opponent almost certainly had better. He made a large bet anyway. In theory-of-mind terms, Straus had decided his only hope was to create the false belief that he had three of a kind. As his opponent deliberated, Straus grew uneasy. He knew he would lose if his opponent called the bet, so, in a move that made poker history, he made a generous offer: for $25, his opponent could choose one of Straus's down cards and turn it face up. His opponent took the offer, threw him a $25 chip, and turned up the 2. After another moment of deliberation, the opponent folded and Straus took the massive pot. What happened? As Straus must have surmised, his opponent thought that to have made such an offer, Straus must have had a pair of 2's down and thus, a full house. Straus was operating on the level of "what they think you think they think."

NOTHING MORE THAN FEELINGS?

I'VE SAID A LOT ABOUT HOW OUR MINDS READ OTHER PEOPLE'S thoughts. But what about reading their feelings? There's an important difference between thoughts and feelings. Thoughts and intentions are invisible. We can infer someone else's thoughts by observing their behavior and by listening to what they say. But emotions hang on our faces and lurk in the sound of our voices. Usually, our inner feelings and outward expressions are consistent: you're feeling elated at winning an Academy Award and you're all smiles as you thank God and your agent. But they can also be out of sync: you feel angry as you realize you've lost another Oscar to Meryl Streep, but you're all smiles as the camera zooms in on your face. We humans have exquisitely sophisticated systems for expressing and detecting emotions in other people. And once again, the face is where the action is.

As I noted in the last chapter, Darwin was the first to claim that facial expressions of emotion are innate, universal, and evolved tools of communication. Nearly a century later, his conjecture was taken up by Silvan Tomkins, a psychologist who claimed that there are nine primary categories of emotion, or affects, and that they are universal, innate, and biologically based. They included the ones you'd probably guess—enjoyment/joy, surprise/startle, fear/terror, shame/humiliation, anger/rage, interest/excitement, distress/anguish, disgust—and one he made up—"dismell." Disgust literally means "bad taste," so you can guess what dismell is.

Tomkins's students Caroll Izard and Paul Ekman later provided crucial evidence for the universality of emotional expressions, showing that they are the same across cultures. People around the world use the same facial expressions when they experience basic emotions like fear, anger, disgust, sadness, surprise, and happiness. Congenitally blind individuals from diverse cultures produce the same emotional expressions as those who can see,[95, 96] supporting the idea that these expressions are innate. Emotional expressions make up a kind of universal vocabulary. People in industrialized and preliterate cultures see the same emotion when they are shown what we would call an angry face or a sad face or any of the other basic emotions.[97] When it comes to signaling our emotional states, facial expressions are the Esperanto of human cultures.

WHY THE LONG FACE?

SO WE USE EMOTIONAL EXPRESSIONS LIKE FEAR, ANGER, AND DIS-gust to communicate our internal states to other people. But why did natural selection favor the particular shapes our faces take on when we're feeling fear or other emotions? For example, why do we open our eyes and flare our noses when we are afraid? If facial expressions are like the words of a social language, are they just arbitrary forms like the words of a spoken language or do they have some inherent meaning?

Darwin speculated that particular emotional expressions evolved for a reason. The curled lip of an angry face, for example, has the intimidating effect of baring one's teeth. But in 2008, researchers at the University of Toronto applied twenty-first-century technology to provide a more scientific test of Darwin's speculation.[98] They used sophisticated computer modeling to map the detailed facial structure involved in a facial expression of fear. Then they ran the computer mapping in reverse—creating a face that had the opposite muscle movements of a fear face. And the result was instantly recognizable—it was the expression of disgust. That seems surprising—why would these two different emotions use the same muscles patterns in reverse?

The answer seems to be that they are not just arbitrary forms—they serve a purpose. Fear spreads the face in ways that enhance our ability to take in sensory information. In contrast, disgust compresses the eyes, nose, and mouth to keep sensory information out. The fear face is about vigilance, taking the environment in; the disgust face is about shutting out or even expelling the environment. This gives us another clue to why natural selection paid so much attention to recognizing emotions in other people. Emotional expressions serve not only as a language of social communication—allowing us to predict behavior and see the effect of our own behavior on other people—they can also save our lives. They can help us take in or keep out danger, and we can use other people's emotional signals to warn us of danger that lurks nearby. By empathizing with someone who is showing signs of fear or disgust, for example, we can prepare ourselves for the worst.

EMPATHY: THE SINCEREST FORM OF FLATTERY?

ADAM SMITH, THE SCOTTISH PHILOSOPHER BEST KNOWN FOR HIS book *The Wealth of Nations,* identified sympathy as the basis of human moral behavior and proposed that we feel another person's feelings by a kind of imitation of his or her mind: " . . . we enter as it were into his body, and become in some measure the same person with him, and thence form some idea of his sensations, and even feel something which, though weaker in degree, is not altogether unlike them" (p. 4).[99]

In a sense, empathy may literally be a form of "inner imitation." When we interact with other people, we tend to unconsciously mimic their behavior, a phenomenon known as the *chameleon* effect.[100] Empathy and its cousins, sympathy and compassion, all involve recognizing the emotional states of another person, but only when we empathize do we actually experience the same emotion as that other person.[101]

The notion that empathy and imitation are biologically two sides of a single coin was bolstered by the discovery of a system of neurons specialized for imitating others. In the early 1990s, a group of Italian neurophysiologists discovered a set of nerve cells in the brains of macaque monkeys that had a very special property. It had already been shown that neurons in a part of the premotor cortex called F5 fire when a monkey engages in goal-directed activity, like using their hands to grasp for food. The Italian scientists were trying to study the properties of these neurons in detail, but serendipity revealed something startling. The F5 neurons also fired when the monkeys were observing an experimenter pick up the food to put it in front of a monkey.[102] The scientists discovered that the same set of neurons that activate when a monkey makes an intentional action also activate when they see someone else make the same action. In fact, these neurons are most active when the monkey imitates the action that they see. These "mirror neurons" seemed to follow the rule "monkey see, monkey do." It wasn't long before scientists claimed to have found a human mirror neuron system (MNS), analogous to the monkey system, distributed in regions of the brain's frontal and parietal cortex.[103, 104]

Here, it seemed, was a neural basis for the chameleon effect: when you watch someone perform an action, brain regions activate as if you yourself were performing the action. Some have claimed that this mirroring may extend beyond the actions of others to their emotions. The MNS is connected to the limbic system (the neural circuits of emotion) through the small strip of cortex called the insula, which you may recall from Chapter 1. So, the theory goes, one biological system for empathy may play out like this: observing the emotional responses of others activates the MNS, which relays information through the insula to the limbic system, triggering our own emotional experience that mirrors the one we observed.

In an fMRI study, a team of American and Italian scientists found that the same network of brain areas, centered on the insula, activates when people are either observing or imitating emotional facial expressions[105] and when they experience pain directly or watch a loved one experience pain.[101] And, as you may recall from Chapter 1, the insula engages when people are exposed to disgusting smells and when they simply watch others express disgust.[106] Rare individuals who have lesions of the insula are unable to recognize facial expressions of disgust and are also unable to experience disgust themselves. Adam Smith wasn't far off: empathy seems to involve a process of enacting the emotional experiences of others in our own brains. We connect with someone else's feelings by simulating them.

But even if the MNS provides a biological mechanism for empathy (and there's plenty of controversy about that), it is likely to be only one part of the puzzle. Neuroscientists have drawn a distinction between "emotional empathy" and "cognitive empathy." Emotional empathy involves the kind of immediate emotional resonance and imitation that mirror neurons might provide—"I feel what you feel"—whereas cognitive empathy involves mind reading or theory-of-mind skills—"I recognize and understand what you feel."[107]

Emotional empathy seems to be the more primitive system. Even newborns display rudimentary types of emotional imitation—for example, smiling in response to their mother's smile. The capacity for this kind of imitation makes infants susceptible to emotional contagion. If you've ever walked into a day care for infants, you've probably witnessed babies doing the emotional equivalent of the wave: one baby's cries will trigger crying in her neighbors, creating a spreading front of wails across the room.* But understanding the mental states of others (cognitive empathy) seems to come later.

Emotional empathy and cognitive empathy seem to rely on different brain circuits. In a study of patients with brain lesions,[107] those with damage to the inferior frontal gyrus, a key node of the mirror neuron network, had severe deficits in emotional empathy but normal cognitive empathy (theory-of-mind abilities). In contrast, those with lesions in the ventromedial prefrontal cortex, a key node in the theory of mind network, had no difficulty empathizing but performed poorly on false belief (Sally-Anne-type) tasks. These studies suggest that there are two independent brain systems in-

---

* Of course, emotional contagion works in adults, too—that's why God made laugh tracks.

volved in what people refer to as empathy. One—the MNS—allows us to simulate and mirror other people's emotional states, while the other—the theory-of-mind network—enables us to appreciate and anticipate what will make someone feel good or bad.

Both systems may be necessary for us to accurately read other people's feelings.[108] The very fact that our brains have multiple systems for detecting and responding to other people's feelings says something about how important empathy is to human nature. Mirroring and emotional contagion are crucial for forming our earliest attachments. Later, as the cognitive empathy system comes on line, we are able to take another's perspective—to see and feel the world through their eyes and to sympathize. Together, these systems help shape our moral behavior. We avoid hurting other people and want to help those in need in large part because we can feel their pain.

LIFE WITHOUT EMPATHY

IF EMPATHY IS A BASIC FUNCTION OF THE NORMAL BRAIN, CAN there be disorders of empathy itself? What would someone look like if he had an intact cognitive theory of mind (an ability to read other people's thoughts, beliefs, and intentions) but an impairment of emotional empathy? This would be a person who would understand what people are thinking and feeling but wouldn't care. That might not be enough to cause problems. But when a lack of empathy is combined with callous and aggressive personality traits, the results can be destructive indeed. There's a name for people like that: they're called psychopaths.

The serial killer Ted Bundy was an exemplar of the psychopathic mind—charming, confident, manipulative, and utterly without empathy or remorse. "I don't feel guilty for anything . . . I feel sorry for people who feel guilt," he said while awaiting execution for more than thirty murders.[109] Recent research suggests that psychopathic individuals have a neurobiologic impairment in the ability to recognize and process fear and sadness in the facial expressions or voices of other people. It's as though they're blind and deaf to the pain of those around them.

That combination of being able to read people but not connect with their fear and pain creates a dangerous mix, especially when someone is motivated—as we all are to some extent—by self-interest. The terms *psychopath* and *sociopath* are essentially synonymous, but contrary to popular belief, neither is a psychiatric diagnosis. There is no category of psychopathy per se in the DSM, psychiatry's official diagnostic manual. The term was coined in the nineteenth century but its modern usage derives from an influential book, *The Mask of Sanity* (1941), by the American psychiatrist Hervey Cleckley,[110] who is also known as the coauthor of *The Three Faces of Eve*, which put multiple-personality disorder on the map. In *The Mask of Sanity*, Cleckley identifies a group of patients whose veneer of sanity, rationality, and even charm covered up a deep disturbance of emotional and social functioning. The psychopath, Cleckley observed, is often outwardly friendly and agreeable, but below the surface, he is insincere, callous, emotionally vacant, pathologically egocentric, given to exploiting others without remorse, "and almost incapable of anxiety." In the realm of psychiatry, psychopathy is most closely tied to the diagnoses of conduct disorder and antisocial personality disorder.

Conduct disorder is a diagnosis made in childhood or adolescence when someone has a persistent pattern of violating rules and victimizing other people. To warrant the diagnosis, a person has to exhibit three or more of the following kinds of behavior: (1) aggression to people or animals, such as physical cruelty, assaults, forced sex, mugging; (2) destruction of property or fire-setting; (3) deceitfulness or theft; and (4) serious violations of rules, such as repeated truancy or running away from home. Antisocial personality disorder is essentially the adult form of conduct disorder.

But you'll notice that these diagnoses are all about behavior and they are not quite the same as psychopathy. It's been estimated that 80 to 90 percent of inmates in maximum-security prisons meet criteria for a diagnosis of antisocial personality disorder, but only 15 to 20 percent qualify as psychopaths. Not all people with psychopathic tendencies have antisocial personality disorder; some are quite high-functioning and successful.

Because their cognitive mind-reading skills are normal, psychopaths are often quite able to keep their true nature well hidden. They can learn to "talk the talk" of normal social relationships. When they do cross the line into criminal behavior, those who know them may be surprised. "He seemed like a regular guy" is the familiar refrain when a reporter interviews a neighbor of the latest serial killer to make the headlines.

The problem with psychopaths, according to James Blair at the National Institute of Mental Health, seems to be a specific impairment in emotional empathy.[111] Most of us are emotionally aroused when we see someone gripped by fear. You can measure this arousal by hooking someone up to a machine that measures skin conductance—how easily an electric current is conducted through electrodes on the skin. Sweating is a sign your emotions have been aroused, and when you sweat, the moisture makes your skin a better electrical conductor. But Blair and his colleagues have shown that when psychopaths are shown faces expressing distress—fear or sadness—they are unmoved. Their skin conductance responses show little or no sign of arousal. In fact, they have trouble even recognizing fear in the faces of others. This is a deficit that's also seen in rare individuals who have damage to the amygdala, that key region of the limbic system that is essential for processing emotional stimuli.

In fact, neuroimaging studies suggest that psychopathy involves a distortion of the brain's fear-processing machinery. Psychopathic individuals have been found to have small amygdalae that have a blunted response when they look at fearful faces or listen to fearful voices.[112–115] And the deficit in appreciating fear seems to develop early: Blair and his team found that children and adolescents who score highly on measures of callous-unemotional traits (a core feature of psychopathy) showed reduced activity of the amygdala in response to seeing fearful faces when compared to healthy children. The more callous they were, the smaller the amygdala response. They also found that callousness was associated with a weaker connection between the amygdala and the ventromedial prefrontal cortex (vmPFC), a brain region involved in moral decision making.[116]

These findings fit with experimental results showing that psychopathic traits are related to problems with recognizing fear and responding empathetically to fear and distress cues from other people. Putting this evidence together, Blair suggests that psychopathy results from a neurobiological disconnect that short-circuits empathy-based learning.[117]

Most of us develop a moral compass in part by learning that exploiting or harming others causes them to feel fear and pain. Because of our capacity for empathy, we find other people's suffering aversive and we learn to avoid doing things that cause it. This emotional learning depends on a circuit involving the amygdala—which detects the other person's distress—and regions of the prefrontal cortex—including the orbitofrontal cortex (OFC)—which registers the connection between their distress and your behavior along the lines of "you just did something that was not good—don't do it again." In tests of empathy and accuracy at judging other people's emotional states, criminal psychopaths show deficits that are similar to those of people with brain damage in the OFC.[118]

Blair's work suggests that psychopathic tendencies result from a dysfunction of this brain circuit: without a capacity to feel other people's fear and distress, the brakes on callousness and antisocial behavior fail to develop normally. When this callousness combines with a predisposition to be impulsive and aggressive—which other research suggests is related to hypersensitive reward circuitry[119]—the seeds of a criminal mind are sown.

The psychopath's disconnection between the amygdala and the OFC may explain why they are so resistant to change or treatment. We rely on this circuit to learn that exploiting other people, causing them pain, is something to avoid. As Blair told me: "It would be very difficult to give them newer value judgments and to really make them care about other people. In order to really feel that hitting another person is wrong, you need to have this basic neural architecture—the amygdala and OFC circuit—intact. It allows you to be able to learn the badness of harming an individual."

In a series of interviews he gave while on death row, Ted Bundy spoke at length about the mental world of someone who could, as he did, brutalize and kill young women. The transcripts are chilling, in part because he tells his story in a detached third-person narrative, which the interviewers allowed because Bundy refused to take responsibility for his actions. Here's Bundy on "Bundy":

> *I think we'd expect a person not to feel much remorse or regret for the actual crime—or guilt in the conventional sense for the harm done to another individual. Because the propriety or impropriety of that kind of act could not be questioned. If it was, then, of course, there would be all sorts of internal turmoil.*
>
> *The guilt or remorse were most prevalent, if they were prevalent at any time, during the period when the individual was uncertain about the results of the police investigation. Once [it] became clear that there was going to be no link made or that he would not become under suspicion, the only thing which appeared relevant was not exposing himself to that kind of risk of harm again.*
>
> *Not thinking about the nature of the act, of the death of the individual herself. The approach is, say, "Don't ever do that again." But as time passes, the emphasis is on "Don't get caught." (p. 96)[109]*

Bundy was capable of feeling the fear of being caught but incapable of feeling his victims' terror as he tortured them. After describing a series of assaults and murders, he said, with evident pride:

> . . . *[I]t's not that I've forgotten anything, or else closed down part of my mind, or compartmentalized. . . . I guess I'm in the enviable position of not having to deal with guilt. (pp. 280–81)*[109]

What causes the brain circuitry differences that nudge some people toward the dark side? The answer, familiar by now, is that it seems to be a combination of genes and life experience. Twin studies suggest that variations in genes account for up to two-thirds of individual differences in psychopathy.[120, 121]

Like other personality traits, psychopathy seems to be a dimension rather than a category. For the most part, people aren't entirely psychopathic or not at all psychopathic. Certainly, criminals score high on measures of psychopathy, and Bundy's case is about as extreme as they get, but studies have shown varying degrees of callousness and psychopathic traits in polite society as well. Up to 30 percent of people in a general population study in Britain exhibited some psychopathic traits (men more than women), though less than 2 percent reached clinical thresholds for psychopathy.[122] At a brain level, many of the regions implicated in psychopathic tendencies overlap with those implicated in studies of empathy, leading some scientists to conclude that callousness and empathic concern are two ends of a spectrum of normal brain function.

COMPASSION FATIGUE

WHAT ABOUT THE OTHER END OF THE EMPATHY SPECTRUM? IS there such a thing as too much sensitivity to other people's feelings? We all know people who are annoyingly touchy-feely, but I mean something else. Imagine how painful or frightening it would be to be constantly tuned into other people's emotional states. In an old episode of *Star Trek*, the crew land on the planet Menarian 2, and are captured by a subterranean race of Vions, who lack all emotion. Captain Kirk is saved by an "empath," a woman who has a special ability to take on and process the painful feelings of others—but at the cost of having to bear the pain herself.

People with Williams syndrome, the genetic disorder we first encountered in Chapter 2, may experience something close to the pain of an empath. You'll recall that Williams syndrome results from a missing sequence of DNA on chromosome 7 and that this specific change produces an extreme interest in other people. Individuals with Williams syndrome are often described as highly empathic and emotionally sensitive.[123, 124] They are happy when you are happy, but feel terrible when you are sad or angry and want to soothe your distress. But their empathy seems to be primarily the automatic, mirroring kind. Like people with autism, they do poorly on theory-of-mind tests.[125] So their attunement to other people is not the result of some keen ability to mentalize, but rather an intuitive sensitivity to distress.

That sensitivity can come at a devastating cost. Despite their lack of social anxiety, people with Williams syndrome often suffer from debilitating generalized anxiety and worry. Compared to the general population, they are about five times more likely to suffer from generalized anxiety disorder.[126] And part of that may be the price of tuning into other people's distress.

The case of Williams syndrome is certainly an extreme. But all of us can experience empathic overarousal at times.[127] You're curled up on the couch, comfortably watching the evening news when they cut to a commercial. You see the sad face of a young girl and a voice says,

This is ten-year-old Maria. She lost her parents when she was only two. She lives in Mozambique with her blind grandmother and her severely disabled sister. As soon as

*the sun rises, Maria is hard at work—gathering firewood,
scavenging for food, caring for her sister, and working in
the fields. Every day is the same. She's tired. Hungry. And
sick. There are millions of children just like Maria who
are hurting, barely surviving . . . One person—just like
you—can help make a difference for one child . . . And all
it takes is one phone call. . . .**

How could you resist such a plea? In his book *The Life You Can
Save*,[128] the philosopher Peter Singer lays out a simple but compelling argument that our failure to donate more to aid agencies is
not only sad but ethically indefensible. There are many reasons for
this moral failure, including our habits of self-interest, the numbing
effect of abstract statistics about human suffering, and an insuf-
ficient capacity for empathy.[129] But in some cases, it might be our
capacity for empathy itself that makes us turn away.[128] When sympathy crosses into empathy, the effect can be paradoxical. Empathizing
means taking on the feelings of someone else. And sometimes that
can be overwhelming, even for the best of us. In these small moments, we are all susceptible to a kind of psychopathic apathy.

What if empathy were your job? Doctors, therapists, aid workers,
and child protective services workers must face the pain and suffering
of strangers day after day. If they were to internalize that pain every
time they were called on to help someone, they would be incapaci-
tated. And some are. They suffer from compassion fatigue, a kind of
emotional exhaustion that can lead to burnout and even posttraumatic
stress disorder. For example, studies have found that 30 to 60 percent
of oncologists experience emotional exhaustion and burnout.[130] In a
study of New York City social workers who were involved in counseling victims of the 9/11 World Trade Center attack, more than a third
of those who had been extensively involved in counseling had symptoms of compassion fatigue and posttraumatic stress.[131]

---

* From a video entitled "One" on the Save the Children website http://www.savethechildren
.org/, accessed December, 23, 2009.

But even when it's not so severe, the costs of caring for the sick, suffering, and dying can have a numbing effect. And there's the irony. We need professional helpers to be empathic but also to maintain emotional distance. I remember well the twin anxieties that gripped me and my fellow medical students as we began our medical training: Will I be able to comfort the anguished without crumpling into a heap of quivering despair or will I lose my humanity and turn into one of those heartless doctors who chirps "How are we feeling today?" when my patients need me the most? Beginning with the emotional trial by fire of dissecting a cadaver on the first day, the process of medical education is carefully calibrated to produce doctors who can both feel and heal. To be effective, those of us whose job it is to care must live in the middle ground between the empath and the psychopath.

---

* *Answers: Malcolm X, Bill Clinton, Scarlett Johansson, and John Wayne*

CHAPTER FIVE

"SOLE MATES": THE BIOLOGY OF ATTACHMENT AND TRUST

*I just wanted to say thank you . . . I was stuck in the same old job,
barely making ends meet . . . I tried Liquid Trust and I finally got a very
nice corporate job . . . While I was using Liquid Trust my relationship
with my girlfriend wasn't going well. Without really knowing what was
going on it went from bad to wonderful. Best of all, she asked me to
marry her! I would highly recommend this product . . .*

—*JOE**

WHAT IF THERE REALLY WAS A POTION THAT COULD make people love you, trust you, believe in you just by applying a perfumelike spray when you got dressed in the morning? That's the promise of Liquid Trust, "the world's first trust-enhancing spray."

Skeptical? Maybe you forgot to spritz yourself this morning.

In a time-honored tradition, marketers have taken a grain of sci-

---

* http://www.verolabs.com/default.asp (accessed January 2, 2010)

ence, mixed it with a serving of hype, and created a product that can "instantly build relationships that were never possible before!" The active ingredient, oxytocin, is a hormone doctors have administered to thousands of women over the past thirty years. Where did the idea that oxytocin could enhance trust and relationships come from?

Research on the biology of trust and relationships has become one of the most provocative frontiers of neuroscience (and may have made the marketing of such potions inevitable). In this chapter, we'll explore the nature of attachment—the social glue that binds us together and the foundation of our capacity to love.

The drive to attach, to affiliate, and to bond with other people begins from the earliest moments of life. It is one of those essential functions that our brains are wired to perform. Our lives are organized around attachments—first with our parents, then friends, lovers, and our children. Recent research shows that all of these bonds may share an underlying biology that involves a symphony of hormones and brain circuits.

THE GLAND THAT ROCKS THE CRADLE

THE STORY OF ATTACHMENT STARTS AS THE MOMENT OF BIRTH APproaches. It's a little like prepping for a blockbuster event—the opening ceremony of the Olympics, a state dinner, the Oscars—a million details have to be coordinated in just the right sequence. Behind the scenes, a cascade of hormones, enzymes, and neurotransmitters have been preparing the mother's body and brain for the remarkable task of extruding another human being and then making sure the helpless newborn survives.

And after eons of experience, natural selection has put a big burden for the event planning on the shoulders of a little peptide called oxytocin. Made from a string of just nine amino acids, oxytocin helps the mother push out the baby, gets the breast milk flowing, and makes the mother's brain see the squirming, crying infant as a bundle of joy. As the due date draws near, a series of chemical brakes come off the mother's pituitary gland, allowing it to make increasing amounts of oxytocin, which are released in pulses into her bloodstream.[1] As she goes into labor, oxytocin stimulates uterine contractions, bringing the painful process of childbirth into full swing. At the same time, oxytocin from the mother's blood (and maybe also from the fetal brain), prepares the baby for its own ordeal. It dials down the metabolic demands of the fetal brain, protecting it from a drop in oxygen and glucose that might occur during the birth process.[2] Meanwhile, other hormones, including estrogen and prolactin, have been stimulating the mother's breasts to fill with milk. After the baby is born and begins to suckle, oxytocin triggers the milk to drop into the canals in the breast and nipple so that it can be delivered to the infant.

Meanwhile, the sight, smell, and touch of the baby stimulate the release of oxytocin in the mother's brain, priming her to love and care for her infant. How? The answer seems to be that oxytocin stimulates the brain's reward circuits that give us feelings of pleasure. One has to marvel at the economy and efficiency of the whole thing.

Scientists and physicians have known for a century oxytocin played a crucial role in delivery and breastfeeding. Since the 1950s, obstetricians have used intravenous oxytocin (a.k.a. Pitocin) to accelerate delivery when labor is dangerously prolonged or to induce labor when a fetus is too far past its due date. But only recently have we learned that oxytocin is a key factor in maternal care, leading to the remarkable discovery that this same hormone may be a key player in how we form attachments, develop trust, and nurture our relationships with our friends, lovers, and spouses.

MOTHER'S LITTLE HELPER

IT TURNS OUT THAT UNTIL THEY HAVE OFFSPRING OF THEIR OWN, many mammals have very little interest in infants. Virgin rats, for example, typically ignore or even reject pups. But in the 1980s researchers found that, by infusing oxytocin into a virgin rat's brain, they could bring out her inner mother, switching on the full spectrum of maternal instincts.[3] Conversely, blocking oxytocin receptors prevents the normal triggering of maternal behavior in rats, while mice in whom the oxytocin gene has been removed don't even recognize their own infants.[4] In the years that followed, increasingly sophisticated experiments have confirmed that oxytocin is a potent "mommy hormone" in species as diverse as sheep, monkeys, and even humans.

How does it work? The evidence suggests that oxytocin participates in at least two chemical shifts that occur in the mother's brain and help her fall in love with her baby. One is a subtle change in her threshold for approach vs. avoidance. Oxytocin binds to receptors in the amygdala and other regions of the brain's limbic (emotional) system to reduce social fears and aversions. It acts like a kind of social Valium—making her more responsive to her infant's distress. Human mothers are acutely attuned to the cries of an infant and respond with approach behaviors whereas nonmothers often find the sound of a baby crying aversive.[5]

But that doesn't account for the passion mothers feel when they see, smell, and touch their newborn infants. Rat studies have shown that, in the presence of other reproductive hormones like estrogen, oxytocin receptors sprout and activate dopamine pathways in the brain's reward circuits. Dopamine is well known as the key neurotransmitter in the brain's pleasure centers—the same chemical responsible for the euphoric effects of cocaine. In essence, the rat pup now acts like a drug, and this "baby buzz" provides a powerful motivator for maternal behavior. Pups are actually more potent than drugs: shortly after giving birth, rat mothers prefer pups to cocaine and even food. Mothers will even cross an electrified grid to get to their pups.[6]

It's not entirely clear if the same story plays out in humans, but there is mounting evidence that it may. They may have been born yesterday, but infants are no dummies when it comes to giving off cues that captivate their moms. As the ethologist Konrad Lorenz observed, a baby's face is a powerful "releaser" of maternal care and nurturing. Lorenz defined the prototype of cute (which he referred to as *Kindchenscema*, or "baby schema"): protruding cheeks, high forehead, large eyes below the midline of the skull, and a small nose and mouth. If that sounds like Mickey Mouse, it's no coincidence.

According to recent studies, reproductive hormones actually bias the female brain to perceive and respond to cuteness. Women are better than men at identifying *Kindenschema* (that is, cuteness) features, and premenopausal women (especially those on oral contraceptives) do best, suggesting that female reproductive hormones prime a woman's brain to respond to cute signals.[7, 8] And, consistent with the animal studies, brain-imaging studies of women have shown that smiling babies and cute babies activate the brain's reward system.[9, 10] In addition, when mothers are shown videotapes of their infant in distress, brain regions involved in mind reading, empathy, and emotional vigilance are also powerfully engaged, reflecting a deep attunement to their baby's feelings and the powerful impulse to comfort and protect.[11]

LOVE POTION #9

SO THE CONNECTION BETWEEN REWARD CIRCUITS AND REPRODUCtive hormones like oxytocin is crucial to bonding mothers and their babies. But that's not all. The same systems seem to help bond us to our soul mates.

Beavers, bats, and marmosets are members of an exclusive clique—"The 3 Percent Club." Only about 3 percent of mammals are monogamous.[12] In modern neuroscience, a rodent known as the prairie vole has become an icon of monogamy. Prairie voles form pair-bonds, and males resemble the ideal husband: they make a lifelong commitment and share equally in parenting the children. If one mate dies, the survivor is unlikely ever to seek a new partner. By comparison, their close relatives, the montane vole and the meadow vole, are cads: males and females are promiscuous, uncommitted, and even abandon their young. What is going on in the brain that makes some animals "true blue" and others hopeless philanderers? The evidence points to oxytocin and to its sister hormone, vasopressin (also known as arginine vasopressin [AVP]).

Oxytocin and vasopressin are both made in and released from the hypothalamus, the master regulator of most hormone systems in the body. The genes for these two peptides sit near each other on chromosomes in all vertebrate species (including us). And that's no coincidence. Somewhere before vertebrates and invertebrates diverged in evolutionary history, an event that scientists refer to as gene duplication happened. In some ancient ancestor of vertebrate animals, a DNA copying error was made that inserted an extra copy of the original gene next door. Over time, mutations in the two sister genes resulted in two distinct hormones. Both oxytocin and vasopressin consist of nine amino acids, but they differ by two, a difference that has had profound effects on animal life.[13]

In the early 1990s Sue Carter at the University of Maryland and scientists at Emory University, including Thomas Insel and Larry Young, began a series of studies that have given scientists the most detailed picture yet of the biology of affiliation and attachment. When they put a male and female vole together and allowed them to cohabitate or mate, they found a remarkable difference between prairie voles and montane voles. For prairie voles, the act of mating or cohabitating triggers a profound bond—the couple will prefer each other to other voles. But to montane voles, the time they spent together and the sex they had means nothing. The reason involves oxytocin and vasopressin—but in ways that depend on the sex of the animal.

For females, the major love potion is oxytocin; for males, it's vasopressin. Injecting oxytocin in the brain of a female prairie vole triggers pair bonding by activating oxytocin receptors in reward centers. In the natural setting, that's essentially what cohabitating and mating do—they cause the release of oxytocin, which binds to receptors in the nucleus accumbens, a key node in the brain's reward circuit where dopamine signals pleasure.[14] Sound familiar? That's essentially the same story I told you about mothers bonding to their infants. In both cases, an event (nursing or mating) stimulates oxytocin release, which in turn stimulates pleasure centers that stamp "joy!" all over the experience of the other (baby or mate). Female prairie voles have more oxytocin receptors in the nucleus accumbens than do nonmonogamous species like meadow voles (and rats and mice),[15] and that may explain why female prairie voles make better partners.

In males, though, vasopressin is the major player in pair-bonding. After a male prairie vole mates with a female, vasopressin induces him to bond with her, fight off other males, and, later, care for their offspring. And the action is, once again, in the receptors. The vasopressin 1a receptor (known as *AVPR1A*) is more abundant in reward circuits of prairie voles than in their promiscuous cousins, the montane and meadow voles. Larry Young's group at Emory was able to turn gigolos into gentlemen by increasing *AVPR1A* in a reward region of the male meadow vole brain. The male meadow voles now bonded to their female partners, just like prairie voles do.

The astounding implication of this work is that a single gene might make the difference between something as complex as whether an animal spends its life in a committed relationship or playing the field. While a postdoctoral fellow in Thomas Insel's lab, Larry Young discovered a subtle but important difference in the gene that makes the *AVPR1A* receptor in prairie voles and montane voles. The part that carries the instructions for making the protein is identical in the two species. The difference lies in the DNA sequence of the gene's promoter—the part of the gene that determines where and when the gene will turn on. The prairie vole gene, they found, has extra repeated sequences in the promoter. When they took the prairie vole gene and inserted it into mice (who are not the most socially engaging of animals), the mice became more interested in other mice. Differences in this promoter sequence of the gene seem to determine how much of the vasopressin receptor is made and correlates with how socially monogamous voles are.*

Research on voles hasn't fully answered the question of how mammals (even voles) end up building a family. But it has given us a fascinating account of how natural selection has engineered a system for solving an adaptive problem—that is, how to bond mammals to their infants and their mates. The answer, admirable in its efficiency, has been to "double-dip." Take a neuropeptide that likely evolved to facilitate female childbirth and lactation (oxytocin) and one that promoted male guarding of offspring and mates (vasopressin) and link them to dopamine-dependent reward systems in the brain. The result: mothers and fathers who care for their young and commit to each other.

But does the vole story tell us anything about human love and attachment? The answer is we don't know yet, but there are certainly clues that it might. Humans have genes for oxytocin, vasopressin, and their receptors that are very similar to those in other mammals. And, like its vole counterpart, the human AVPR1A gene also has repeated DNA sequences in the promoter region that differ among people. Could these genetic variations affect male pair-bonding in humans?

Researchers at the Karolinska Institutet in Sweden tested that hypothesis in a study of more than 550 couples.[18] They had the couples complete a questionnaire that measured partner bonding, with questions like: How often do you kiss your mate? How often are you and your partner involved in common interests outside the family? Have you discussed a divorce or separation with a close friend?

---

* Of course, in the wild, it's not quite so simple. Some prairie voles have been found to sneak sex with females on the side.[16] And some studies have found that in their natural habitat, the AVPR1A promoter sequence doesn't always predict monogamy.[17]

Men who carried one variant of the *AVPR1A* gene reported lower partner-bonding scores and were more likely to be unmarried or to have major marital problems.

Oxytocin also seems to affect how couples interact. In one study, researchers asked forty-seven couples to pick two areas of their relationship that were a source of conflict.[19] They then had the couples inhale a nasal spray that contained either oxytocin or a placebo and asked them to discuss the conflict issues for ten minutes while trained observers rated their verbal and nonverbal behavior using standardized scales. The couples who had inhaled oxytocin had more positive interactions and lower levels of the stress hormone cortisol. Other studies have found that variation in the oxytocin receptor gene are associated with partner-bonding and romantic closeness among couples.[20]

Brain-imaging studies support the idea that romantic love and mother-infant bonding rely on shared brain biology. In a British study, volunteers who claimed to be "truly, deeply, and madly in love" underwent fMRI scans while they looked at pictures of their beloved.[21] When the results were compared to scans of women looking at pictures of their babies, the overlap was striking. Both romantic love and maternal love activated brain regions rich in oxytocin and vasopressin receptors and those involved in reward circuitry.[22]

Larry Young has even speculated that if the neural foundations of romantic love involve tweaking systems that evolved for maternal-infant bonding, we might need a new perspective on human love and sexuality. "The stimulation of the cervix and nipples during sexual intimacy are potent releasers of brain oxytocin," he points out, "and may function to strengthen the emotional tie between partners." Is this why some men are so ga-ga over women's breasts? Could foreplay be about turning on those "ancient maternal bonding systems"?"[23] If you need proof that there's some kind of link, consider this: according to an unscientific search of Billboard.com, there are more than forty thousand songs with the word *baby* in their title—and most of them are not about infancy.

Let's just say that "Hit Me Baby One More Time!" is not an ode to maternal masochism.

In a funny way, modern neuroscience may be turning Freud's ideas on their head. Freud's claim was that infants' relationships to their mothers are based on sexual desires; neuroscience suggests that mothers' sexual relationships are based on their love for their infants. Either way, we're probably going to need a lot of therapy.

So it may well be that natural selection used this basic blueprint for wiring both maternal love and pair-bonding in mammals as diverse as voles and humans. Of course, no one would claim that that's all there is to human relationships. Human love is undoubtedly much more "splendored" than neuropeptides and dopamine. But it may have roots in the bond between a mother and her baby.

ATTACHING IMPORTANCE

IN THE PAST 150 YEARS, PSYCHOLOGISTS AND PSYCHIATRISTS have proposed innumerable theories of how the human mind works. Most of them are born and die in obscurity. But every so often, one of these ideas takes hold and changes the way we view ourselves in some fundamental way—shifting paradigms and launching whole new directions for research and, sometimes, clinical practice. The most powerful of these ideas move beyond the rarified world of science to shape popular conceptions of human nature. Those ideas that survive usually have two features: a persuasive champion and an explanatory model that seems at once contrarian and profoundly clarifying, and then self-evident. Often there is a third element: the idea seems to arrive at a moment when it resonates with the social politics of its age. Sigmund Freud was a prolific and brilliant communicator of his psychoanalytic theories. His claim that we are endowed from infancy with sexual and aggressive drives that explain our dreams, desires, fears, and neuroses at first seemed absurd and even scandalous. But the explanatory power of psychoanalytic theory eventually triumphed and became a dominant paradigm by which both professionals and later the public came to see human behavior.

Behaviorism, the main competitor to Freudian psychoanalytic models, grew out of the work of John Watson, B. F. Skinner, and others who claimed that the human mind is a blank slate on which experience writes the scripts for our behavior. The idea that all behavior is learned—by conditioning, reinforcement, and punishment—and therefore is eternally malleable fit well with the American ideals of equality, pragmatism, and opportunism. The appeal of learning theory broadened with the backlash against biological determinism that developed in the aftermath of the racist eugenics of Hitler's Germany.

But in the 1950s another paradigm-shifting idea about human development began to take hold. Its leading proponents were John Bowlby, a British psychiatrist and psychoanalyst, and his colleague Mary Ainsworth, an American psychologist. In the 1940s, Bowlby wanted to understand the effect of early maternal separation on child development. It was a question that had become particularly poignant in the wake of World War II after untold numbers of children were left motherless. In fact, Bowlby had been commissioned by the World Health Organization to prepare a report about the outcomes of children without families and as he researched the issue, he found psychoanalytic concepts wanting.

In 1952 Bowlby's colleague Jimmy Robertson made a powerful but low-budget film called *A Two-Year-Old Goes to the Hospital*, about a healthy two-and-a-half-year-old little girl named Laura who is admitted to the hospital for a minor operation. As was the custom, her mother leaves the child in the care of the hospital team. Using a handheld camera, Robertson simply documented the eight days of her stay without narration. But no narration was needed. As the days go by, Laura becomes visibly shaken and frightened, with plaintive cries of "I want my mummy" and "I want to go home." She begins to be more withdrawn at her mother's visits. On the eighth morning, "she is shaken by sobs."[24] The film was instrumental in changing hospital policies to allow visiting and overnight stays by parents to minimize separation distress.

Psychoanalytic theory and behaviorism could scarcely have had a more different view of child development, but neither of them showed much interest in concepts like love and attachment per se. In fact, their explanation for the infant's bond to its mother was ultimately similar: it was a side effect of gratification. To the psychoanalyst, it was about satisfying the child's oral desires for the breast; to the behaviorist, the child's apparent affection was simply behavior reinforced by feeding and other pleasurable stimuli. But those accounts seemed incomplete to Bowlby. They didn't capture something fundamental about the child's attachment to its mother, and he began to look for other explanations.

He found them in part in the work of ethologists who had been writing about their observations of animal behavior. He knew of Konrad Lorenz's work showing that goslings had an innate capacity to "imprint" on their mothers. But even more relevant was the work of an American psychologist named Harry Harlow, who had been studying maternal deprivation in rhesus monkeys. Harlow's experiments have since become part of the canon of developmental psychology—familiar to almost anyone who has taken an introductory psychology course in the past fifty years.

A Mother Out of Whole Cloth

Harlow was trying to identify the active ingredient in maternal-infant bonding. Where Freud famously asked "What does a woman want?" Harlow wanted to know "What does a baby want?" What was the infant seeking by attaching to its mother? To answer that, he took sixty newborn rhesus macaque monkeys away from their mothers on the first day of life and raised them in a laboratory. Harlow and his team soon noticed that the infants developed a strong connection with the cheesecloth blankets that lined the floors of their wire cages.[25]

When the blankets were taken away, the monkeys became violently emotional and oppositional. Was there something about the contact with the cheesecloths they craved? Were these motherless babies findings some "mother-ness" in the softness of the blankets? Harlow decided to create two kinds of "mother surrogates" and mount them in the infant's cages to see what effect they would have. The cloth mother was a cylinder of wood covered in terry cloth, and the wire mother was a wire-mesh cylinder of the same size. Bottle holders were installed in the upper part of the "mother" to provide milk. The only real difference between the cloth mother and the wire mother was a layer of cloth—a soft exterior that would provide some comfort to the touch. But that difference was anything but small.

In Harlow's classic experiments, half the monkeys were fed by the cloth mother and the other half by the wire mother. He observed the infants with their surrogate mothers for about six months. Regardless of which mother fed them, the infants spent almost all of their time holding on to the cloth mother. When Harlow introduced a series of fear-evoking stimuli (e.g., a moving toy bear), infants raised on the cloth mother ran and clung tightly to her. After a few minutes, they were soothed and relaxed and start exploring. But the infants raised by the wire mother clutched themselves, rocked back and forth, and cried out, unable to settle.

In another set of experiments, he gave the infants a choice between a wire mother with a milk bottle and a cloth mother without. Once again, they preferred the cloth mother and clung to her when they were afraid. The implication was clear: baby monkeys would rather have a mother who could comfort them than a mother who could feed them.

Harlow's studies seemed to rebuke the prevailing idea that infants bond to their mothers because mom is associated with satisfying basic needs like thirst or hunger. Rather, they supported a conclusion that John Bowlby had already begun to draw: infants have an innate need to attach. Attachment isn't a by-product of ful-filling basic drives like hunger—it's an end in itself. As Harlow put it, "man cannot live by milk alone" (p. 677).[26]

Bowlby's view of attachment was influenced greatly by an evolutionary perspective. The central idea was that natural selection has helped animals develop an attachment behavioral system for ensuring an infant's safety and survival. We don't know much about the details of life in our hunter-gatherer past, but it's pretty clear that safety was an issue. There were predators, rival social groups, and the ever-present risk of malnutrition and dehydration. Staying close to and bonding with your parents would have had obvious survival value. The attachment behavioral system would motivate infants to stay close to their caregivers and establish a relationship that would allow them to safely explore the environment, knowing that someone had their back.

One of Bowlby's most powerful insights might seem paradoxical: attachment is liberating. With a secure attachment in place, you are freed to go out into the world and learn what you need to learn. Without it, you need to either expend a lot of energy managing your relationship to your caregivers or go it alone and deal with your own distress. The instinctive nature of attachment means that children will try to form attachments regardless of how responsive or unresponsive their caregivers are. The poignant fact is infants will even become attached to abusive caregivers.

MOMMIE DEAREST

WHY? ANIMAL STUDIES SUGGEST THAT PART OF THE REASON babies will attach even to a hostile caregiver has to do with a biological switch in the infant brain.

If you're a newborn mammal, say a rat pup, bonding with your mom requires connecting to her, staying close, suckling, and letting her do what she needs to do to protect you. But moms have their own challenges. They have to drag their infants around, make sure they eat, keep them warm, and sometimes hide them from predators. So, if you're a helpless newborn, sometimes you're going to be jostled, dragged, stepped on, pinched, or have any of a variety of other unpleasant experiences. And that creates a problem. Normally you want to steer clear of situations and individuals that cause pain. As we know, rats, humans, and other mammals have fear circuits that are dedicated to helping us avoid pain and discomfort. When something threatens or hurts us, these circuits can be life-saving by helping us learn to avoid the danger: once bitten, twice shy. But forming an attachment to your mother is also lifesaving when you're a totally dependent infant. And you don't want to avoid her even if she tramples or bites you from time to time. So how is a baby supposed to act?

Research by Dr. Regina Sullivan and her colleagues at New York University suggests that natural selection helped mammals handle this dilemma by creating a sensitive period in which the infant's fear circuits are dialed down while circuits that drive approach behaviors are allowed to fire on all cylinders. Sullivan and her colleagues found that shortly after a rat pup is born, Mom's presence has a soothing effect that lowers the infant's levels of a key stress hormone, corticosterone. The low stress hormone levels keep the infants' fear circuits in the off position,[27] and the rat pup will even endure electric shocks to get to its mother. This shift in how the brain learns creates a protected time for bonding and "allows the infant to attach to the caregiver at all costs."[28]

Separation from Mom, which normally happens as weaning begins, triggers a rise in stress hormone levels and the fear conditioning system switches on, allowing the infant to begin learning to avoid danger as it ventures out into the world.

There are two unsettling implications of Sullivan's work. First, early-life stress can disrupt the biology that normally bonds mother and infant. As long as Mom is around, she provides a buffer that keeps the corticosterone level low and lets the attachment process proceed normally. But prolonged separation or a chronically stressed and unavailable mother can increase this stress hormone level, prematurely closing the window for forming stable attachments and disrupting the mother-infant bonding process.[29]

The second implication is that the "attachment neural circuit" is basically blind to the quality of care. That means the infant will bond to a caregiver regardless of how nurturing she is. A wide variety of mammals—from rats to dogs to monkeys and humans—appear to have this kind of innate push to approach their caregivers with a blind trust, to give them the benefit of the doubt no matter how they are treated in return.

In humans, the strange phenomenon of attachment even in the face of danger has provided some of the oddest and most compelling stories of the past thirty years. In a comment on Sullivan's findings, the biologist Robert Sapolsky drew a parallel to the ordeal of Jaycee Dugard, the girl whose story captivated the nation when she was freed from eighteen years of brutal captivity. She had been abducted at age eleven and kept by her deranged captor, Phillip Garrido, in a hidden backyard compound where she was raped and gave birth to two of his children. And yet, she seems to have had many opportunities to flee and chose not to. The "family" traveled in public and in later years she worked as the graphic artist in Garrido's print shop, interacting with customers, making phone calls, and writing e-mails. When she was first interviewed by Garrido's parole officers, she described him "as a 'great person' who was 'good with her kids.' "[30] In her private diaries she expressed her anguished yearning for freedom, but also wrote, "I don't want to hurt him, sometimes I think my very presence hurts him . . . I will never cause him pain if it's in my power to prevent it."[31]

The stories of Jaycee Dugard, Patricia Hearst, and other high-profile abductees who seemed to form a bond with their captors are sometimes labeled examples of Stockholm syndrome, a phenomenon that was described following a six-day hostage ordeal in a Swedish bank in 1973. When the hostages were released, they kissed and hugged their captors.[32] Actions like these suggest that our need for attachment might, in some cases, be more powerful than reason.

Attachment in human infants may not have the distinct sensitive periods that Sullivan finds in her rats, but it does involve a kind of "perceptual narrowing" that we've seen with other sensitive periods. Recall that with vision, language, and emotion perception, young children's minds go from a state of being broadly receptive to more narrowly committed. Attachment in infancy follows a similar path. For the first few months of life, newborns can be comforted by a variety of people. By seven or eight months, they begin to discriminate, focusing their attachment behaviors on their primary caregivers. They can tell the difference between "mommy" and a stranger, and they want mommy. And over the next several months, they show more and more distress on separation. After all, their sense of time and the future is fuzzy—it's not clear that when mommy leaves, she's ever coming back. There are two key signs that an infant is attached: he becomes fearful around strangers (stranger anxiety) and he shows distress when his caregiver leaves (separation protest and separation anxiety). And so, over the first year of its life, the baby narrows its attachment system to focus on its primary caregiver.

And then, in the two to three years that follow, attachment behaviors come into full focus and center on what John Bowlby identified as three key functions: (1) maintaining proximity and avoiding separation; (2) using the caregiver as a secure base from which to explore; and (3) using the caregiver as a safe haven to which the child can turn for comfort and support.

STRANGER IN A STRANGE ROOM

DEVELOPMENTAL PSYCHOLOGISTS WHO STUDY ATTACHMENT HAVE relied heavily on an experimental paradigm called the Strange Situation that was developed in the 1960s by Mary Ainsworth, a long-time colleague of Bowlby's. It typically lasts about twenty minutes and takes place in a room with toys and two chairs. The mother and her twelve- to eighteen-month-old toddler go through a series of episodes that are designed to be mildly and progressively stressful for the child. The idea is that threatening situations—encountering strangers, being abandoned, being injured or ill—activate the child's attachment system. The goal of the Strange Situation is to create an environment in which the nature of the child's attachment is revealed. And it includes two scenarios that ought to powerfully trigger our evolved attachment systems: being left alone in an unfamiliar place and being left with a stranger.[33]

The Strange Situation involves eight standardized periods, each lasting one to three minutes, in which the child is observed interacting with her mother (or other primary caregiver) and a stranger who is part of the research team. Twice during the procedure, the mother leaves the room. For the first separation, the mother leaves the child with the stranger and, for the second, the child is left alone. Later, the mother returns, and the researchers gauge how the child responds to reunions.

Based on this simple but evocative series of episodes, Ainsworth and her colleagues were able to classify the child's behavior into one of three main attachment patterns.[34, 35] Most children (about 55 to 60 percent) fell into the "secure attachment" group. These children use their mothers as a secure base from which they can explore the environment. When the mother leaves the room, the child may become distressed, but when she returns the child is happy to see her or is easily comforted and quickly returns to playing and exploring.

The remaining groups are classified as "insecure attachments." About 20 percent of children exhibit avoidant attachment—they readily explore the room and don't show much interest in Mom. When left alone, they don't cry or seem distressed, and when she returns, they don't engage with her and even actively avoid her. If Mom picks the child up, the child may stiffen and pull away, seeming to want to keep his distance and go back to the toys. And finally, about 10 percent of children are classified as ambivalent. They are wary about exploring and often have a good deal of stranger anxiety. They're extremely upset when Mom leaves, and are ambivalent at the reunion—seeking comfort from her but continuing to be distressed and angry with her.

In later years, it became clear that about 15 percent of infants had a profile that didn't fit well into these groups, prompting the creation of a fourth "disorganized/disoriented" category.[36] These were children who seemed to lack any coherent strategy for responding to separation or reunion with their caregiver. They seemed overwhelmed and frightened, and their responses often seemed contradictory or idiosyncratic. They might scream for Mom to come back only to freeze or drop to their knees and rock back and forth when she appeared. Or cling to her while also pulling away.

Although the details of this attachment typology have been debated, it's proven remarkably durable. These same basic attachment patterns appear in Europe, America, Africa, and Asia, even though the cultural traditions of how caregivers and children interact may be quite different.

But even if attachment behaviors in infancy are innate and universal, the quality of the attachment depends crucially on the nature of nurturing. Given the chance, all infants will form attachments, but the particular style of the attachment will be determined by how the caregiver and the infant interact. Caregivers who are sensitive and responsive to their baby's needs are likely to foster secure attachments in their children. They offer consistency and predictability in their interactions with their children. And bolstered by that steady presence, the child learns to regulate his own emotional states, when to seek comfort and when he can handle things on his own.

As we saw with temperament in Chapter 2, though, attachments depend on the dynamic interplay between caregivers and their children—the goodness of fit. A child whose nervous system tends to be highly excitable or who is temperamentally needy or fearful will challenge the patience and nurturing capacities of her parents. And if the mother is stressed, lonely, or excitable herself, the result may be an insecure attachment.

John Bowlby argued that a central component of the attachment process is the development of what he called "internal working models." Through their daily interactions with their caregivers, infants and children develop a mental representation of their primary relationship and their place in it. Is Mother* trustworthy and responsive? Is she erratic, frightening, or unpredictable? What effect do I have on her? Am I lovable? In the reflection of this relationship, we begin to discern who we are and what we can expect of others. As we grow, this template may be updated and revised, but it provides the scaffold on which we build our other attachments throughout life. It shapes whom we seek for friends and lovers, and creates a lens through which we interpret their behavior. It is the root of our assumptions about the comfort or the pain that attachments entail.

Secure attachment is the norm, but all these organized attachment patterns are normal—that is, they are adaptive responses to different kinds of care. Rather than attachment gone wrong, insecure attachment behaviors might be strategic compromises. One solution to life with an unpredictable or erratic caregiver might be the alternating attachment behaviors that characterize the ambivalent pattern: be vigilant for signals of abandonment, be dramatic in demanding attention, resist intrusive behavior. Or, if you're born to parents who are unavailable or rejecting, the self-reliant style of avoidant attachment might be your best bet: don't make too many demands on a standoffish parent, keep your distance, find other outlets for stimulation. So natural selection may have equipped the infant mind with a menu of attachment strategies that would be sensitive to cues about the circumstances of our family environment.[33]

---

* I am using "Mother" as a shorthand for caregiver. While early attachment theorists focused predominantly on the mother-child relationships, subsequent research has shown that fathers and other caregivers can also be the object of our primary attachments.

Insecure attachment is not a disorder, but understanding the psychobiology of childhood attachment helps us understand how things can go awry. Almost all conditions that we recognize as psychiatric disorders involve relationships and attachments in some way. But are there disorders of attachment per se? That is, are there disorders in which we can say that the fundamental problem is one of attachment? It seems so.

When the process of attachment is catastrophically fractured, the result can be a devastating inability to bond and form relationships. In the most extreme case, some children are simply never given the opportunity to form an attachment. We enter life biologically prepared and primed to form an attachment to a caregiver. Our brains expect some kind of caregiver with whom we can identify and bond. But what if that caregiver never comes? Without the opportunity to attach, children can be profoundly disturbed.

The diagnosis of "reactive attachment disorder" (RAD) is reserved for children whose early caregiving environment was one of persistent neglect, abuse, or chaos and who have gross distortions in their ability to form social bonds. Much of what's known about the disorder emerged from studies of children who spent their infancies in Romanian orphanages and other bleak institutional settings. RAD is relatively uncommon, affecting less than 1 percent of children; but up to 30 to 40 percent of institutionalized children or children who have landed in foster care because of abuse or neglect show signs of the disorder.[37]

Children with RAD can be emotionally withdrawn. Often, they are unable or even afraid to seek comfort from other people when they are upset. Or they may be indiscriminately sociable with no primary attachment but "seemingly willing to seek and accept comfort from almost anyone, including strangers."[37]

While RAD exemplifies a tragedy of the human experience, it also underscores the remarkable resilience of the human mind. Despite the profound impairments that children with this condition suffer, their situation is not hopeless if they are given a chance. Most children with RAD who are adopted out of institutions into supportive homes no longer meet criteria for RAD after time has passed, although they may still bear scars.

But as we'll see, the dramatic deprivation that causes RAD is not the only way disruptions in early attachment can cause long-lasting pain.

"WHY SHOULD I TRUST ANYONE?"

THE BEEP! BEEP! BEEP! WOKE ME UP WITH MY HEART POUNDING. I looked at the clock. 3:30 A.M. I groped for the pager on my bedside table and looked at the message. There was a phone number and the message "Sandra—emergency—please call." It was the third time in the past three weeks that an emergency page had come from that number. The pounding was subsiding as I began to gather my thoughts and dialed, not sure what to expect.

"Hello, it's Dr. Smoller answering a page."

Sandra was crying on the other end of the line. "I don't know what to do." I'd been Sandra's psychopharmacologist for several years.

"Can you tell me what's going on?"

"I don't have a therapist!" she said angrily.

A month earlier, Sandra had told me she'd fired her therapist of three years because she felt the therapist wasn't listening to her and was cold and unfeeling.

"I know, and I know that's been very difficult for you." But I calmly reminded her that we had agreed that when these panicky feelings came on, she would write them down and we would talk about them at our next meeting.

"I don't know if I'll be there," she said.

"What do you mean?"

"Oh nothing. Fine—have a nice vacation!" she said sarcastically and hung up. Our next appointment was two weeks away, after my return from a ten-day vacation.

Two weeks later, I found myself sitting in my office waiting for Sandra. The clock was winding down on our appointment and I began to worry. She had told me she might not make the appointment. But what did she mean by that? Several months earlier, I'd gotten a page from a colleague in the emergency room. Sandra had made a suicidal gesture by taking a week's worth of her antidepressant and antipsychotic medication after a disagreement with her therapist. As she would have known, it was a nonlethal overdose. But that incident ran through my mind as I tried to figure out how to respond to her absence. Had she done something self-destructive? Or worse? With ten minutes left in the appointment, Sandra arrived, looking annoyed. "The bus was late," she said with exasperation. I was relieved.

In the weeks that followed, we talked about the events around my vacation. She acknowledged that she had felt abandoned—by her therapist, by me, by everyone. Over time, the rift healed. But she remained on guard and fragile. "Why should I trust you? Why should I trust anyone?"

Sandra had survived a chaotic childhood. Her mother was unpredictably moody and suffered from depression on and off for most of Sandra's early years. In Sandra's childhood memories, her mother was often in bed, waking up to bark orders or asking Sandra to comfort her. She recalled her father, a successful businessman, as aloof and incessantly critical. He had a drinking problem, and when he drank he was loud and scary. In elementary school, she had problems sleeping and frequent nightmares. When the nightmares were bad, she would go to her parents' room and ask to get into bed with them, but her mother always ordered her back to her room. In high school, she began to drink, sometimes ending up having casual sex after a bout of binge drinking. She had a series of painful relationships that often began with sex. She would fall desperately in love with boys who expressed interest in her, idealizing them and then feeling devastatingly abandoned and betrayed when it became clear that they were just interested in a good time. The agonizing sense of mistrust and loneliness that she had begun to feel as a child grew more and more entrenched. What had been a wound became a scar that kept reopening.

Sandra suffered from a condition that psychiatrists and psychologists call borderline personality disorder (BPD). The term *borderline* was first coined in the mid-twentieth century, when psychoanalysis dominated psychiatric practice and the main diagnostic question was "Who should be analyzed?" In the Freudian tradition, psychoanalysis developed as a treatment for neuroses. Patients with psychoses, like schizophrenia, were considered poor candidates for analysis. John Gunderson, a pioneer in the classification and treatment of BPD, noted that the borderline label was originally a term of art, loosely used to describe patients who were at the boundary of neurosis and psychosis and who were liable to regress into psychotic states during analysis.[38] But beginning in the 1960s and 1970s, psychiatrists began to see borderline personality as a syndrome in and of itself, and in 1980 it achieved the status of an official diagnosis, borderline personality disorder, in the DSM-III.

Unfortunately for many people, the iconic example of BPD is *Fatal Attraction*'s notorious Alex Forrest, played by Glenn Close, who seduces and then terrorizes Michael Douglas's character. As a rule, people with BPD are not bunny-boiling stalkers, but then Hollywood tends to favor boffo box office over nuanced narratives. Still, the signs and symptoms of BPD can be dramatic. Those affected are prone to torrents of emotional pain that are often expressed as rage, panic, and self-destructive behavior. They may experience intolerable feelings of emptiness and have a hypersensitivity to perceived rejection, betrayal, and abandonment—feelings that can trigger emotional storms and frantic efforts to avoid being alone. Sometimes these episodes culminate in self-injury, including cutting themselves or attempting suicide. One in ten succeed.

A hallmark of the disorder is a pattern of unstable and intense personal relationships. People with BPD may alternate between idealizing ("you're the only one who understands me") and angrily devaluing caregivers ("you never cared about me!"). These shifts are typically triggered by separation or perceived abandonment. For friends and family, being on the other end of a relationship with someone with BPD can be a bewildering roller coaster, leaving them walking on eggshells to avoid provoking another crisis.

BPD has had a certain notoriety in clinical circles as well. It's sometimes used as a synonym for "the difficult patient." And that can be an obstacle to compassionate care. In 1978, an influential paper by James Groves, entitled "Taking Care of the Hateful Patient," appeared in the *New England Journal of Medicine* and challenged physicians to recognize the ways in which patients with borderline and related personality traits trigger doctors' own unconscious negative impulses (or, as psychoanalysts would say, negative countertransference). Because of their hostile dependency and emotional volatility, these were patients "whom most physicians dread to treat."

That image has begun to change for at least two reasons. First, newer, more effective psychotherapies have been developed specifically for BPD, providing an antidote to therapeutic hopelessness for patients and clinicians. And second, long-term follow-up studies have shown that the prognosis for BPD is much better than anyone had thought, with improvement rates of up to 85 percent by ten years of follow-up.[38, 39]

Still, there is controversy over the disorder. Some have argued that it's simply a misnomer for unrecognized depression, bipolar disorder, or posttraumatic stress disorder. Others have gone even further to claim that the whole notion of personality disorders is a muddled concept without a clear scientific or medical basis.

Regardless, it's clear that the diagnosis of borderline personality disorder captures some kind of enduring pattern of unstable emotions and relationships that can wreak havoc on affected individuals and those around them. And, as with most psychiatric disorders, genes play a role. Based on twin studies, variations in genes account for anywhere from 35 to 70 percent of BPD risk in the population.[40-42]

In some ways, BPD is a perfect storm of many of the phenomena we've talked about so far. The vulnerable child may start life with a genetic endowment that creates a temperamental bias toward emotional reactivity. Then major adversity or a hostile or erratic home environment may program the stress hormone system in ways that make it even harder for the child to regulate her emotions. They may also bias her brain toward perceiving and feeling negative emotional states. These perceptions in turn affect the development of theory of mind and empathic skills resulting in a tendency to misread other people's intentions and feelings and a hypersensitivity to signs of threat and loss. But there is increasing evidence that much of this plays out in the process of attachment.

Studies have consistently found a strong association between insecure attachment and BPD.[43] It's been said that "BPD is typically not a disorder of the unloved but a disorder of those who were loved inconsistently."[44] In one of the few prospective studies that have followed children from infancy to adulthood, Harvard psychologist Karlen Lyons-Ruth and her colleagues identified several steps along the trajectory from insecure attachment to adult borderline personality. Initially, researchers observed twelve-month-old children in the Strange Situation procedure. As Lyons-Ruth's team coded the videotaped interactions, they found an interesting pattern. Children who went on to develop borderline traits tended to have mothers who displayed a pattern of withdrawal and emotional distance when they reunited with their infants after a separation. The children also tended to seek more contact than the average child. That combination—a child with a heightened need for contact and a mother who, out of fear or trouble handling her own distress, tends to withdraw when her infant expresses a need to be close—spelled trouble.

In one typical videotaped interaction, we see the mother come back into the room and stand still rather than approach her crying child. The child runs over to his mother who picks him up and begins to comfort him. She kisses his cheek and walks him over to some toys on the floor and stays by his side while he begins to play. So far, so good. But after a few moments, she moves away and sits on a seat several feet away. He seems confused and toddles over to her. She offers him a toy. The little boy's distress begins to show, but instead of comforting him, she puts the toy in his little hand. He begins to cry and slowly moves away. And then something odd happens. He freezes. He seems bewildered and dazed, almost as though he's retreated into his own world, as he stares at the toy. This goes on for nearly thirty seconds until he brings the toy to his mouth and begins to soothe himself. The interaction is subtle and perhaps unremarkable—but for a developmental psychologist, it's revealing. Over countless small daily interactions, this combination of maternal withdrawal and disorganized attachment behavior seemed to nudge children in Lyons-Ruth's study along a path toward borderline traits later in life. But whether they continue on that path depends in part on what happens next.

In the face of an inconsistent or withholding caregiver, some children develop strategies to control the relationship. They may become caretaking themselves, trying to smooth things over and make sure mother isn't upset or angry. Or they may become more dramatic in their demands for caretaking, responding to distance or separation with emotional tantrums as if to say "I will not be ignored!" And, when the subjects in the Lyons-Ruth study reached middle childhood, this kind of controlling behavior on the part of the child proved to be another step along the trajectory toward borderline traits and a tendency to engage in self-injury later in life.

Through a collision of genetic vulnerability and inconsistent caregiving, a child at risk for BPD may form an "internal working model" of attachment figures as untrustworthy and liable to disappoint or harm them. Instead of developing an ability to regulate her own emotions and needs, she becomes preoccupied with controlling the emotional states of others.

The need to monitor and control the mental states of others may reinforce the hypersensitivity to emotional signals that's characteristic of BPD.[45] Brain-imaging studies find that individuals with BPD have a hypersensitive response of the amygdala and other limbic regions when they're shown emotional faces.[46, 47] This exquisite sensitivity may have enduring effects on the development of their theory of mind and empathic capacities. On the one hand, some studies have shown an enhanced ability to read the mental states of other people. One study used a test developed by Simon Baron-Cohen called the Reading the Mind in the Eyes Test.[48] It works like this: subjects are shown a series of pictures of faces that include only the eyes and the regions right around them. They are then asked to choose which of four words correctly describes what the person in the photograph is thinking or feeling: "panicked," "cautious," "friendly," "regretful," and so on. Compared to healthy controls, subjects with BPD were significantly more accurate at reading mental states from the photographs.[49]

On the other hand, studies have shown that those with BPD have a subtle bias toward reading negative affect in others—especially anger and fear. It as though the gain is turned up on their affect detectors—they see anger and other negative emotions faster than other people, but they are also prone to see them when they're not there.[46] As we saw in Chapter 3, victims of child abuse and neglect show this same hypersensitivity to reading negative emotions; remarkably, childhood histories of abuse and neglect are reported by up to 90 percent of individuals with BPD.[47, 50]

TRUST ME

ALL OF US ARE BORN WITH BRAINS THAT EXPECT TO ENCOUNTER a caregiver and are prepared to form an attachment to that person (usually Mom). Barring a catastrophe—severe deprivation, for example—we will attach over the course of the next two or three

years. But normal variations in attachment behavior and temperament, coupled with particular styles of caregiving can set a trajectory toward healthy relationships on the one hand or unstable ones
on the other.

The key ingredient underlying all of this is the establishment of
a sense of trust. Trust is the glue that bonds relationships—from
personal friendships to geopolitical alliances. Check out your favorite source of news, and you'll realize that it's at the core of most
stories that capture our attention: sex scandals (from Tiger Woods
to John Edwards), finance (from Bernie Madoff to the Wall Street
backlash), and global conflict (from the brinksmanship with Iran
to the Arab-Israeli conflict). Of course, trust is also fundamental
to forming attachments. Our basic sense of trust in others emerges
early in life, as infants experience their caregivers and come to see
the world as basically stable and safe or precarious and threatening.

The influential psychoanalyst Erik Erickson (1968) put the development of a basic sense of trust versus mistrust at the very foundation of personal development, calling trust "the cornerstone of a
vital personality." As Erikson put it, "Mothers create a sense of trust
in their children by that kind of administration which in its quality
combines sensitive care of the baby's individual needs and a firm
sense of personal trustworthiness within the trusted framework of
their community's lifestyle. This forms the very basis in the child
for a component of the sense of identity which will later combine
a sense of being 'all right,' of being oneself, and of becoming what
other people trust one will become" (pp. 103–104).[51]

BRAIN TRUST

LATER IN LIFE, THE MENTAL MACHINERY OF TRUST BECOMES MORE
complex—we use theory-of-mind skills and emotion-detection skills
to gauge another person's intentions and to see if they are deceiving
or cheating us. But only recently has trust itself become the subject

of scientific study. And research on the biology of trust is once more implicating our old friend oxytocin.

In a groundbreaking experiment, a team of scientists at the University of Zurich had volunteers play a simple two-person game.[52] Both players start out with twelve units of money. One player, the investor, makes the first move by giving none, some, or all of his money to the other player (the trustee). The trustee can give the investor anything from zero to all of his money back. The catch is that, going into the game, the researchers tell the players that they will triple any money the investor gives the trustee. The more money the investor gives, the more money there is for the two players to split. If the trustee can be trusted to give some money back, it's a win-win situation.

Before the game began, players were given a single dose of a nasal spray that contained either oxytocin or placebo. Investors who inhaled the oxytocin became much more trusting—that is, more likely to give others money in the belief that their generosity would be reciprocated. Interestingly, oxytocin only worked when subjects were interacting with other people. When the oxytocin-snorting investors played against a computer algorithm, there was no increase in their trust behavior. It also wasn't a matter of their just being nicer (what psychologists call *prosocial*), because trustees who were given oxytocin didn't increase the amount they gave back to an investor.

The study spawned a wave of research that suggested oxytocin is central to the biology of trust and attachment. Later studies found that inhaling oxytocin can increase positive communication between couples,[19] enhance generosity,[53, 54] and make hostile faces seem more familiar.[55] Now you see how the business model for Liquid Trust was born.

THE FEEL-GOOD HORMONE OF THE YEAR?

BUT HOW EXACTLY COULD OXYTOCIN—JUST A LITTLE STRING OF nine amino acids—tweak our brains to see others as more trustworthy and make us more cooperative and even loving? The answer seems to involve three related brain systems that are familiar by now: maternal care, social cognition, and harm avoidance.

First, as the work with rats and voles suggested, natural selection made use of the maternal functions of oxytocin and expanded them to create a mechanism for bonding to others. Start with a peptide hormone—oxytocin—that kicks in when a newborn arrives—just when nurturing needs to start. Now hitch that peptide up to the brain's emotion and reward circuits and you can coordinate child*birth* with child*care*. The motivation and reinforcement for caring for an infant was in place. In humans, with our complex cognitive and emotional capacities, that bonding became something more profound—love. And so it may have been a relatively small evolutionary step to translate a system for maternal love into other bonds and attachments, including romantic love.

But that's not the whole story of how oxytocin connects us to other people. It also seems to help us tune into the thoughts and feelings of other people by enhancing mind reading and empathy. After inhaling oxytocin nasal spray, people are more likely to fixate on the eye region of faces[56] and do better on tests of reading emotional cues.[57] People who carry a particular variant of the oxytocin receptor gene have a similar advantage in reading emotional expressions.[58] And perhaps the most striking evidence that oxytocin augments mind reading are studies showing that oxytocin can help people with autism spectrum disorders become better at recognizing emotional cues from other people's faces and voices.[59, 60]

The third element of oxytocin's effect may be the most fundamental to trust, and it has to do with fear. It turns out that the amygdala is loaded with oxytocin and vasopressin receptors and that they have opposite effects on fear behavior.[61] Oxytocin turns down the amygdala's fear response and vasopressin turns it up. Together, they appear to create a balance that helps determine where an animal sits on the approach/avoidance continuum. Dialing up oxytocin nudges the brain toward approach by dialing down fears and inhibitions. But oxytocin seems to have its greatest effect on reducing fears of other people. Several studies have now shown that oxytocin dampens the amygdala's reactions to people's faces[62-64] and biases us to see positive aspects of other people.[65, 66] In some ways, the effect of oxytocin on how we see others is the opposite of the effects of trauma and early adversity. Earlier we saw that children who have suffered abuse and neglect are biased to see negative emotions.[67] If early trauma puts dark shades over our mind's eyes, oxytocin seems to give us rose-colored glasses.

The ability to reduce fear is crucial, because if oxytocin merely enhanced our mind reading, that might not do much for trust—it might even make us less trusting. After all, if you are more sensitive to what people are up to, you might be more vigilant about getting scammed. But once you add in the attachment-promoting and fear-reducing effects, now you've got something that just might add up to trust. In other words, one interpretation of oxytocin's trust-enhancing power is that it helps us tune in to other people while bathing them in the warm glow of goodness. It's like what some people describe when they take Ecstasy ("I love you, man")—and yes, Ecstasy stimulates oxytocin.[68, 69]

So oxytocin reduces our fears about other people and pushes us to give them the benefit of the doubt. It's a kind of anticynicism effect—we expect the best of other people. And at the same time, it makes us less sensitive to betrayals of trust. In an elegant study, the University of Zurich team took the trust game one step further. Their original study of oxytocin's effect on investors in the trust game was missing a key element of how we decide to trust in real life: we develop trust over a number of interactions by seeing what people do and learning how trustworthy they are. In the new study, the researchers went one more round. After inhaling oxytocin or placebo, subjects played the role of the investor in the trust game with a human partner. After the initial round, the investors were given feedback about how their investments fared: they were told that the investee gave back money only about 50 percent of the time.

Now what would you do if you were the investor in the trust game and you were given another opportunity to play the game? You trusted some guy with your money, hoping he'd reciprocate and make you both richer. But half the time, the selfish jerk kept the money for himself. You'd probably be much less willing to trust him again. And that's exactly what happened in the experiment for those given placebo—on the second round, they offered their trustees much less money. But those who were given oxytocin had no drop-off in their trust behavior after being betrayed—that is, oxytocin seemed to make people insensitive to betrayal.

While the subjects played the game, their brains were being scanned by an fMRI. After learning that their trust had been betrayed, subjects who got placebo had a much stronger response in the amygdala and related fear circuits compared to those who got oxytocin. Previous studies had shown that the amygdala is, among other things, a trust sensor: it lights up when people are shown faces that look untrustworthy[70] and patients with amygdala damage tend to view others as more trustworthy and approachable.[71] The placebo group also had activation of the caudate nucleus, a region shown to be involved in adapting to another person's behavior in trust games.[72] But the oxytocin group didn't have these activations of the amygdala and caudate. By damping down the brain's social fear and social judgment circuits, people with oxytocin on board continued to see their beneficiaries as trustworthy even after they'd been double-crossed.

Survival of the Skittish

Like attachment, mind reading, and harm avoidance, trust has obvious implications for survival. Could our minds have evolved specialized mechanisms for determining how much we trust others? After all, trust is essential to the success of our relationships and to the scores of social encounters we participate in every day. Every time you buy something from a salesperson, confide in a colleague, or make an investment, you are choosing to trust someone— someone who might take advantage of you. The consequences of being cheated could be devastating, depending on when and in whom you placed your trust.

That situation was no different for our hominin ancestors. In evolutionary terms, a person's survival (and opportunity to pass on their genes) could easily depend on being right about whom he trusted. For example, females, who need to commit to investing lots of resources in bearing and caring for offspring, would have been strongly motivated to select mates who could be trusted to stick around and contribute. Males, who can never be certain about the paternity of their offspring, needed to know when a female was unfaithful. And any time you take a risk by cooperating in a dangerous situation or sharing your resources, your very survival may depend on the trustworthiness of your collaborators. Given how high the stakes are, it might be worth dedicating some of the brain's operating system to the task of detecting trustworthiness and monitoring the outcomes of social exchanges.

And that seems to be the case. Our brains are exquisitely adept at detecting violations of trust. We make judgments about trustworthiness at warp speed and with very little information. For example, we can detect when someone is being authentic or insincere just based on how they smile.[73]

In an elegant series of experiments, the evolutionary psychologists Leda Cosmides and John Tooby showed that people are remarkably good at "cheater detection." Most of us are not very facile with formal logic or reasoning. In fact, when presented with a standard logical problem in which people are asked to decide what information is needed to determine whether a logical rule has been violated, the vast majority of people get it wrong. But when the problem involves determining whether someone has violated a rule by cheating in a social exchange, we suddenly become expert logicians—60 to 85 percent of people get it right. Our cheater detection skills, like our theory of mind skills, have the hallmarks of an evolved mental mechanism, and Cosmides and Tooby have argued that these skills are part of our "universal human nature."[74] By age three or four, children from industrialized Europe to rural Nepal understand when someone has violated a social contract,[75] and hunter-gatherers from a remote region of the Ecuadorean Amazon perform just as well on cheater detection tasks as Harvard undergraduates.[76]

There is also at least some evidence that the human mind's ability to detect cheaters is rooted in particular brain areas. In one study, psychologist Valerie Stone, along with Cosmides and Tooby gave reasoning tests to R.M., a patient who had been in a bicycle accident that extensively damaged his limbic system, including the orbitofrontal cortex and both his right and left amygdalae.[77] They found that compared to healthy controls and other patients with brain damage, R.M. reasoned just fine about most kinds of rule violations, but when it came to detecting social contract violations, he was lost.

The evidence so far suggests that trusting and avoiding betrayal involve two mental components. One is more about emotional judgments and relies on subcortical brain regions like the amygdala and striatum, where oxytocin and vasopressin help calibrate our willingness to trust, our expectations about other people's behavior, and our wariness of being betrayed.[64, 72, 78] The other is more about reasoning and seems to involve higher cortical circuits,[77, 78] providing an innate set of mental algorithms that determine when someone has broken a social contract.

Dys-Trust

"Why should I trust anyone?" Sandra's question became a *focus of our discussions over the next several months. I encouraged her to reengage in therapy and had given her several referrals, but she never followed through. What was standing in the way? I asked. "I've had it," she said. She had come to see her relationships, including those with her therapists, as a series of disappointments and betrayals. In her early twenties, she had hooked up with a man named Mark, whom she'd met at a club. She was wary of getting involved, and the more time they spent together, the more fearful she became. She was almost paralyzed by a sense that she was going to do something that he'd use as an excuse to leave her.*

But it didn't happen and "after a while, I let my guard down and I let myself fall in love." Everything was fine at first. He seemed genuinely interested and would listen and comfort her when she was upset. One day they came up with a business idea—he was a talented chef and she knew some people in the food industry, and they decided to launch a business catering for parties. Unfortunately, that was a catalyst for conflict. They argued and fought about almost everything, with Sandra often erupting into angry outbursts, accusing him of devaluing her ideas and treating her like a child. Finally, Mark told her he couldn't take it anymore and ended the relationship. She was devastated and her anguish took the form of relentlessly hounding him and demanding he explain why he had lied to her and used her.

In Sandra's mind, Mark had made promises he never intended to keep. When he cut off all communication with her, she upped the ante. She sent him an angry e-mail that included a threat to sue him. And then she took a knife and cut up her forearms. She called Mark to tell him what he'd made her do, but he didn't answer the phone. And so she paged her therapist, who convinced her to go to the emergency room.

The next crisis happened a few years later. Her therapist at the time had apparently tried to make Sandra feel secure by encouraging her to call anytime if she needed to. She had even given Sandra her home phone number. As Sandra told it, the therapist became more of a friend than a therapist. Sandra did feel a sense of calm knowing that her therapist was always available. But one night, Sandra was feeling lonely and was ruminating about whether "alone" would be her fate forever. She began to feel panicky. She picked up the phone and called her therapist. But there was no answer. She left a distraught message, asking her therapist to call immediately. As one hour and then another passed, Sandra became overwhelmed. She began drinking and then began cutting. Finally, she called an ambulance and by the next morning, she was in the hospital. "Now do you see why I don't exactly feel like I should trust you?"

I've argued that BPD is a disorder of attachment. But from another angle, it's a disorder of trust. Individuals with BPD are exquisitely sensitive to signs of betrayal, insincerity, and abandonment—and, as we've seen, neuroimaging studies suggest that, like victims of child abuse more generally, there is a perceptual bias toward seeing other people as hostile and threatening. The residue of their insecure attachments and the experience of inconsistent love would only fuel the presumption that relationships are fraught with danger. They yearn to connect but have learned to beware.

And since trust is the foundation of cooperation, maintaining stable relationships is a challenge. What would you see if you asked individuals with BPD to play the trust game? Read Montague and his colleagues at Baylor College of Medicine asked this question by having healthy volunteers play a multiround trust game with trustees who were either other healthy volunteers or individuals with BPD. Using fMRI, they were able to capture brain images of the players in action. The healthy pairs got into a rhythm of cooperation and trust: investors transferred money and trustees reciprocated. If an investor made a low offer, signaling a possible loss of faith in his partner, the trustees often increased the amount they returned as if to coax the investors to trust them again. These gestures of goodwill usually worked, restoring the cooperative relationship and the flow of money. But when the trustee was someone with BPD, the social exchange often broke down. The BPD partners didn't try to repair the relationships with gestures of goodwill, and soon the exchange collapsed.

What happened? When asked, the BPD subjects reported much lower trust than the healthy volunteers. When healthy trustees were faced with a low offer, brain scans revealed heightened activity in the insula, a region of the limbic cortex previously implicated in violations of social norms and the sense of being exploited. But the BPD trustees didn't have the same insula response. It was as though they weren't surprised by a low offer. They expected to be screwed. Of course, one study doesn't prove the case, but it does fit with the idea that BPD involves a neural bias to mistrust and to expect the worst from relationships.

ONLY CONNECT

WE'VE SEEN THAT THERE ARE EVOLUTIONARY AND BIOLOGICAL IN-fluences on human relationships: from maternal love and attachment to romantic love and trust. And as varied and complex as these phenomena are, there are unifying threads running through them. At one level, they all reflect one of the most profoundly enriching features of the human mind: our ability to connect. We not only cooperate, we care for each other. Our connections to other people begin from the moment of birth and set the trajectory of whom we become and how we live. Love, commitment, cooperation, estrangement, and grief: these are the chapter headings for the story of our lives. They clearly cannot be reduced to biology. But it's also clear that biology is at work.

In a circuitous way, our capacity to love is the reward we get for having a brain that does much of its developing after we're born. The mature human brain is big. Literally too big to bear. To deliver a head that contained a full-grown human brain, the female pelvis would have to be radically restructured. You might say there was a "goodness of fit" problem. So human infants have relatively small heads that grow over a period of years to accommodate the developing brain. We are an altricial species—born helpless and vulnerable. We need nurturing and protection. Our ancestors undoubtedly faced strong selection pressures to find a way to keep infants alive until they could fend for themselves. The solution was twofold: (1) use hormone systems and reward circuits to motivate maternal nurturing and cement pair-bonds between parents so they protect and support their infants; and (2) endow the infant with an innate drive to attach and make use of that caregiving. And, happily, those same systems provide a biological foundation for connecting with one another throughout our lives.

So here's the remarkable thing: our relationships are always, directly or indirectly, under the spell of these two biological imperatives—parent-child bonding and child-parent attachment. The connections we later forge are not only psychologically but even biologically rooted in that first bond. The very circuitry and chemistry of parenting, attachment, trust, and romance are linked.

We are all endowed with the biological drive and equipment to connect. Even in the face of life's challenges, we usually manage to attach and find love. But the manifold variety of how it all plays out depends on the particulars of who and what we find along the way: the goodness of fit in our earliest attachment and the catalogue of triumphs and traumas that we encounter in childhood and beyond. As these accumulate, they rework our models and bias our perceptions and interpretations of other people. Variations in our genes may influence how we respond to these experiences. But for some of us, adversity takes a toll that is hard to overcome, distorting the trajectory of our capacity to love and to trust. Sometimes the result is a disorder of attachment: reactive attachment disorder in children or borderline personality disorder in adults.

Oxytocin seems to be one key player throughout the life cycle, promoting rewarding relationships, enhancing trust, and buffering our social fears. Maybe the idea of bottling oxytocin as liquid trust isn't so far-fetched after all. Except, of course, there's a problem. It's hard to see how spritzing oxytocin on yourself each morning would have the intended effect of "making the people around you have a strong feeling of trust," as the website claims. Wouldn't it be the reverse? You spray on oxytocin on your shirt and go on a date—presumably you're the one who's more likely to be the easy mark. You'd have to wonder, as Aretha Franklin once poetically put it, "Who's zoomin' who"?

There's another wrinkle in the story that makes the prospects of giving people oxytocin more complicated than we might hope: its effects may depend on your attachment history. For example, in one study, people with borderline personality disorder who were given oxytocin before playing a version of the trust game actually became *less* trusting and cooperative.[79] It may be that, by making people more attuned to social cues, oxytocin can heighten concerns about trust and intimacy in those who've had a history of insecure and painful attachments.

And yet, oxytocin's use as a therapeutic aid may not be far off. Larry Young, the neuroscientist whose work has helped establish oxytocin and vasopressin as social peptides, thinks so: "I think that we probably will see a time when some target, some drug, is used to stimulate the oxytocin system . . . to increase social perception or social cognition."

The brightest hope right now has been oxytocin's potential to enhance social cognition in people with autism spectrum disorders. In small short-term studies, oxytocin has been shown to enhance their ability to read mental states, make eye contact, experience trust, understand social cues, and engage in social interactions.[59, 60, 80] There are few options available to ameliorate autism, so this seems like a welcome advance, even though much work remains to be done.

And what about more mundane problems? Studies suggesting that oxytocin can reduce marital conflict are also intriguing. As

Young commented, "People go to a marital therapist to try to solve the problems that they have in a relationship. If oxytocin is helping you tune in to other people, making you more empathetic, better able to perceive the emotions of the other person, it seems like it might be useful" for couples therapy. And, as we will see in Chapter 7, there is a precedent for using drugs to tweak the brain in ways that make therapy more effective.

How far can we go with this? The limitations of Liquid Trust notwithstanding, what would it mean if there were a drug that could be used to make us more loving and trusting? Who knows—it might be the solution to bipartisan gridlock. And yet the question can't help but evoke Orwellian fantasies. An opiate for the masses? A weapon of war?

If only we could trust each other not to go there.

CHAPTER SIX

THE BRAIN OF THE BEHOLDER: BEAUTY AND SEXUAL ATTRACTION

IN EARLY OCTOBER 2000, TWO TECH-SAVVY TWENTYSOME-things named James Hong and Jim Young were sitting around drinking beer. Hong was an unemployed computer engineer, and Young was a Berkeley grad student getting a joint degree in electrical engineering and computer science. The moment of inspiration came "when a comment Jim made about a woman he had seen at a party made us think, wouldn't it be cool if there was a website where you could tell if a girl was a perfect ten?" Hong later recalled.

It might have stopped there, but these were Bay Area computer geeks in the dot-com age. Within a few hours, they had mapped out the idea and Young began writing code for a site that let users post photos and rate other people's looks. Before a week had passed, they took their site live, with ten photos of friends and a catchy name: Am I Hot or Not?[1]

When word got out online, things went viral and the response was overwhelming. Within twenty-four hours, the site had more than 150,000 views, and Hong and Young panicked and shut it down. But realizing they had a runaway hit on their hands, they quickly leased additional servers and reopened. By December, less than two months after their beer-soaked epiphany, the site was getting nearly 15 million hits per day.[1]

The concept of attractiveness ratings had clearly struck a chord. Predictably, others jumped in the game and Am I Hot or Not spawned a series of copycats, though none were as successful. In 2003 Mark Zuckerberg, a sophomore at Harvard, got into trouble for creating a "hot or not"–style site, posting students' ID photos so that other students could rate their "hotness." Within days, amid overwhelming traffic and complaints from fellow students, he had to close it down.[2] But it turned out okay for him. Three months later he tried again, launching another website from his dorm room. It was called Facebook.

What was it about judging people's appearance that was so irresistible? As the critic David Denby noted, it may tap into our culture of "snark," where snide invective has become the dominant idiom.[3] But it may also have tapped into something more enduring about the human mind. Converging evidence from evolutionary biology, psychology, and neuroscience suggests that our brains have a biological interest in assessing the attractiveness, beauty, and sexual appeal of other people. Even without the facile clickability of a website, we're constantly making judgments about who's hot and who's not.

In this chapter, we'll explore what makes us attracted to each other and what the challenge of choosing a mate tells us about the brain and its vices. Along the way we'll encounter some intriguing questions: What's the difference between a beautiful face and one that only a mother could love? Could supermodels owe their careers to parasites? Why do gay men have more older brothers than heterosexual men do? Can there be disorders of desire?

HOT TOPICS

WE ALL KNOW THAT SEX SELLS, AND ADVERTISERS HAVE CREATED something like a nation-state of beautiful, sexy people who inhabit the two-dimensional terrain of magazines and television. The "beauty industry" offers eager consumers an endless supply of new solutions to the scourges of imperfection and aging. People are willing to endure an astonishing battery of physical insults—skin exfoliation, teeth bleaching, toxin injections, and surgical procedures—to defend their ideals.

Where do these ideals come from? Perhaps the most fashionable answer is to blame that ubiquitous villain, the media. But even if it were true that a cabal of media types conspire to impose their will on a helpless public, they would fail without mass collusion. How did these images become so powerful? Why exactly *does* sex sell? Are ideals of beauty and sexual attractiveness based in the biology of the human mind? Rather than defining standards of beauty, could advertisers and plastic surgeons be tapping into a well of aesthetic preferences that evolution prepared our brains to favor?

SKIN-DEEP OR HARDWIRED

IN 1991 NAOMI WOLF'S BESTSELLER *THE BEAUTY MYTH* ARGUED that modern Western conceptions of female beauty and sexuality are the creation of a male-dominated power structure determined to "mount a counteroffensive against women." She tried to debunk the notion that biology plays a role in standards of attractiveness, calling that notion a "beauty myth" that was born with the Industrial Revolution and disseminated by the unprecedented reach of male-dominated mass media. As she put it:

> The beauty myth tells a story: The quality called "beauty" objectively and universally exists. Women must want to embody it and men must want to possess women who embody it. This embodiment is imperative for women and not for men, which situation is necessary and natural because it is biological, sexual, and evolutionary: Strong

men battle for beautiful women, and beautiful women are
more reproductively successful. Women's beauty must cor-
relate to their fertility, and since this system is based on
sexual selection, it is inevitable and changeless. None of
this is true. (p. 12)[4]

One problem with Wolf's argument is that it's a bit of a false dichotomy. Like the outmoded nature vs. nurture argument, Wolf makes it seem as if it's all biology or all politics. She is surely right that ideas about beauty or sexuality are not objective or changeless, but it's hard to imagine a mainstream biologist making such a claim.

Clearly, ideals of beauty have varied over human history and across human societies. Even within developed Western nations, iconic images of female and male attractiveness have shifted substantially. Between 1953 and 2003, the average *Playboy* centerfold gradually grew thinner (although the very thinnest centerfolds were actually more common in the 1950s).[5]

Increasingly, images of beauty have become . . . well, unnatural. On the occasion of Barbie's fiftieth birthday, the BBC News Magazine applied Barbie's dimensions to a real-life woman. Their volunteer was a twenty-seven-year-old woman named Libby. Holding her height constant at 5'6" and giving her Barbie's dimensions turned Libby into a tiny-waisted waif. But if her waist was held constant, the only way to achieve Barbie's proportions was for her to sprout to an Amazonian height of 7'6". Male action figures experienced a similar transformation. From the 1960s to the 1990s, GI Joe, Luke Skywalker, Batman, and the like have practically exploded into heaps of muscularity. Harrison Pope, the psychiatrist we met in Chapter 1, has been studying the vagaries of body image and its cultural transformations and found that if you extrapolated the dimensions of modern action figures to the size of a 5'10" man, they would have physiques "far exceeding the outer limits of actual human attainment" (p. 70).[6]

Even if there is no media conspiracy, let's just say the weight of the evidence does support an effect of media exposure on our internalized ideals of attractiveness. Among women, exposure to mass media portrayals of thinness as a female ideal are associated with an internalization of the thin ideal and greater dissatisfaction with their own bodies.[7]

So is there any reason to believe that our responses to sexual attractiveness and beauty have a biological component? There are two lines of evidence to consider when we answer that question. One is about what scientists call "proximate" causes of behavior and the other is about "ultimate" causes. Proximate causes are the specific brain and behavioral mechanisms that control our behavior. Ultimate causation has to do with why and how these biological mechanisms might have arisen in the first place. And "ultimately," it's about evolution.

IS BEAUTY A THING OF THE PAST?

WHAT'S THE EVOLUTIONARY ARGUMENT FOR WHY OUR MINDS AND brains are biologically attuned to beauty and sexual attraction? It boils down to the fact that sexual attractiveness is related to reproductive success and that makes it susceptible to natural selection. Now, the field of evolutionary psychology is sometimes accused of spinning "just-so stories": begin with an observation about human behavior and come up with a story that relates it to our Pleistocene past. And critics have challenged its assumptions about the relevance of our ancestral past to our modern minds.[8]

But let's start with a few notions that ought to be uncontroversial (unless you're a Creationist). Humans are the product of natural selection, and natural selection favors traits and behaviors that increase the likelihood of transmission of an animal's genes. To reproduce, you need to find and attract a mate. If other humans of your sex are around, you may need to compete for mates. And if your goal is to produce offspring who have the best shot at reproducing themselves, you'd do well to be able to discriminate among potential mates. Basically, you want to choose a mate who maximizes the probability that your offspring will survive and reproduce.

So, from an evolutionary standpoint, the argument is that we will be attracted to features that advertise that a potential mate is a good bet for successfully transmitting our genes into the next generation. Natural selection would have promoted physical and behavioral traits that are effective signals of a high-quality mate. The more effective we are at convincing a potential mate that we have desirable traits, the more desirable and therefore reproductively successful we would be.

Ideas about the evolution of sexual attractiveness have undergone their own evolution, beginning, appropriately, with Darwin. In 1871, with the publication of *The Descent of Man, and Selection in Relation to Sex*, Charles Darwin "dropped the other shoe," as the biologist E. O. Wilson put it. Twelve years after the publication of *On the Origin of Species*, which laid out the argument for evolution by natural selection, he introduced the concept of "sexual selection." The idea emerged from Darwin's recognition that many traits and behaviors in the animal kingdom do not have obvious advantages for survival. From the dramatic plumage on male peacocks and the euphonious song of the nightingale to the elaborately branched antlers of the red deer and the brilliantly colored face of the mandrill— Darwin saw these traits as puzzles to be solved. They require a lot of energy to build and maintain, creating an opportunity cost that must have been outweighed by some kind of benefit. Darwin's solution was that these ornaments and displays were an expression of sexual competition. They were the outcome of a process of sexual selection that favors traits that improve the odds of mating.

For the most part, Darwin observed, the pressure was on males, who needed to compete with one another for access to females. That may have sparked a sexual arms race that made males bigger, meaner, and tricked-out with weaponry like big muscles, fangs, and horns. They also needed to impress the ladies, which is where peacock tails come in—they're part of a charms race.

In 1972, evolutionary biologist Robert Trivers contributed another fundamental insight. His parental investment theory argued that mating strategies are shaped by the relative amount of investment a male or female has to make in order to increase their offspring's chance of surviving and reproducing. He wrote: " . . . the sex whose typical parental investment is greater than that of the opposite sex will become a limiting resource for that sex" (p. 141).[9] In most species, including humans, females make the greater investment and males often have to compete for mating rights and guard their mates so that they're not cuckolded. Under some circumstances, males will do best by inseminating multiple females and investing little. But females can even the score by favoring males who are willing to stick around and provide resources for their young. So sexual selection should impel females to choose mates whose appearance and behavior signal both "good genes" and "good parent." Of course, under circumstances where offspring have a much better chance at surviving with two parents around, both males and females will have a vested interest in long-term pair bonding, also known as monogamy.[10]

Decades of field research have largely upheld the original predictions of sexual selection and parental investment theory, but it's also clear that the variety of mating behavior in the animal kingdom is more complex than first thought.[11] Today, most evolutionary biologists believe that sexual selection has shaped male and female mating strategies—that is, how, when, and with whom we have sex. Among primates, mating strategies range from monogamy (in gibbons, for example) to polygyny (as in a male gorilla and his harem). Polyandry—that is, females with male harems—is relatively rare. Humans, of course, practice a wide range of mating behaviors— across and within cultures and even within a given lifetime. Sometimes we're monogamous and sometimes we're polygamous.*

---

* Polygamy is the broader term that includes polygyny (one male mating with more than one female) and polyandry (one female to several males). Polyandry, as a cultural practice, is quite rare. Less than 1 percent of preindustrial cultures practice it. Another 16 percent have monogamous marriage systems. But the clear winner is polygyny—occurring in more than 80 percent.[12]

Humans have a menu of mating strategies that we can and do pursue—depending on what our options are.[12] Sometimes it pays to pursue "short-term" mating and sometimes it pays to commit. And that has implications for what's hot and what's not.

MIND YOUR PLEASING CUES

IF WE ACCEPT THAT OUR ANCESTORS HAD REASON TO DISCRIMI- nate among potential mates, the next question is what cues did they use to do it? Unfortunately, we can't ask them, so this is where things get more theoretical (and more controversial). But the answer would have depended on what strategy they were pursuing. Since the costs of bearing a child are generally greater for females, we'd expect them to have looked for longer-term mates—males who have resources to invest in their offspring and a disposition that suggests they're willing to be a provider and a partner. Being the physically weaker sex, females might also favor males who show signs that they can defend a family. For shorter-term trysts, women would presum- ably look for a male who looks like he's got good genes—that is, a guy who's high-status, strong, and healthy.

For males, the costs of child bearing are lower and having more sex partners could mean having more offspring. If the goal is maxi- mizing your reproductive success, the most important cues you should respond to are those signaling fertility. That means youth, good health, and sexual characteristics that reflect reproductive hor- mones like estrogen. Under some circumstances, the evolutionary view predicts that males should be more likely than females to seek short-term sex and multiple partners. Of course, in difficult environ- ments, where two-parent children do better, men would do better looking for good mothers and partners.

These are the kind of predictions that set many people on edge, in part because they are often oversimplified and taken as justifi- cation for sexist behavior. The pop culture version of evolutionary psychology's view is that men have a biological imperative to "spread their seed" and women are innately gold-diggers who just want to seduce high-status men. But that's a pretty poor rendering of the argument. What's worse is that it's often accompanied by the classic "naturalistic fallacy": the idea that what's natural is right. But just because we have some kind of evolved predisposition doesn't mean that that's how we ought to behave.

The reality is a bit more nuanced. Evolutionary theory gives us a framework for understanding how the human mind developed, what its biases are, and why it doesn't behave like a blank slate. But nowhere does evolutionary theory claim that our mental mechanisms are independent of context or insensitive to environment or culture, or that there isn't a range of normal individual differences. We don't live in the Pleistocene anymore. The idea that our brains evolved in response to conditions that held in our evolutionary past doesn't mean that they are well-adapted to the modern world or that we can't use them differently in our enlightened present.

So let's look at the evidence. Are men really more inclined to promiscuity? Many studies supporting the evolutionary account are based on asking people in Western cultures (often college students) about their sexual behavior, a rather narrow test of the theory.* But if standards of attractiveness and mate choice are simply the creation of a male-dominated media culture, we'd expect to see lots of variation in different parts of the world. Yet the data are fairly consistent.

A survey of more than fourteen thousand people from forty-eight countries on six continents found that men reported more promiscuity than women in every country. In cultures where men outnumbered women, monogamous behavior was more common. This may be because in those situations women are able to dictate the terms of mating when demand for women outstrips supply.[15] Mo-

---

* Though, admittedly, some of these surveys are pretty striking. In one widely cited study, male and female student researchers "of average attractiveness" asked strangers on a college campus if they would go to bed with them. None of the women asked by males agreed, but 75 percent of men asked by women said yes.[13]

nogamy was also more common in cultures where rearing offspring was precarious and the need for parental investment was higher. These were countries where rates of malnutrition, low birth weight, and infant mortality were high. In nations with the highest levels of gender equity (e.g., more equal wages, more women in government), women were more promiscuous, but they were still well behind the men. Of course, the problem with studies like these is that they are based on correlations and self-report. They can't tell us much about why men and women report different sexual behaviors or how much of that has to do with biology or culture. But there are other reasons to think that evolution has had a hand in how we judge attractiveness and in shaping our sexual desires.

HOT GENES

PEOPLE FROM DIVERSE CULTURES SEEM TO AGREE ON CERTAIN elements of attractiveness in male and female faces, and preferences for attractive faces develop as early as infancy, well before we are exposed to the world of supermodels.[15] From an evolutionary standpoint, being attracted to a face is like tasting sweetness. We taste sweetness, not because there is something inherently "sweet" about a strawberry or a candy bar, but because natural selection shaped our brains to experience something pleasing when our taste receptors encounter sugars. Animals whose brains responded in this way would have been drawn to foods that provided ready sources of energy. In this same way, there is nothing inherently beautiful about a face, but beauty acts as a cue about the value of a potential mate. So what are we responding to when we see beauty?

Each of our bodies is a vessel for the genes we inherit and transmit. They pass through us in a relentless quest to carry on and live to see another generation. Genes that enhance successful reproduction will out-compete those that don't. For evolutionary biologists, there's a clear implication: we'd expect animals to have developed mechanisms for evaluating the quality of potential mates.

If we accept that natural selection acts at the level of genes, desirable mates would be those that have good genes.* So what are good genes? There might be many answers to that question, but a few seem particularly relevant to reproductive success—starting with genes that promote healthy development and disease resistance. And, in species where parental investment is an issue, the sex that has to invest more resources would also do well to pick up on signs that a potential mate is a good investor to avoid being "left holding the baby."

Researchers have tried to test these predictions by looking at what people and other animals find attractive and seeing how well they line up with expectations. Most studies have highlighted three features that would have ranked high on a Pleistocene version of Am I Hot or Not. And each of them could be an advertisement for "good genes": averageness, symmetry, and sexual dimorphism.

AVERAGE IN THE EXTREME

WE TEND TO EQUATE BEING "AVERAGE" WITH MEDIOCRITY. BUT when it comes to physical traits, the average can be extraordinary.

By many accounts, the greatest racehorse of all time was a chestnut-colored thoroughbred named Eclipse. Born in Berkshire, England, on April 1, 1764, the day of a great solar eclipse, he began racing in 1769. For the next seventeen months he was undefeated in race after race. He didn't just win—he crushed the competition. In fact, in 1771, after an unblemished record of victory, he was retired, mainly because of a lack of competition. He was so revered as a

---

* There are alternatives to the idea that attractive features signal good genes. For a more extensive account of these issues, see Matt Ridley's marvelous book, *The Red Queen*.

champion that after his death, his skeleton was preserved in the hope that veterinarians could one day divine the secret of his perfection. In 2004 Dr. Alan Wilson, an expert in biomechanics and veterinary medicine, led a team of scientists at the Royal Veterinary College in an effort to decipher the mystery of Eclipse's magic. Starting with precise measurements of his skeleton, they used computer modeling to reconstruct the dynamics and mechanics of his movement. And they found something remarkable.

Eclipse was perfectly average. The shape and length of his legs were at the midpoint of the range of modern horses. And "being right in the middle of normal," Wilson surmised, may have been the secret of his success: the very averageness of his proportions gave him just the right balance of flexibility and strength. As Wilson put it, "When they all come out optimum, it looks like average."[16]

But Eclipse was not only great on the track—he was also great in the sack. After retiring from racing, he began a long and legendary career as one of the top sires of all time. He fathered 344 winning horses and it's been estimated that 80 to 95 percent of all living thoroughbreds have traces of Eclipse in their bloodline.[17, 18]

The link between averageness and good genes has implications for what we find attractive in each other. The surprising fact is that beautiful people are more likely to have average than unusual features. Actually, average in this context doesn't mean "ordinary," but rather features that are roughly in the statistical middle of the features seen in a population.

It was Francis Galton, Darwin's half-cousin, who first discovered the beauty of the average. In an effort to study the essential features of different "types" of people (criminals, consumptives, Englishmen), he developed a technique he called "composite portraiture" in which he aligned and superimposed photographic plates. By doing this, he was able to create a composite, statistically averaged face which he could then take as the exemplar of the group of faces he had combined. In 1881, while describing his technique before the Photographic Society, he commented on the surprising image he got when he averaged the faces of a group of consumptive men whom he had selected for their ill-appearance: "The result is a very striking face, thoroughly ideal and artistic, and singularly beautiful. It is, indeed, most notable how beautiful all composites are"(p. 272).[19]

In fact, Galton saw a business opportunity for the Society's members: make a composite family portrait. People would love it. "The result is sure to be artistic in expression and flatteringly handsome. . . . Young and old, and persons of both sexes can be combined into one ideal face. I can well imagine a fashion setting in to have these pictures" (p. 273). Galton's technique never caught on, but his insight about the beauty of averaged faces had legs. Over the past two decades, researchers have used more sophisticated digital technology to create composite faces, and the results are in line with Galton's original observation: the average is beautiful.[15]

So what's so great about being average?

From an evolutionary perspective, averageness may be a signal of good genes. For one thing, genetic mutations and chromosomal abnormalities that interfere with development are likely to create an appearance that is unusual when compared to the average features of a given population. In fact, facial averageness has been correlated with better health.[15] Average features may signal what geneticists call *heterozygosity*, which means there is a greater diversity of genetic variants within an animal's genome. A face that approximates the average features of a group may reflect more diversity among the variants (alleles) in a person's genes. Inbreeding, which is known to increase the risk of genetic disease, homogenizes an animal's genome and makes it less varied. Many mutations cause problems only when an individual carries a double dose (that is, they are recessive), so having a mixture of alleles across the genome would reduce the risk of genetic disease. And like averageness, greater genetic heterozygosity (diversity) has been associated with better health.[20]

A blending of alleles could also reduce the risk of infection by parasites and other pathogens. The major histocompatibility complex (MHC) is a region of the genome that's packed with genes that control our immune responses. Over evolutionary time, human genomes have accumulated a tremendous variety of MHC alleles or variations to help us combat the endless number of parasites and other pathogens that threaten us. We have to be nimble to keep up with these continually evolving pathogens, and having a variety of MHC immune genes can be lifesaving. More diversity in the MHC translates to more options for resisting disease (up to a point). If averageness is a reflection of genetic blending, it may act as an advertisement to potential mates for a more optimal set of immunity genes.

But does MHC diversity make you hot or not? To address that question, one group of researchers took photographs of men's and women's faces and analyzed their DNA.[21] Then a separate group of men and women rated the faces on overall attractiveness as well as averageness and symmetry. In males, MHC heterozygosity was indeed associated with higher ratings of both averageness and attractiveness.

If part of our attraction to potential mates is driven by a search for more diverse MHC genes for our offspring, maybe we should be drawn to people whose MHC genes are different from ours. The child of parents who have dissimilar MHC genes would have more MHC diversity. When lovers say "you complete me," maybe they're not kidding: when it comes to resisting disease, "good genes" really means "complementary genes." Indeed, Oxford scientists found that married couples were much more likely to have complementary (dissimilar) MHC genes than random pairs of people, although, for unclear reasons, that was only true for European-American and not West African couples.[22]

I assume you don't have a genotyping lab in your pocket when you're out on a date, so how are you supposed to know if a potential mate has MHC genes that are different from yours? You might be able to smell it on them: it turns out that MHC genes seem to influence body odor. It's not clear exactly how this happens, but species as diverse as fish, lizards, mice, and humans seem to be able to distinguish MHC similarity by smell. To see whether MHC similarity influences sexual attractiveness in humans, most scientists have turned to the "smelly T-shirt" test. In a typical version of the study, men are asked to wear a T-shirt for a few days to capture their body odor. Women are then asked to rate the T-shirts according to how attractive or pleasing they are. For the most part, women in these studies seem to prefer mates who have dissimilar MHC alleles.[23, 24]

One study actually addressed a more direct question about the MHC and mate choice: does MHC similarity predict sexual desire among couples? Christine Garver-Apgar and her colleagues at the University of New Mexico recruited forty-eight romantically involved couples and asked them a battery of questions about their sexual feelings for each other and for other people.[25] Consistent with the evolutionary hypothesis, women in relationships where the couples had similar MHC genes were less sexually aroused by their male partners, and their partners also rated those women as less sexually adventurous with them. The women in MHC-similar couples also reported cheating more on the partners in their current relationship, but not in past relationships, which rules out the possibility that these women are just more promiscuous.

When it comes to immunity genes, opposites attract.

TURN THE OTHER CHIC

AVERAGENESS HAS ITS LIMITS. THE WEIGHT OF THE EVIDENCE SUGgests that although average faces are usually more attractive than unusual faces, they are not necessarily the most attractive. There's another beauty trait that, like averageness, may have evolved to signal disease resistance—symmetry. In general, more symmetrical faces and bodies are more attractive.[15, 26]

Our bodies are billboards that advertise our genetic idiosyncrasies and the diary of insults and injuries we've endured: our limps, our scars, our sun-weathered complexions. But our evolutionary ancestors were even more likely to show their cards than we are. In the days before modern medicine and cosmetics, parasites and infections like leprosy might cost you an eye, take a patch out of your skin, or a chunk out of your nose. Asymmetries in the shape of your face or body might also signal a genetic problem in the development of body parts. Biologists refer to these small random differences between the left and right side of the face and body as *fluctuating asymmetries*. A potential mate who has fluctuating asymmetries might be one who has less desirable genes: more mutations that interfere with normal development and an immune system that's more susceptible to disease. If our brains read symmetry as attractive because it's a sign of disease resistance, supermodels may owe their careers in some small way to the parasites that long-ago shaped our biological "hot or not" meters.

Still, the effect of symmetry on attractiveness is small,[27] and sometimes asymmetry can be beautiful. Iconic beauties like Marilyn Monroe and Cindy Crawford are well known for their "beauty spot"—a mole on one cheek that many people find attractive.

Proving that nothing is beyond the reach of scientific study, researchers set out to answer that age-old question: Where's the best place to have a mole? They showed images of women's faces with various configurations of beauty spots (technically *melanocytic nevi*) to a panel of 250 male and female judges that included physicians and artists.[28] When it comes to facial moles, it turns out that location matters. The most attractive moles are unilateral and off to the side. The closer it is to the midline (like on your chin or the bridge of your nose), the less attractive it is. And symmetry fared worst of all—women with a mole symmetrically placed on each side of the face were rated least attractive.

STUDS AND BETTYS

PERHAPS THE LEAST SURPRISING OF THE BEAUTY FEATURES IS sexual dimorphism, which basically means we tend to like faces that look most like the sex they represent. Males are attracted to feminine-looking females and females are attracted to masculine-looking males. Highly feminine features like full lips, high cheekbones, and a small chin, are strongly preferred by males across races and cultures.[15] The same goes for hormone-related signs of female fertility: larger breasts, smaller waist-to-hip ratios. The reason may have to do with the influence of reproductive hormones—estrogen in females and testosterone in males—that cause these sex differences during puberty. A distinctly feminine face may act as a fertility signal for men, telling them that this woman has reached sexual maturity and has estrogen on board.

And, in fact, women with larger breasts and an hourglass shape have higher reproductive potential as reflected in higher estrogen and progesterone levels during their menstrual cycles.[29] For males, the square jaw of a masculine face signals high testosterone and, by implication, sexual potency and perhaps social dominance. Since testosterone can also suppress immune function, a very masculine face on a healthy male may also be saying: "My genes are so good, I can handle loads of testosterone and still be healthy."

POST-MENSTRUAL SYNDROME?

ON THE OTHER HAND, RUGGED HANDSOMENESS MIGHT BE A double-edged sword. A guy who's swimming in testosterone might be good for breeding strong, healthy offspring, but he might also be aggressive and unfaithful, which is not the kind of guy you want raising your kids. This might explain an intriguing observation about women's preferences, namely that women change their minds about male attractiveness depending on where they are in their menstrual cycle.

In one of the first studies to show this, women were presented with a series of male faces and asked to pick the one they thought was the most physically attractive.[30] Unbeknownst to the women, the researchers had digitally manipulated the faces to make them more masculine or more feminine. They categorized the women based on the phase of their menstrual cycle. The group categorized as "high conception risk" were in the follicular phase (the first half of the menstrual cycle, before ovulation and therefore able to conceive during that cycle) and those categorized as "low conception risk" were in the luteal phase (after ovulation has occurred, and thus unlikely to conceive). The high-conception-risk women were more attracted to the more masculine faces while the low-conception-risk women preferred the less masculine faces. In a second experiment, the researchers allowed women to morph male faces to be more masculine or feminine and asked them to pick the face they find most attractive for either a short-term relationship or a long-term relationship. When women in the high conception risk group chose a short-term mate, they made the face more masculine. Other studies have shown that during their most fertile phase, women report being attracted to more masculine bodies and voices and having more sexual interest in men other than their partners.[31]

Why would a woman's menstrual cycle affect how she views a man's sex appeal? One explanation is that it's an adaptation that allowed ancestral women to optimize their options. Recall that more masculine features are thought to be one index of "good genes" and less masculine features suggest a more cooperative (better parent) kind of guy. Women who cheat on their long-term partners are more likely to do so during the follicular phase, when they are most likely to conceive. Since men can never be certain that a child is theirs, a woman might be trading up for "good genes" by having a tryst with the hunk just before she ovulates while holding on to Mr. Mom for the long term. It's the best of both worlds—she gets genetic benefits from her lover and material benefits from her partner.

If mate preferences change with hormonal shifts (in estrogen and progesterone) during the menstrual cycle, could you alter women's preferences by manipulating their hormones? This sounds like the diabolical plan of an evil (male) scientist, but, in fact, tens of millions of women enroll in this experiment every month. They take oral contraceptives.

The pill works by suppressing estrogen and keeping progesterone levels high enough to block ovulation. The usual cyclic shifts in these hormones are flattened out. Could taking the pill change how women feel about men? Indeed, several studies have now reported that pill-users don't show the preferences for more masculine or symmetrical faces, or for the sweaty T-shirts of MHC dissimilar men that other women show around the middle of their menstrual cycles. It's as though the pill has not only flattened their hormonal cycles but their sexual desires as well.

There's an interesting flip side to this story. These same hormone cycles also affect how men look at women. Around the time of ovulation, just as a woman is beginning to tune into more masculine men, she actually becomes more sexually attractive herself. At their peak of fertility, women's faces, voices, and odor change in subtle ways that make them more attractive to males.[32-34] Some studies show that women dress more provocatively during these periods,[35-38] and are more likely to flirt with and fantasize about sex with men other than their primary partner.[37, 39] But the pill appears to mute this surge in sex appeal that normally happens around ovulation.[40]

At this point, you're probably thinking, that's interesting, but there's only one way to really study the pill's effect on sexual attractiveness. That's right: lap-dancing. So, as a service to science, psychologists at the University of New Mexico went to "gentleman's clubs" in the Albuquerque area to recruit lap dancers for a study of hormonal effects on female attractiveness. As you may or may not know, lap dancers perform topless dances on the laps of men who have to remain seated and can't touch them. The dancers make their money through the tips they get for each dance. The more alluring they are and the more dances they do per shift, the more money they make.

Every day for a two-month period, the dancers were asked to log onto a website and report on where they were in their menstrual cycle and how much they earned in tips. By measuring the dancers' income, the researchers could get a direct, numerical estimate of how attractive each dancer was. So they could now ask: What effect do ovulatory changes have on male tipping? The results were clear: earnings shot up during the peak days of fertility (the week before ovulation). On average, the women made nearly twenty dollars an hour more during their fertility peak compared to the last ten days of their cycle (when they would have been unable to conceive).

But women taking the pill didn't get that earnings boost. The authors concluded that the pill's steady dose of estrogen and progesterone had quashed the subtle shifts in female attractiveness that occur across the menstrual cycle. The evidence that hormones enhance physical and behavioral cues of fertility suggests that natural selection may have promoted "come hither" cues that attract men when conception is most likely. But the pill, by artificially blocking the hormone surges that trigger ovulation, may short-circuit those cues.

These studies raise questions that might have implications beyond the economics of lap dancing. Most of the studies have been small and have limitations, but they suggest that men and women look at one another differently when a woman is ovulating, and the pill blocks this natural cycle. Could the fact that millions of women take hormones that might interfere with normal mate preferences be affecting how and with whom they form relationships?[40] Might they have chosen other mates? Would a woman who met her mate while she was on the pill find him less attractive when she comes off?

A 2011 study of more than 2,500 women suggests that the pill could indeed be affecting women's desires and mate choices. In the study, each woman was asked about her relationship with the biological father of her first child.[41] Women who met their male partners while taking the pill reported being less sexually attracted to and aroused by their mates than those who hadn't been on the pill. On the other hand, the pill-taking women were significantly *more* satisfied with nonsexual aspects of the relationship, such as how successful the men were as financial providers. And, overall, women who had been using the pill were more likely to stay with their partners over the long term. In other words, women who chose their mates while on the pill went on to have relationships that were less sexually satisfying but more durable. That fits with the idea that by blunting hormonal cycles oral contraceptives could bias women to choose the stable partner over the "hot" guy.

GLOBAL AVERAGES

AS INTRIGUING AS AN EVOLUTIONARY ACCOUNT OF BEAUTY AND sexual attraction may be, there are plenty of ambiguities. No one has convincingly shown that preferences for averageness or other beauty indicators of "good genes" are actually associated with greater reproductive success in humans. The most obvious limitation is that we can't directly study what our hominin ancestors did or how they behaved. Most of the studies I've discussed so far have drawn their subjects from industrialized cultures. And, as we know, standards of beauty in those cultures have been changing. The fact that other animals have developed strategies for mate choice that resonate with our own certainly supports the evolutionary theory, but other animals aren't subjected to the powerful and shifting influence of cultural conventions that we are. They've never had to face the judgmental standards of religious law or fashion magazines. There is a thriving industry of animal magazines—*Cat Fancy*, *Birdtalk*, *Modern Dog*, *Reptiles*, and the like—but animals don't read them.

The other kind of evidence we'd like to see if the evolutionary arguments hold water is some degree of cross-cultural consistency. If our standards of beauty are embedded in a universal human nature, they ought to apply beyond New York, Paris, and London. In particular, it would be helpful to know what goes on in today's preindustrialized societies, especially hunter-gatherer groups that might more closely resemble our own ancestral roots.

One such group, the Hadza, are a nomadic hunter-gatherer society who live in remote savannah-woodland areas in North Tanzania. There are only about one thousand Hadza living today. The men hunt and collect honey while the women dig for wild tubers and gather berries, and collect baobab fruit. They are a predominantly monogamous society and less than 5 percent of Hadza women marry outside the group. They also have almost no access to "the media."

Several years ago, while a graduate student in anthropology at Harvard, Coren Apicella traveled to Tanzania to see whether beauty signals found in industrialized countries would appeal to the Hadza.[42] She began with forty photographs of young Hadza individuals (twenty men, twenty women) and an equal number of photographs of young, white adult British men and women. For each group, she used computer morphing software to create two types of composites with different numbers of faces in each. The "more-average" faces were made by blending all twenty (male or female) faces in each group and the "less-average" faces combined five of the twenty faces. She then asked the Hadza and the British subjects which faces they found more attractive. If natural selection has biased us to see averageness as beautiful, we'd expect the more-average faces to get higher ratings even in the media-naive Hadza society.

Have a look at the figure. Which set of faces do you think are more attractive—the top row or the bottom?

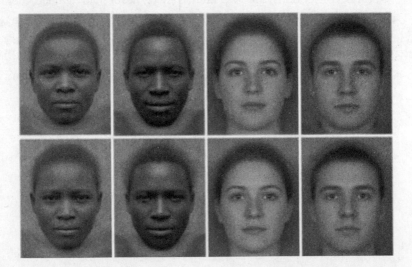

The more-average faces are in the bottom row. Both the Hadza and the British rated the twenty-face Hadza composite as more attractive than any of the five-face Hadza composites. The European and hunter-gatherer groups agreed: the more averaged, the more attractive.

Apicella's results seem to support the hypothesis that at least one of the alleged beauty signals that media-soaked Westerners find attractive is also attractive to a society that, culturally speaking, is worlds away. But that wasn't the whole story. When Apicella showed the Hadza the European composites, they had no preference for the twenty-face images.

I asked Apicella how she made sense of this. She noted that Westerners have been exposed to a lot of African faces, but the Hadza have not been exposed to Western faces. They have no mental template for an attractive white face. "We might have this biological or universal preference for what is average," she theorized, "but it's the environment and the faces that we're exposed to that will shape our prototype of what is average and . . . which faces we find attractive."

Perhaps there's a basis for reconciliation here between the Darwinists and the beauty-mythers. Apicella's view at least opens the door to the idea that cultural fashions (including the Western fashion industry) might indeed shape our standards of beauty, even if some of the basic parameters are innate. Our minds may be primed to perceive some physical features as beautiful, but they may need to be calibrated by exposure to the facts on the ground. If we are bombarded with enough images of "waifer"-thin models, our prototype of average may shift.

The environment may tweak our biological templates of attractiveness in other ways as well. In another study, Apicella along with psychologist Anthony Little and anthropologist Frank Marlowe, compared the appeal of symmetry in Hadza and British adults.[43] If symmetry is attractive because it signals disease resistance, the effect might be stronger in the Hadza who sleep on the ground, are exposed to wild animals and plants, and have no access to modern medical care. In other words, for the Hadza, symmetry might be a more reliable signal of good disease resistance genes. And, in fact, both the Hadza and the British rated symmetrical faces of their own group as more attractive, but the effect was stronger for the Hadza. Once again, these data suggest a degree of universality in our criteria for beauty, but the details of the environment we live in can blunt or enhance their salience.

We've seen that evolutionary theory and a growing body of data across cultures and species suggest that our sexual behavior and our desires have been shaped at least in part by sexual selection. But it's important not to overstate the power of these ancestral influences. To the extent that these influences operate on us today, they merely bias our brains toward certain information and behavioral strategies. They are not determinative; instead, they provide an envelope within which we operate. How they play out depends on the local circumstances of our lives: our social and physical environment, competing needs, and even chance. And as I mentioned earlier there are a variety of facts about our sexual preferences and behavior that aren't easily explained by an evolutionary account. There have clearly been cultural fluctuations in beauty standards (e.g., the recent Western emphasis on thinness in women) that might have overridden or distorted any biases evolution gave us in judging attractiveness. But this doesn't mean our modern cultural preoccupations were born in a Madison Avenue meeting. If there's a middle ground between the two poles of social constructionism and evolutionary determinism, it may be that Western culture has discovered the power of supersizing features that we are biologically prepared to find attractive.

Could our modern media culture be feeding us souped-up versions of what we are instinctively drawn to? In the late 1940s, the ethologist Niko Tinbergen noticed that he could elicit powerful instinctual responses from animals by using stimuli that exaggerate the natural triggers of innate behavior patterns. For example, herring gull hatchlings peck at their parent's bill to beg for regurgitated food. Tinbergen found that he could get hatchlings to peck like mad—beyond their normal rate—when he presented them with a phony model of a bill that exaggerated its natural shape, angle, and color.[44] In the same way, parental birds will ignore their own eggs in favor of fake super eggs that are much bigger than anything they would encounter in nature.[45] By pushing an animal's biological buttons, these "supernormal stimuli" essentially trick the brain into finding them irresistibly attractive.

Drawing parallels between animal and human behavior is always a tricky business, but the concept of supernormal stimuli does resonate with some familiar features of modern Western culture.[46] Consider our eating habits. Natural selection prepared us to enjoy foods high in fats and sugar because they're energy-rich and would have had obvious survival value for our hominin forebears. Our ancestors (like millions around the world today) were much more likely to die of malnutrition than complications resulting from obesity. But in the modern industrialized world, high-calorie foods have become cheap and plentiful. We now crave combinations of fat and sugar that never existed in the natural world, including the fast-food trifecta of megaburgers, large fries, and a shake. No one in the Pleistocene ever tasted ice cream (much less a fudge sundae), but a lot of us would eat it before anything else in the four food groups. Exaggerate and they will come.

It's tempting to speculate that something like this accounts for the potency of the increasingly extreme fashions in male and female beauty. From the unusually large breasts and thin waists of Victoria's Secret models to the broad shoulders and washboard abs of Abercrombie and Fitch hunks, we are surrounded by images of attractiveness that may have taken our biologically prepared preferences to an unnatural extreme. We've gone beyond cosmetics that accentuate full, red lips and smooth skin to cosmetic surgeries and Botox and collagen injections that literally reshape our bodies to maximize signals of youth, fertility, and health. It's as though our conceptions of attractiveness have been torn free of their biological roots by a flood of images that push our mental buttons.

BEYOND THE STRAIGHT AND NARROW

YOU MIGHT HAVE NOTICED A BIG HOLE IN OUR DISCUSSION OF sexual preferences and mate choice. After all, not everyone follows the "boy meets girl" story line. Since perhaps the beginning of human existence, same-sex behavior has been a part of our experience. If natural and sexual selection have driven human mate preferences to maximize reproductive success, homosexuality would seem to present a puzzle. Assuming most homosexual couples don't reproduce, wouldn't genetic variants that promote homosexuality be disadvantaged and fade away in the evolutionary contest to leave descendants?

In American life and politics, few debates are as charged as the one surrounding sexual orientation. The nature of sexual attraction and sexual orientation has always been more than a scientific issue. One of the most contentious battles in modern politics centers on whether people who are primarily attracted to individuals of their own sex should be allowed to marry. Beliefs about the origin of homosexual behavior have been used to justify social policy, religious proscriptions, and legal decisions—mostly in the direction of discrimination against gay people.

The debate has often come down to three explanations: homosexuality is a choice, homosexuality is learned behavior, or homosexuality is innate. The first two are often seen as compatible and in opposition to the third, and people usually pick sides. In a 2009 national survey of Americans, 47 percent said that homosexuality is "something people choose to be" while 34 percent said it is "something people are born with." Only 19 percent said they didn't know.[47] Those who believe that it's biologically innate tend to have more favorable views toward homosexuality, but pro- and antigay activists don't necessarily endorse a nature or nurture position. As the sad history of discrimination based on skin color makes clear, the fact that something is innate or immutable doesn't preclude its being the target of prejudice and social exclusion. And, indeed, in the late nineteenth and early twentieth centuries, biological theories of homosexuality were used to justify persecution and bizarre attempts to "cure" the condition—including testicle transplants.[48] On the other hand, there are modern conservative groups like the Family Research Council, which emphasizes the role of choice in sexual behavior and asserts not only that "homosexual conduct is harmful" and "unnatural" but also that "there is no convincing evidence that a homosexual identity is ever something genetic or inborn"[49, 50]

The complexity of talking about sexual orientation is obvious even in the question of how common it is. If you were going to do a survey to determine the prevalence of homosexuality, what would you ask people? Are you attracted to members of your own sex? Are you *only* attracted to members of your own sex? Have you had a same-sex sexual partner? Do you consider yourself homosexual? The answers you get would depend a lot on which questions you ask.

In a comprehensive survey of sex in the United States in the 1990s, only 1.4 percent of women and 2.8 percent of men said they thought of themselves as homosexual or bisexual, but about three times as many said they'd had a same-sex sexual partner at some point in their lives.[51] A large survey by the National Center for Health Statistics in 2002 found that 49 percent of men and 65 percent of women who'd had a same-sex partner considered themselves heterosexual.[52]

So what's the evidence that same-sex behavior is part of the "biology of normal"? First of all, hundreds of animal species engage in same-sex sexual behavior: insects, snails, birds, dolphins, sheep, and the list goes on.[53] While same-sex pairing is not uncommon in the animal kingdom, the kind of exclusive homosexuality that we recognize in humans is quite rare. However, the long history of human homosexual behavior and the fact that it exists in cultures around the world certainly suggest that it's a part of human nature and might even have an adaptive function.* And evolution-minded scientists have proposed a variety of possible explanations.

For example, there's the gay uncle hypothesis, a spin-off of the established evolutionary idea of kin selection. It's obvious that re-producing is a good way to pass on your genes. But there are other, more indirect ways as well. Behaviors that increase the reproduc-tive fitness of your close relatives (kin)—protecting them, nurturing them, and so on—could also help transmit your genes (since the more closely related you are, the more genes you share). My niece shares 25 percent of my genes, so helping her survive and reproduce helps send my own genes into the next generation.

The key implication is that genes promoting kin-directed altru-ism could be maintained in a population if they offset the cost of not having your own offspring. By extension, some argue, natural selection might maintain the frequency of genes predisposing to ho-mosexuality if they also promote kin-directed altruism toward close relatives.

It's a coherent hypothesis, but if it were true, it would require a very large effect since, for example, gay men have only a fifth the number of children that heterosexual men do.[55] To compensate for that difference, gay men would need to be some kind of superuncles. And the few studies that have tested the theory have found little evidence that gay men provide enhanced caregiving for their nieces and nephews—at least in Western cultures.[55, 56]

On the other hand, Canadian psychologist Paul Vasey has been able to find some support for the gay uncle hypothesis in his stud-ies of sexual behavior in the South Pacific islands of Independent Samoa. In addition to men and women, Samoan culture has defined a third gender referred to as *fa'afafine* (literally, "in the manner of a woman").[57] The *fa'afafine* are biologically male but tend to be ef-

---

* Actually, there is a debate among historians about just how old the concept of sexual ori-entation really is. Although same-sex sexual behavior is clearly ancient, social construc-tionists argue that the idea of classifying people as exclusively heterosexual or homosexual didn't appear until the mid-nineteenth century.[54]

feminate and their sexual relationships are with other males (though not other *fa'afafine*). In a series of studies, Vasey and his colleagues found that *fa'afafine* invest much more time and effort in caring for their nieces and nephews than do Samoan women and heterosexual men.[58] What's more, their avuncular tendencies are not related to a more general altruism toward unrelated children.

Why hasn't this been seen in Western cultures? Vasey argues it's possible that modern industrialized cultures—where families are dispersed and homophobic attitudes are common—are too removed from the conditions of our ancestral environment and simply don't allow the adaptation to be expressed. But that explanation is hard to prove or disprove.

But there's another observation about the *fa'afafine* that does match up with Western studies. Their mothers have more children than the mothers of heterosexual Samoan men.[59] The same phenomenon has been reported in studies of Italian and British families—mothers and aunts of gay men have more children than those of straight men.[60, 61] Based on these data, the "fertile female" hypothesis claims that homosexuality in men is in part a by-product of genes that enhance the fecundity of females. But again, it's only a hypothesis at this point.

The notion that homosexual behavior is related to natural selection presupposes that genes are involved. But none of the theories I discussed above really provide direct evidence for this. So, is there any convincing evidence that a gene affects same-sex sexual behavior? The answer is clearly yes . . . if you're asking about fruit flies. The fruit fly *Drosophila melanogaster* has been a staple of genetic research for decades, and its courtship behaviors have been meticulously catalogued and dissected.

For nearly forty years, scientists have known that a single fruit fly gene—known as *fruitless*—controls courtship behavior in males and females.[62] In 2005 researchers showed that *fruitless* can act like a genetic master switch—capable of turning male courtship and mating behavior on and off.[63, 64] Males and females express different versions of the *fruitless* gene, the result of a difference in how the gene's message is spliced together after it's transcribed from the gene itself. Using genetic engineering, females can be forced to express the male version, and when they do, something dramatic happens: they stop mating with males and start courting females. When it was reported, this discovery was stunning. Here was a single gene that could trigger a highly complex behavior pattern (what we might metaphorically call "sexual orientation") in animals.* *Fruitless* does this by encoding a transcription factor that turns on or off a whole set of other downstream genes, which in turn encode elements of sexual behavior.

The fruit fly provides a kind of "proof-of-principle" example that genes can regulate sexual behavior and mate choice. But let's face it: people are a little more complicated than fruit flies. The flies' entire lives last a month and their brains have one millionth the number of neurons ours have.

So what do we know about genes and human sexual orientation? For one thing, several family studies suggest that homosexuality runs in families: in those studies, siblings of gay men or women were about two to five times more likely to be gay than were the siblings of straight individuals.[65-67] But because family members share both genes and environments, family studies by themselves can't prove or disprove that a trait is genetic. It's conceivable that a child learns to be gay from his parents or siblings for reasons that don't involve genes. So far, though, there's no convincing evidence of differences in gender identity or sexual orientation between children raised by lesbian or gay parents and children of straight parents.[68]

Studies of twins have consistently found that sexual orientation is heritable, meaning that variations in genes account for a proportion of differences in sexual orientation across the population.[69-72] In the largest study so far, including more than seventy-five hundred twins, the heritability of same-sex sexual behavior was higher for

---

* It has since become clear that the full range of male courtship behavior involves other genes in addition to *fruitless*, including, most importantly, a gene known as *doublesex*.

males (34 to 39 percent) than females (18 to 19 percent),[71] suggesting that genetic differences affect sexual orientation in humans and the effect may be stronger for gay men compared to lesbian women. On the other hand, since the heritability seems to be quite a bit less than 100 percent, these data also imply that most of the difference among people in sexual orientation is not due to genetic variations.

But if genes are involved, which genes are they? The few studies that have attempted to map genes related to sexual orientation have had conflicting results.[73-77] The bottom line is that, to date, no specific genes influencing homosexual behavior in humans have been identified. Which is not to say that they don't exist—there's really been no large-scale effort to find them.

BROTHERLY LOVE?

ONE SURPRISING PREDICTOR OF MALE HOMOSEXUALITY HAS TO do with families but only indirectly with genes. And no one is quite sure what to make of it. It's called the "fraternal birth order effect": having older brothers increases the likelihood of homosexuality in men. This effect has been seen in studies from Canada, Italy, Britain, the United States, and even among the *fa'afafine* of Samoa.[59, 60, 78-80] Estimates from this research indicate that each additional older brother increases the odds of male homosexuality for the younger brother by 33 percent* and that 15 to 30 percent of men can attribute their sexual orientation to their fraternal birth order.[79, 81, 82]

How could having older brothers impact your sexuality? The leading theory points to the immune system. Canadian researchers Ray Blanchard and Anthony Bogaert have proposed that while a mother is carrying a male child, she may be exposed to male-specific

---

* This is a *relative* increase in the odds. So if the baseline odds of a male being gay is, say, 2 percent, this means that having an older brother would increase the odds to 2.66 percent (i.e., 2 percent + [.33 x 2 percent]).

proteins from the fetus. Because her immune system sees these proteins or antigens as alien, she develops an immune response— essentially, antimale antibodies (possibly directed against proteins made by genes on the Y chromosome). With each successive male child, the mother's immune response may grow stronger. In effect, her body remembers how many male offspring she has carried. When these antibodies encounter the fetal brain, they might alter the function of neural circuits involved in sexual differentiation and, as a result, sexual preferences later in life.

The idea of mothers developing antibodies to fetal proteins is not a new one. The most familiar example is Rh incompatibility. Pregnant women are routinely screened to see whether they carry Rh factor, a blood group protein that's absent in a small propor- tion of women. When an Rh negative mother carries an Rh positive child, she may develop anti-Rh antibodies when she is exposed to the fetus's Rh factor during delivery. This could be a problem if she has another Rh positive child: her antibodies may attack the fetus's red blood cells, causing a severe anemia.

While there is no direct evidence that an immune reaction really affects sexual orientation, there is some intriguing circumstantial evidence.[79, 82] First, the birth order effect is only seen for males with older brothers. Females with older brothers and boys with older sis- ters do not have an increased likelihood of being gay. Second, males who have nonbiological (adopted or step-siblings) older brothers don't show the effect, suggesting that a younger brother's homo- sexual orientation doesn't just stem from some psychological effect of having older brothers. Rather, it seems that the brothers have to have shared the same womb. Third, it's the *number* of biologi- cal older brothers—and not the amount of time reared with older brothers—that predicts sexual orientation. Finally, there are known male-specific proteins that are expressed early in development and are found on the surface of brain cells, providing a possible target for an antibody response. Clearly, more direct evidence is needed before this hypothesis can be fully evaluated, and no one has suggested that the fraternal birth order effect explains most of male homosexuality. For one thing, most gay men have no older brothers.

But if this "maternal immune hypothesis" is right, it demonstrates an important point about how nature and nurture can produce behavioral traits. As we saw, twin studies indicate that genetic variation contributes to sexual orientation, but also that most of the differences in gay vs. straight preferences are accounted for by environmental (nongenetic) factors. But even if environmental factors have a larger effect on same-sex preferences, this still doesn't mean that a gay person would have a choice about their orientation. The maternal immune hypothesis would be an example of an environmental factor that has nothing to do with choosing or learning—in this case, it's the environment of the womb.

The prenatal environment has been considered in other ways as a biological contributor to sexual orientation. Testosterone and estrogen have powerful "organizing" effects on brain development that begin in the womb. Testosterone "masculinizes" the fetal brain, directing development toward male sexual behavior and gender identity.[83] One line of evidence suggests that lesbianism might be related to exposure to excess testosterone during fetal development. And studies suggest that a simple index of how much testosterone you saw as a fetus may be readily at hand.

Try this: straighten your right hand, put your fingers together, and look at how they line up. Males and females tend to have a difference in the length of their second (index finger) and fourth fingers (ring finger)—known as the second digit to fourth digit ratio (2D:4D ratio). In males, the second finger (2D) tends to be slightly shorter than the fourth (4D), but in females the second and fourth fingers are about the same length. It's thought that the 2D:4D ratio is pretty much determined by how much testosterone you were exposed to in the first trimester of pregnancy. The theory goes that girls with the male pattern (a smaller 2D:4D) ratio were exposed to more testosterone than girls with the more typical female pattern. A recent analysis of a large number of studies that have looked at this found that lesbian women do indeed have a smaller average 2D:4D ratio than heterosexual women, though the effect was small.[84]

So what can we say at this point about the biology of same-sex sexual behavior? We know that it is widespread among animal species and that in at least one species (fruit flies), specific genes and neural circuits have been identified. The possibility that it has adaptive functions in animals and humans is certainly plausible, but so far the evidence is unconvincing. Twin studies suggest that sexual orientation is partly heritable, but there are plenty of unresolved questions. In humans, no specific genes influencing homosexuality have been identified. On the other hand, there is no credible evidence that same-sex orientation is either a choice or learned behavior for most gay or lesbian people. Like everything else we've discussed, sexual orientation is a complex and multidimensional part of our lives. It can't be reduced to nature or nurture and there's no reason to think that a single cause is necessary or sufficient.

A BEAUTIFUL MIND

REMEMBER THAT, FROM AN EVOLUTIONARY STANDPOINT, PERCEIVing beauty is kind of like tasting sweetness—an enticement to desire, an experience that became rewarding because it was tied to something that enhanced fitness. Just as a sweet taste might signal a valuable source of energy, an attractive face or body might signal a high-value mate. If that's right, it makes a strong prediction about where we might find biological responses to beauty and sexual attractiveness: the brain's reward system. And if the idea is that we are "wired" to respond to certain features of attractiveness, it would be reassuring to see the wiring. But now we've crossed over into the realm of proximate causes—the here-and-now workings of the brain.

In 2001 a group of scientists at Massachusetts General Hospital and MIT in Boston tested that hypothesis by showing pictures of four sets of faces to groups of heterosexual men: beautiful women, average-looking women, beautiful men, and average-looking men. The men gave the beautiful women and the beautiful men high attractiveness ratings and were even willing to work to keep looking at them (by pressing keys on a keypad), but their brains were more discriminating. Only the beautiful female faces activated the nucleus accumbens, a key reward center that is also turned on by all manner of guilty pleasures: cocaine, speed, nicotine, money, and, yes, sweet tastes.[85] In other words, the men appreciated the aesthetic appeal of beautiful people of both sexes, but only the beautiful female turned on the brain's pleasure centers. Since that original study, others have found that looking at beautiful faces engages reward circuitry in the brain, particularly the nucleus accumbens and the orbitofrontal cortex (OFC),[85–88] an area involved in tracking whether experiences are rewarding or aversive.[89, 90]

As you might expect, sexual preferences also matter. When asked to give attractiveness ratings, men and women generally agree on how beautiful other men and women are.[90] Even heterosexual men can appreciate the appeal of a beautiful man. But what we say and how our brains respond may be quite different. In one study, men and women, regardless of their sexual orientation, gave virtually identical answers when asked to rate the beauty of photographs of other men and women. But fMRIs of their brains revealed a hidden signature of sexual preferences: in heterosexual women and gay men, brain reward centers (including the OFC) were more strongly activated by attractive male faces, while in lesbian women and heterosexual men, these centers lit up for attractive female faces.[91]

Physical features beyond facial attractiveness also appear to have this rewarding effect on the brain. Men seem to be drawn to the hourglass shape of a woman's body.[92, 93] The average waist:hip ratio of a man tends to be in the range of 0.8 to 0.95, while in women the average is in the range of 0.67 to 0.79.[94] In the 1990s psychologist Devendra Singh suggested in a number of studies that American men are most attracted to women whose waist:hip ratio hovers around 0.7.[95] He also found that while *Playboy* centerfolds and Miss America winners have grown slimmer over the years, their waist:hip ratios remained relatively constant within the range of .68 to .72. Studies in several other countries have largely supported the idea that men are typically attracted to women with an hourglass shape, although the favored waist:hip ratio does vary across cultures.[92, 96–98]

Men seem to make these judgments in the blink of an eye. In one study, researchers tracked men's eye movements using an infrared camera while they were shown pictures of a naked woman whose body was digitally morphed to vary her waist:hip ratio (0.7 vs. 0.9) and breast size. Within 200 milliseconds, their eyes registered her waist/hip area and they judged the low waist:hip ratio (0.7) as more attractive.[99] Singh interprets this phenomenon from an evolutionary perspective: a low waist:hip ratio has been associated with fertility, youth, and health and so, like facial cues of good genes, an optimal waist:hip ratio might have served as a signal of mate quality. It's certainly possible that these preferences are driven by media exposure, but even men who were blind from birth rated mannequins with a waist:hip ratio of 0.7 as more attractive than those with a larger ratio when they were asked to feel and touch them.[100]

For many years in Western cultures, women have gone to great lengths to maintain and exaggerate an hourglass figure—from corsets and girdles to tummy tucks and liposuction. Perhaps it's all an attempt to push buttons in the brain of the beholder. In one study, Singh showed men "before and after" pictures of naked women who underwent a cosmetic procedure that involved removing belly fat by liposuction and grafting it to the buttocks, effectively creating a surgically enhanced hourglass. The men rated women who achieved a waist-hip ratio of about .70 as most attractive[94] and their reward centers (including the OFC and nucleus accumbens) lit up for the new and surgically improved waist:hip ratios.[101]

But—shock!—men and women are different: men appear to have a stronger OFC response to physical attractiveness,[88] supporting a whole lot of psychological research showing that physical appearance seems to matter more to men than women. To take one example, a massive BBC Internet survey of more than two hundred thousand men and women spanning fifty-three nations asked about traits that people desired in a mate. "Good looks" were in the top three traits for 43 percent of the men but only 17 percent of the women.[102]

Of course, none of these studies really answer the question of whether our neural reactions to human beauty and sex appeal are innate or acquired. That's because they are simply looking at patterns of brain activity among people who live in a particular (modern, Western) culture. Also, the fact that reward circuits turn on when we look at attractive people doesn't necessarily mean that these responses are innate—it's still possible that our brains have been culturally conditioned to find certain features rewarding.

However, the available data sketch a plausible picture about the biology of attraction that ties the ultimate and proximate mechanisms together. Evolution has primed us to recognize certain signals of mate quality and sexuality in the faces and bodies of other people and added desire by linking these perceptions to reward. In doing so, we become attracted to them.

But again, we're talking about a mental bias here. Some things may be biologically more likely to turn us on, but our specific preferences are also experience-dependent. All kinds of experiences—the crush we had in grade school, the fashion trends of our time, and, yes, the sexual politics of our culture—get piled on top of the biological foundation that we bring to the world. Our brain circuits are undoubtedly shaped by the portfolio of associations we acquire through the particular trajectory of our social and sexual lives. There's clearly more to being "hot or not" in the twenty-first century than signaling good genes, and one man's (or woman's) sexy may be another's "yuck!"

So there is some persuasive evidence that our minds are attuned to evaluating sexual attractiveness and that our brains get a buzz from sensing hotness. But does our understanding of normal tell us about how things can go awry? Are there disorders of sexual desire?

DANGEROUS LIAISONS

PSYCHIATRIST AVIEL GOODMAN OFFERED THIS CASE OF A MAN whose sexual desires got the best of him:

> *An executive in his midthirties, Harold would say with a smile that his Achilles' heel was his "weakness for the fair sex." When an attractive woman indicated to Harold that she was interested in him sexually, he found himself unable to resist, or more accurately, he found himself unable to want to resist. He experienced himself almost as a victim, sexually drawn to women against his will. Harold's fiancée ended their engagement after he repeatedly broke promises to her that he would stop sleeping with other women. When Harold began to use his apartment in the city for midday sexual liaisons, his lunch breaks stretched longer and longer. His formerly superior work performance began to slacken and he did not receive an expected promotion. Harold's boss warned him that he could lose his job if he was unable to keep business and pleasure separate in his life. Harold resolved that he would turn over a new leaf and for six weeks he kept his sexual behavior in check. Then, when he was out of town on business and had just finished dinner with his work team, he commented that his neck and back were tight. His secretary offered to give him a back rub, and he accepted the offer without a moment's thought. The back rub resulted in a sexual encounter. Upon returning to his office, Harold continued to engage in sexual activity with his secretary. Soon she began to*

pressure him for an exclusive relationship. When he re-
buffed her, she filed a suit against him for sexual harass-
ment. He was fired immediately.[103]

Many of the conditions that fall under the heading of Sexual
and Gender Identity Disorders in psychiatry's diagnostic manual
(the DSM) have less to do with sexual attraction than with sexual
function: female sexual arousal disorder, male erectile disorder,
premature ejaculation, and so on. But then there's the group of
conditions in a category known as *paraphilias*. That's a term that
few outside of the mental health (and perhaps legal) professions
have probably heard, but some of the syndromes may ring a bell:
exhibitionism, fetishism, voyeurism, sexual sadism, pedophilia,
and frotteurism. Okay, maybe *frotteurism* isn't a household word.*
The thing that ties these disorders together is what people often
call *deviant sexual arousal*, or more specifically, a pattern of intense sexually arousing fantasies, urges, or behaviors that typically
involve nonhuman objects, sexual humiliation, or nonconsenting
people.

Paraphilia is one of the most interesting examples of how fuzzy
the line between normal and abnormal can be. Having an odd sexual
interest—say a fetish for rubber dolls—doesn't buy you a diagnosis
of a psychiatric disorder. Remember that psychiatry has a standard
for when a set of symptoms or behaviors crosses over into the land
of disorder: it has to "cause clinically significant distress or impairment in social, occupational, or other important areas of functioning."[104] The goal here is to avoid pathologizing normal behavior—a
frequent criticism leveled at psychiatry. Without the "distress or impairment" standard, the risk is that we might too easily label someone with a disorder whether or not it causes them problems. But in
the case of paraphilias, that leaves the door open to some awkward
scenarios.

---

* Frotteurism (from the French word meaning "one who rubs") is a disorder characterized by intense sexual urges or fantasies about rubbing oneself against another person.

Imagine a man, John, who is powerfully aroused by sadomasochistic pornography. He buys pornographic literature and movies, and after his wife, Connie, goes to bed, he spends four hours per night online looking at S&M porn sites and online chat rooms. He's not bothered by it, and his wife is unaware. Technically, we wouldn't say that he has a disorder yet. But one night, Connie finds him cruising the Internet. She is alarmed and disgusted and a major conflict ensues. The marriage soon begins to deteriorate, and within a year, despite counseling, the couple divorce.

At this point, we'd say John has a disorder: his behavior has now caused marked distress and impairment. But it wasn't until his wife discovered and objected to his interests that his behavior became a disorder. Had she slept soundly that night and never discovered his secret world, he would not have met the criteria for an illness. Here's a case, then, where the diagnosis of a disorder boils down to another person's sensitivities.

Although the precise causes of paraphilia are not well understood, there is one unequivocal genetic risk factor: carrying a Y chromosome. The proportion of paraphiliacs who are women is vanishingly small compared to men. The Canadian researcher Ray Blanchard told me, "A lot of paraphilias are so rare for females that you could probably write a case for each one you saw. And I think that speaks to the biology. I think it speaks to the fragile nature of the male developing brain compared to the female brain—it more easily goes awry."

How common are paraphilias? We don't really know. As Blanchard told me, "If you go knocking on doors, and say, how's your sexual appetite for . . . let's just say eight-year-olds, nobody is going to tell you that. They don't tell you that after they've been arrested. Your average sex offender will deny paraphilic interests even when the guy has done so much of this stuff that there's no other explanation. Trying to do an epidemiological study, I think, would be almost impossible."

On the other hand, we know something about the relative frequency of different fetishes, one form of paraphilia. The Internet is buzzing with social networks of fetishists. There are hundreds—if not thousands—of Internet groups that like-minded people can join to discuss and share their lust for inanimate objects and body parts. In one study, researchers trying to figure out which fetishes are most popular used Yahoo to scour the Internet for fetish-related discussion groups and found nearly four hundred groups composed of thousands of members.[105] And there was a clear winner. Foot fetishes trampled the competition, accounting for 47 percent of those with body-related fetishes, with second place going to bodily fluids at a mere 9 percent. For those who were partial to inanimate objects, footwear accounted for nearly a third of group members.

A SEMINAL EVENT

UNFORTUNATELY, WE DON'T KNOW MUCH ABOUT THE CAUSES OR biology of paraphilia, and it's not at all clear whether the kind of sexual attraction and arousal that occur in paraphilias involves the same mental or brain systems that we've discussed when it comes to regular heterosexual and homosexual attraction. However, the connection may be closer for conditions that involve the extremes of sexual interests.

In the last years of the twentieth century, an event occurred that led to an unprecedented change in human sexual experience. For the first time in history, millions of people were able to watch other people have sex. The Internet had arrived. The rise of Internet pornography is only the latest chapter in the codependent history of technology and sexual stimulation. Indeed, the very existence of some paraphilias has risen and fallen with advances in technology. Take the case of telephone scatalogia, better known as obscene phone calling. Here's a psychiatric disorder that only became possible with the invention of the telephone. As communication technology has evolved, the telephone is becoming passé, and there are indications that telephone scatalogia is becoming less common. But now we have new media to take its place. Who knows—in the coming years it may be replaced entirely by "sexting."

But the coevolution of sexual behavior and technology has a much longer history. The invention of the printing press in the fifteenth century enabled the spread of "obscene" books and pamphlets. In the nineteenth century photography arrived and flooded the world with a new kind of sexual imagery. And of course, twentieth-century motion pictures, television, and home videos created a full-blown pornography industry that reached into our homes. But for sheer scope, volume, and variety, the World Wide Web is unparalleled as a medium for the dissemination of porn.

Just in case you had any doubts about how mainstream Internet porn has become, consider a few numbers.[106] A new pornography video is made every thirty-nine minutes—more than thirteen thousand per year—and every second, more than twenty-eight thousand people are viewing porn online. There are more than four million pornographic websites and four hundred million pornographic Web pages on the Internet. By 2006 worldwide pornography revenues topped $97 billion annually (with China and South Korea accounting for the majority), a figure that exceeded the revenues of Microsoft, Google, Amazon, eBay, Yahoo, Apple, and Netflix combined. A survey of American college students published in 2008 found that 87 percent of the male students used pornography,* almost three times the rate among the females. On the other hand, nearly 50 percent of the female college students felt that viewing pornography is an acceptable way to express one's sexuality,[107] and an estimated 9 million U.S. women access porn in a given month.[108] Okay, okay, you get the point. People like porn.

---

* One expert told me that this means that 13 percent of male students are liars.

YOUR BRAIN ON SEX

WHOSE BRAINS ARE MORE AROUSED BY EROTIC IMAGES AND porn—men's or women's? If you guessed men, you'd be buying into an age-old stereotype . . . and you'd be right. Brain-imaging studies suggest that even though men and women report similar levels of sexual arousal when viewing couples having sex, emotion (limbic) circuits in men's brains are more "turned on."[108] Heterosexual men are much more aroused by sexually explicit images of women than they are by images of men. While both sexes show increased activity in reward regions when they look at sexual images,[109, 110] women's brains are less discriminating about what turns them on. They are equally aroused by sexual images of men and women, despite the fact that they *say* they are more aroused by looking at men. How much of these sex differences are due to innate biology or cultural learning is unclear.

The accelerating reach of the Internet has clearly made porn consumption a common practice. In recent years clinicians have seen the emergence of what has been called "Internet pornography addiction" and "compulsive cybersex." The idea that people can be addicted to porn makes sense if we consider the evidence that sexual imagery stimulates reward circuits. These are the same circuits that become trapped in the grip of addiction to street drugs. And just like some people can experiment with cocaine or speed and not get hooked, some people can use pornography recreationally. But for others, the pull is too great.

In his confessional book *Porn Nation*, Michael Leahy describes the Internet as the "rocket fuel" that drove his journey from recreational user to full-blown addict. Before he encountered the bottomless well of sexual images available on the 'net, he was limited by "lack of availability (or accessibility) and anonymity. But the Internet smashed through both of those barriers." Soon enough, Leahy realized he could get ahold of any image he wanted, "tapping into new genres or categories of porn that I never knew existed before. And I could do it all instantaneously and anonymously" (pp. 57–58).[111]

TOO MUCH OF A GOOD THING?

LEAHY'S INTERNET PORN ADDICTION TURNED OUT TO BE ONLY A way-station on the road to a larger problem. In recent years, this problem has become a staple of celebrity news and scandal journalism. I'm talking about sex addiction. With the 2009 debut of *Sex Rehab with Dr. Drew,* sex addiction achieved the ultimate in iconic status—its own reality TV show.

With all this, you might be surprised to learn that sex addiction is not officially a disorder according to mainstream psychiatry. At least for now. Right now, sex addiction is not in the DSM, but those responsible for defining sexual disorders for the next edition (DSM–5) have proposed a diagnosis of "hypersexual disorder," a condition that in many respects captures what people mean when they say "sex addiction." The man who gave it that name and has spearheaded the definition of hypersexual disorder is an affable and energetic psychiatrist at McLean Hospital in Belmont, Massachusetts, named Martin Kafka.

Kafka has been studying paraphilias and sexual disorders for nearly twenty-five years. As he acknowledges, the idea that some people engage in excessive sexual behavior is not a new one. In the nineteenth and early twentieth centuries, it went by different names: Don Juanism or satyriasis (for men) and nymphomania (for women). The new diagnosis of hypersexual disorder would be made for recurrent, intense, normophilic* sexual fantasies, arousal, urges, and behaviors that are excessive and lead to significant distress or impairment.[112] The diagnosis includes a variety of subtypes depending on

---

* *Normophilic* refers to normal, or conventional, sexual interests as opposed to paraphilic, or deviant sexual interests.

how and where the hypersexuality is expressed: pornography, cyber-sex, masturbation, or—the tabloids' favorite—"sexual behavior with consenting adults," also known as "protracted promiscuity." And, as you might expect from what we've seen about the biology of sexual attraction and mating, men are more susceptible than women.

But there's a basic dilemma involved in defining the diagnosis: How much sexual desire is too much? Where should we draw the line between a robust sexual appetite and a disorder? Kafka's answer has less to do with quantity than consequences: an inability to control thoughts or behaviors, repetitively engaging in fantasies and behaviors to the exclusion of other important activities or obligations, using sex or sexual thoughts to alleviate mood problems, or pursuing sexual fantasies despite risk or harm to oneself or others. And, of course, being diagnosed with hypersexual disorder would mean having significant distress or impairment. Kafka is convinced that at the extremes, an excess of sexual appetite can be harmful and worthy of treatment:

> *I think there's no question that this condition exists and that it causes very significant impairment and that it's also associated with the spread of STDs as well as very significant pair-bond impairments: separations, divorces, and so on. People get fired from work for looking at pornography. When you have this condition it causes very serious adverse consequences. So, in my opinion, this doesn't stigmatize this group, it actually destigmatizes this group. It says, it really is a condition. There is research supporting this condition, it has criteria, and we need to explore what are the best treatments of this condition.*

Why not just call it sex addiction? Kafka argues that using the word *addiction* would be claiming something about the causes and biology of the condition that we just don't know yet. Is it really an addiction? Or is it more like a compulsion?

What's at heart here is a debate among psychiatrists and others about whether a normal appetitive behavior can ever become an addiction. Appetitive behaviors are those that aim to satisfy a basic need: food, water, sleep, sex. So far, no one has proposed the idea of water addiction or sleep addiction.* Traditionally, the idea of addiction has been reserved for out-of-control behaviors that focus on getting and consuming something we don't normally need: illicit drugs, alcohol, and perhaps gambling. These are things that become addictive because they hijack our brain's reward mechanisms. And those mechanisms were presumably designed to help ensure that we would be motivated to seek out the things that we do need to live and reproduce—that is, food, water, sleep, sex, and, as we saw in Chapter 5, attachment and love.

Addictions tap into the same reward circuitry as those basic needs in ways that are more direct and potent than the experiences that the circuitry was designed to find desirable in the first place. That's why they are so powerful and dangerous. Rats will press a lever to get cocaine until they die of starvation or dehydration. And for some people, cocaine gives the reward system a direct chemical jolt that even attachment or sex can't match. But at this point, it's still unclear whether or not overexposure to porn or sex itself can make someone a sex "addict."

So we've seen two ways that the human capacity for sexual attraction might go awry. One, paraphilia, occurs when the object of the attraction is unconventional—that is, deviant. The other, hypersexuality, is said to occur when sexual interests are excessive. But what about the opposite end of the spectrum? If there's such a thing as too much sexual attraction and drive, can there be too little? The DSM would call for a diagnosis of hypoactive sexual desire disorder when a lack of sexual fantasy or desire causes problems for a person. Like hypersexual disorder, this diagnosis implies that there

---

* Food is another matter. Recent studies in rodents and humans do suggest that food can activate reward circuits in ways that are similar to that seen with drugs of abuse.[112, 113]

is a normal range of sexual attraction and that being outside that range can be a source of distress and impairment.

About 1 percent of the population say that they are "asexual"—that is, they have never really been sexually attracted to another person.[115] They appear to have a higher threshold—perhaps biologically—for sexual attraction,[116] but they don't have a disorder. Perhaps more than any other aspect of human behavior, separating normal from abnormal sexual attraction is fraught with value judgments.

But research on the psychological and biological roots of sexual attraction has given us the outlines of an explanation for how we judge who's hot and who's not. From our ancestral past, natural selection has endowed us with mental biases that shape our mate choices. Those biases operate, at least in part, by tuning brain circuits involved in reward and emotion processing into beauty and sexual signals and creating desire. And yet there are clearly individual differences in what turns us on. That's because those same systems are plastic—they can be tweaked, retuned, or even hijacked by experience, conditioning, supernormal stimuli, and even the sexual politics of our culture. But at the end of the day, these things probably draw their power from the fact that we have mental machinery and neural mechanisms—software and hardware—that care about seeking and choosing mates. There may be a beauty myth, but it's based on a true story.

CHAPTER SEVEN

REMEMBERING TO FORGET

The Biology of Fear and Emotional Memory

ELEN ANTHONY WAS IN HIGH SCHOOL WHEN IT HAP-
pened. "I kept on saying, 'Where are we going to go?
What can we do? What difference does it make whether
we get killed now or later?' I was really hysterical," she later recalled.
"My two girlfriends and I were crying and holding each other and
everything seemed so unimportant in the face of death. We felt it
was terrible we should die so young" (p. 50).

Sylvia Holmes had a more practical thought: "I looked in the
icebox and saw some chicken left from Sunday dinner that I was
saving for Monday night dinner. I said to my nephew, 'We may as
well eat this chicken—we won't be here in the morning'" (p. 54).[1]

These women, like millions of other Americans, were glued to
the news reports. Only miles away, terror had struck out of the blue
in the form of a devastating attack on American soil. According
to initial estimates, at least forty people lay dead near the site of
the attack, "their bodies burned and distorted beyond all possible
recognition." A federal official urged calm and tried to reassure the
nation that a military counterstrike was under way. But that was

doubtless little comfort as word came that highways were clogged with vehicles fleeing the fiery explosions and poisonous black smoke.

Many Americans had gathered with their loved ones, preparing to meet death together, by the time the final announcement came:

> *Tonight the Columbia Broadcasting System, and its affili-ated stations coast-to-coast, has brought you* War of the Worlds *by H. G. Wells . . . the seventeenth in its weekly series of dramatic broadcasts featuring Orson Welles and the Mercury Theatre on the Air.*

The *War of the Worlds* broadcast, on Halloween night 1938, has become a legendary example of how fear and panic can be contagious on a large scale. As newspaperwoman Dorothy Thompson wrote in an editorial two days later, the most frightening thing about the show was the public's irrational response. The broadcast was filled with patently implausible details: Martians had attacked New Jersey with death rays; millions of New Yorkers had reportedly fled the city minutes after the attack was announced; and other events that would have spanned hours were condensed into the brief time-scale of a radio drama. Not to mention the fact that "the public was told at the beginning, at the end, and during the course of the drama that it *was* a drama." But the really terrifying fact, she wrote, was that Welles and his colleagues had "uncovered the primeval fears lying under the thinnest surface of the so-called civilized man. . . . If people can be frightened out of their wits by mythical men from Mars, they can be frightened into fanaticism by the fear of Reds, or convinced that America is in the hands of sixty families, or aroused to revenge against any minority, or terrorized into subservience to leadership because of any imaginable menace."

Sadly, Thompson's analysis was amply confirmed in the years that followed with the frenzied anti-Semitism of Hitler's Germany, the internment of Japanese Americans, McCarthyism, the Cold War, recurrent ethnic violence, and more recently, the anti-Muslim and anti-immigrant fearmongering of right-wing pundits. Fear has been a versatile tool for mobilizing opinion and action in the service of all sorts of agendas.

In a much more benign way, the attention-grabbing function of fear has not escaped marketers and media moguls. Just think of the "watch or die" promos that compel us to watch the nightly news:

> *"The popular toy that could kill your child!"*
> *"Shocking new research: Is your shower giving you cancer?"*

In the past few years, we've ridden waves of panic as apocalyptic warnings have ebbed and flowed: mad cow disease, bird and swine flu, global warming, economic collapse, and cyber-terror. As Robert Brockway lays out in his book *Everything Is Going to Kill Everybody*, there is no shortage of impending disasters you probably don't even realize you should be worrying about. In a lighthearted entry on the end-of-days potential of genetically modified foods, he reassures us that "All joking aside, though: Plants are going to murder your family."

All joking aside, though, we do seem to be living in an era of escalating fear. During his first inaugural address in 1933, Franklin Delano Roosevelt tried to calm a nation under economic duress by declaring that "the only thing we have to fear is fear itself." But more than sixty years later, fear itself became a weapon of war. The "war on terror" marked the first time in history that the named enemy was not a nation state but an emotional state. In the years since 9/11, threat levels have risen and fallen, dragging our nervous systems along with them.

Why are we so susceptible to the power of fear? There is substantial evidence that fear occupies a privileged place in our minds. Natural selection has wired our brains to feel it. Our most basic fears are echoes of the threats our ancestors were likely to face. But the neural machinery that evolved to sense danger has left us vulnerable to fears and anxieties our ancestors never imagined.

With advances in neuroscience and molecular biology, we now have a much more nuanced picture of how fears arise, recede, and sometimes hijack our lives. In many ways, research on the biology of fear and anxiety has provided the best example of how a detailed understanding of the biology of normal can reveal both fundamental features of our everyday lives and how things can go awry to create disorder. And, as we'll see in this chapter, scientists are beginning to exploit that same biology to tame terror.

THE ORIGINS OF FEAR

WHAT IS FEAR? A SIMPLE, BUT USEFUL, DEFINITION IS THAT FEAR is an emotional response to a perceived threat. Gordon Gekko, the fictional tycoon from *Wall Street*, famously said, "Greed is good." The same could be said of fear. We need it to survive. In the face of danger, fear mobilizes us to fight or flee. When our ancestors faced the proverbial saber-toothed tiger, those who were afraid lived to tell the tale—and had children to pass it on to.

So it should come as no surprise that our minds are not blank slates when it comes to fear. Avoiding harm is so fundamental to survival that fear was the first emotional system to be wired into the animal nervous system. We all have fears, and some of them are almost universal. Polls typically find that Americans' top five fears are heavily weighted toward things that would have been universal threats in our evolutionary history: (1) public speaking; (2) snakes; (3) confined spaces; (4) heights; and (5) spiders.[2] You might be wondering what public speaking has to do with our ancestral past. Obviously early hunter-gatherers weren't giving PowerPoint lectures on the finer points of taking down a wildebeest or collecting berries. But the fear of public speaking is really about (real or imagined) social threat. What we fear when we speak in front of a group of people is that we'll show signs of fear or embarrassment and be judged harshly. Evolutionary theories about social and performance anxiety have suggested that being stared at by strangers would have signaled a dangerous situation in our ancestral past and that blushing and other signs of social anxiety might have developed as appeasement displays that communicate submissiveness in an attempt to avoid attack by strangers.[3, 4] Public speaking fears are powerful and ubiquitous. As Jerry Seinfeld once joked, "According to most studies, people's number-one fear is public speaking. 'Death' is number two! Now, this means to the average person, if you have to go to a funeral, you're better off in the casket than doing the eulogy."[5]

BE AFRAID, BE VERY AFRAID

THE WORLD IS A DANGEROUS PLACE. IN CHAPTER 2 WE LEARNED that our brains are tuned to sense threat and avoid harm from the very start of our lives. We are born with rudimentary neural circuitry dedicated to keeping us safe by allowing us to experience fear in the face of threat. But this early sensitivity to signs of danger can only get us so far. Natural selection has wired us to detect and avoid badness, but the infant brain can't anticipate the endless variety of life's possible dangers. We need a mechanism for reacting to the particulars of the environment we are born into—what specifically we should approach because it is safe or nurturing, and what we should avoid because it is dangerous or harmful. Fortunately, natural selection has given us that mechanism: we can *learn* to fear.

In the past two decades, research on the biology of fear has been extraordinarily productive. Electrophysiologic and neuroimaging studies have mapped the basic anatomy of fear circuitry. Neuroscientists have drilled down to the level of synapses and cells to identify the specific chemical events that create and modify emotional memories. And in just the past few years, researchers have used these discoveries to understand the biology of pathologic anxiety and develop promising methods that may relieve the suffering of the millions of people afflicted with disorders of fear and anxiety.

A Dog and His Boy

If a friend asked you to name the most famous dogs of the twentieth century, you might first wonder if your friend has too much time on his hands and then you might come up with names like Lassie, Rin Tin Tin, and Checkers, or maybe even Scooby Doo and Spuds Mackenzie. Chances are the names Beck, Milkah, Ikar, Ruslan, and Toi would not spring to mind. But in the canon of canines, they clearly deserve more credit.[6] Along with a few dozen other dogs (whose names you also didn't know), they helped lay the foundation for modern psychology and neuroscience. You probably know them by the more popular but sadly generic moniker they've been given: Pavlov's dog. Their individual names may not ring a bell, but bell ringing was ironically the basis of their claim to fame.

In the late nineteenth century the Russian physiologist Ivan Pavlov was making major strides in unraveling the physiology of the digestive system. Pavlov discovered that much of how we digest food depends on neural reflexes that coordinate the secretion of "digestive juices" from the salivary glands, stomach, and pancreas. In 1904 he was awarded a Nobel Prize in Physiology or Medicine for his breakthrough experiments.

In the course of this work, Pavlov made another discovery that made his name a household word. He found that placing food in the mouth of a dog would reliably trigger the dog to secrete saliva and gastric acid. But he also found that if he rang a bell each time he fed the dog, the animal would begin to salivate at the sound of the bell—whether or not it was actually followed by food. Pavlov realized there were two kinds of reflexes at work here. The first, triggered by the real food, he called an "unconditioned reflex," and the second, triggered by a stimulus that had merely been associated with the real thing, he called a "conditioned reflex."

That simple observation, which seems almost self-evident today, was revolutionary. It illuminated something much more than the physiology of salivation. It revealed a fundamental mechanism behind learning and memory. It soon became clear that essentially any stimulus could be associated with a conditioned reflex. The implication was that animals can learn about how the world works and the complicated contingencies of their environment by this simple process of association. We can predict the future by registering how things occurred together in the past.

Today, Pavlov's model of learning is known as *classical conditioning*. Several years after Pavlov reported his findings, an American psychologist named John B. Watson took Pavlov's idea and ran much further with it. Watson founded an entire branch of psychology, which he called "behaviorism," on the idea that both animal and human behavior can be explained by stimulus and response learning. He believed that psychology should be a science of observable behavior. In staking that claim, he challenged his colleagues to sweep aside the muddled and untestable jargon of studying the mind as a hidden entity, approachable only by introspection. As he put it in a 1920 article that later became a behaviorist manifesto:

> *I believe we can write a psychology . . . [and] . . . never use the terms consciousness, mental states, mind, content, introspectively verifiable, imagery, and the like. I believe that we can do it . . . in terms of stimulus and response, in terms of habit formation, habit integrations and the like.* (pp. 166–67)[7]

When it came to the nature vs. nurture debate, Watson and his behaviorist disciples were squarely in the nurture camp. The birth and rise of behaviorism ushered in the ascendancy of the "blank slate" view of the human mind. Conditioning was the quill with which any life story could be written. Watson famously wrote:

> *Give me a dozen healthy infants, well-formed, and my own specified world to bring them up in and I'll guarantee to take any one at random and train him to become any type of specialist I might select—doctor, lawyer, artist, merchant-chief and, yes, even beggar-man and thief, regardless of his talents, penchants, tendencies, abilities, vocations, and race of his ancestors. I am going beyond my facts and I admit it, but so have the advocates of the contrary and they have been doing it for many thousands of years. (p. 82)[8]*

Watson never attempted the dozen healthy infants experiment, but in 1920 he reported another experiment on infant learning that was profoundly influential.[9] He wanted to see whether he could use conditioning to create long-lasting emotional reactions in a human being. To test this, he performed an experiment on the infant son of a wet nurse in the Harriet Lane Home for Invalid Children. Albert B. (or, as he became known to generations of psychologists, "Little Albert") was a healthy nine-month-old baby when Watson began a series of conditioning trials that seem remarkably cruel in retrospect. In the modern era of ethical review boards and research oversight, it's clear that Watson's study would never be allowed today.

The research began by documenting that Albert had never been seen to express fear. One day, Albert was placed near a four-foot steel bar that was suspended in the air. While one experimenter distracted him, another suddenly struck the bar with a hammer, creating a loud, frightening noise. Albert was startled and began flailing. The hammer blow was repeated a second and then a third time. "On the third stimulation," Watson reported, "the child broke into a sudden crying fit. This was the first time an emotional situation in the laboratory has produced any fear or even crying in Albert." Watson notes that he considered the possibility that maybe this wasn't such a good thing to do to an infant, but he quickly dismissed the ethical issues, "comforting ourselves that such attachments would arise anyway as soon as the child left the sheltered environment of the nursery for the rough and tumble of the home." (One has to wonder what kind of home Watson was raised in.)

About two months later, Watson began to try conditioning Little Albert. In this case, making a loud noise was analogous to holding out the meat that caused Pavlov's dogs to salivate. Modern psychologists call this an "unconditioned stimulus" that triggers an innate "unconditioned response" (e.g., salivating or, in this case, a fear reflex). Now Watson's task was to see whether pairing this unconditioned stimulus with something else, a "conditioned stimulus," could make Albert learn to fear something new. When Albert was about eleven months old, Watson brought him back to the laboratory and pulled a white rat out of a basket. Albert was curious but "just as his hand touched the animal," Watson reported, "the bar was struck immediately behind his head. The infant jumped violently and fell forward, burying his face in the mattress." A moment later, as Albert again tried to touch the rat, the bar was struck again. Albert "jumped violently, fell forward and began to whimper." Over the next two weeks, the scenario was repeated several times. And after the seventh trial, Watson presented the rat alone. The laboratory notes read that *The instant the rat was shown the baby began to cry. Almost instantly he turned sharply to the left, fell over on left side, raised himself on all fours and began to crawl away so rapidly that he was caught with difficulty before reaching the edge of the table"* (Watson's italics).

Watson wrote, exultantly, "This was as convincing a case of a completely conditioned fear response as could have been theoretically pictured."

But it gets worse. Over the next two months Watson tested whether Little Albert's fear would be transferable to objects that resembled the rat to varying degrees: a rabbit, a dog, a fur coat, cotton wool, a Santa Claus mask, a set of blocks. The results confirmed Watson's suspicions. Albert had paroxysms of fear when presented with the similar stimuli—especially the bunny—but had no fear when given the blocks.

In summing up his findings, Watson noted, "These experiments would seem to show conclusively that directly conditioned emotional responses as well as those conditioned by transfer persist, although with certain loss in the intensity of the reaction, for a longer period than one month. Our view is that they persist and modify personality throughout life." In other words, if all went according to plan, Little Albert would be screwed up for the rest of his life. At the conclusion of his case study, Watson suggested, prophetically, "It is probable that many of the phobias in psychopathology are true conditioned emotional reactions either of the direct or the transferred type."

With the case of Little Albert, Watson established the experimental paradigm of "fear conditioning" that is still used in laboratories throughout the world and that has allowed neuroscientists to discover the neural and molecular basis of normal fear and anxiety disorders.

FEELING THE PAST

Where were you on September 7, 2001? How about November 15, 2001? Chances are, you can't recall. But if I asked you where you were on the morning of September 11, 2001, you probably have a pretty good idea. By stamping the events of that morning with powerful feelings, your brain formed an emotional memory that's lasted more than a decade.

Emotions are the brain's way of attaching salience to our experience. By adding feeling to fact, they help us pick out the important signals from the infinite noise of the world around us. They focus our attention on potential rewards and looming threats. And they help us learn, remember, and then anticipate events that matter. In a sense, then, memory itself is more about the future than the past. We hold on to some experiences and associations because they help us predict what might happen again. We remember so that we can be prepared. And events and situations that elicit feelings get a privileged place in our minds. Fear learning and fear memories are a subset of emotional memory and we now know in great detail how they work, thanks to the legacy of Pavlov and Watson.

CIRCUIT TRAINING: THE ANATOMY OF FEAR

WE'RE ALL FAMILIAR WITH THE PHENOMENON OF CONDITIONED FEAR. Perhaps you were bitten by a dog when you were twelve and since then the sight of an approaching dog makes you sweat. Or maybe you fumbled a big presentation at work last week and now the thought of tomorrow's presentation to the client has you panicking. But how do we learn to fear and how do emotional memories shape our lives?

We've already talked about the hub of the brain's fear system— the amygdala—and we've seen that it has a role in everything from temperament and attachment to trust and empathy. In simple terms, you can think of the basic wiring diagram of our fear circuitry as an alarm system with three major nodes. The amygdala receives information about threats, stamps them with fear, and alerts the rest of the brain to focus attention and trigger stress hormones and fight-or-flight responses. The prefrontal cortex hears about the threat from the thalamus and the amygdala and has a calming, inhibiting effect on the amygdala. The cortex also generates the cognitive experience of fear and worry. And finally the hippocampus processes the context of the threat and also helps us remember the fear experience.

To give you a sense of how the circuit works, let's consider a situation you might have experienced. You're on an airplane en route from New York to Boston. All of a sudden, the seat belt light turns on, the pilot comes over the loudspeaker and says (in that nonchalant pilot way), "Ladies and gentlemen, we're heading into a little bit of turbulence, and I've gone ahead and turned on the seat belt light. Please remain in your seats. Flight attendants, please be seated." Next thing you know, the plane is shaking and you feel a sudden queasiness as the plane seems to drop and then bounce on a pocket of air. For fifteen long seconds, it's nothing but shaking and shuddering, dropping and bumping. And then, after what feels like hours, the plane is once again humming along. You're alive. As you peel your fingers off the armrest, you feel your heart pounding and beads of sweat on your brow. You grab a copy of SkyMall and try to get your mind off what just happened by studying the design features of an award-winning electronic mosquito trap.

What just happened? We can think of the frightening sensations of the turbulence as the unconditioned fear stimulus and the seat belt light and pilot's voice as conditioned stimuli. As your senses took in these stimuli, they relayed the information to the thalamus, a structure deep in the brain that sends sensory information to the amygdala along two pathways. The first is a direct line from the thalamus to the amygdala and has been dubbed "the low road" by neuroscientist Joseph LeDoux, who helped define much of this circuitry.[10] The low road is fast but crude, telegraphing a rough but instantaneous version of the fear stimulus ("DANGER! THREAT!") into the amygdala and triggering an immediate fear response. The second route from the thalamus, which LeDoux calls "the high road," runs up to the prefrontal cortex and then back down to the amygdala. The high road is longer and slower, but conveys more detailed information about the threat ("The plane may be going down! We're shaking and falling!" "The pilot's saying something!").

As the plane was shaking up and down, the two paths converged on the front gate of the amygdala, a collection of neurons known as the basolateral amygdala. These neurons do the work of associating the unconditioned stimulus (the plunging plane) with the conditioned stimuli (seat belt light and pilot's voice). Meanwhile, your hippocampus, which sits in the temporal lobe, fed your amygdala information about the context of the danger situation (the airplane in midflight). Neurons in the lateral amygdala then pass the information to the output side of the amygdala, known as the central nucleus. This part of the fear center alerts other brain regions that trigger stress responses (the hypothalamus), fearful thoughts and feelings (the prefrontal cortex), the fight-or-flight response (the brain stem), and memories of the fearful situation (the hippocampus).

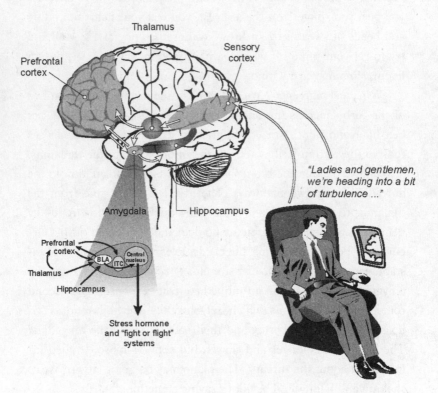

Fear circuitry including detail of the amygdala and its connections to other fear-processing regions. BLA: basolateral amygdala; ITC: intercalated cells.

By the time you were in Boston, waiting at the luggage carousel, your plane ordeal had begun to fade from your mind. But in the deep recesses of your fear circuitry, a memory formed through a process known as consolidation, ready to reawaken if the conditions are right.

Now it's a week later and you're on a flight back to New York. As the plane approaches the airport, the fasten seat belt light comes on and the pilot begins to speak, "Ladies and gentleman . . ." All of a sudden, you freeze. Your heart is pounding and your grip tightens on the armrest. Your mind races back to the terrifying turbulence of your last flight and you're too preoccupied to hear the pilot finish his sentence, "we're approaching New York and we should be on the ground in about five minutes." The conditioned fear memory has been retrieved.

In recent years, neuroscientists have been able to go beyond this basic map of our fear circuits and dissect the chemical and cellular events that generate our emotional memories.

To understand how our memories are formed, you should understand a little bit about the dynamic connections in our brains. Recall that a brain circuit is made up of a series of neurons that communicate with each other across synapses—tiny gaps of about 20 nanometers between neurons. When one neuron is stimulated, an electric charge flows down the neuron and causes the release of packets of neurotransmitters into the synapse. The neurotransmitters cross the synapse and bind to receptors on adjacent neurons, causing chemical or electrical changes that propagate an electrical signal along that neuron, and the chain continues from neuron to neuron.

Contrary to popular notions, memories are not sitting in some compartment of the brain waiting to be called up. Rather, they are stored in these dynamic synapses. In large part, laying down memories involves strengthening the biochemical connections within a circuit of neurons. When experience excites a circuit of neurons, it literally boosts the strength of the connection between the neurons in that circuit and enhances the transmission of signals. That process, known as "long-term potentiation" is a fundamental mechanism of learning and memory throughout the brain, and one of the best-understood examples of experience-dependent plasticity—that is, how experience remodels our brains (see Chapter 3).

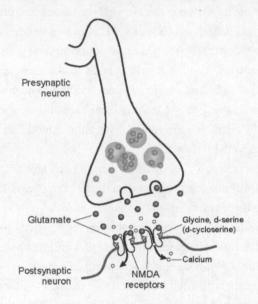

Presynaptic
neuron

Glutamate

Glycine, d-serine
(d-cycloserine)

Calcium

Postsynaptic
neuron

NMDA
receptors

A simplified picture of a glutamate synapse. Glutamate released from a presynaptic neuron binds to NMDA receptors on a postsynaptic neuron. Cofactors (glycine or d-serine) boost the effect of glutamate; the drug d-cycloserine (discussed later in this chapter) can also act as a cofactor that enhances the effect of glutamate.

That's what appears to happen in the amygdala when fear conditioning occurs. Information about the unconditioned and conditioned fear stimuli is carried by neurons from the thalamus, which release a neurotransmitter called glutamate into synapses within the basolateral amygdala. The glutamate crosses the synapse to bind to a type of glutamate receptor called the NMDA receptor on the amygdala neurons. These synapses register the coincidence between the unconditioned fear stimuli (the plunging plane) and the conditioned stimuli (the seat belt light and pilot's voice). Once the glutamate is bound, it opens a channel that allows calcium to rush into and stimulate the amygdala neurons.[11] The calcium also triggers a cascade of chemical events within the neuron that drive the formation of new proteins, including more NMDA receptors that make the neuron more sensitive to triggering a fear response the next time we encounter the conditioned stimuli. When two neurons communicate like this, the connection between them gets stronger by the process of long-term potentiation. The next time glutamate is released, the neural connection fires more easily. And it's this strengthening of connections that encodes information as memories.

A whole set of other brain chemicals can dial up or down the intensity of this fear learning by changing our level of emotional arousal. In general, we pick up fears more easily when we're more emotionally aroused. For example, emotional stress unleashes the hormones cortisol and norepinephrine, which enhance fear learning in the amygdala by stoking our arousal level.[12] The complex balance of these and other neurotransmitters, neuropeptides, and hormones determines how intensely we respond to threats and learn to fear them. New proteins and receptors are made that literally remodel neurons and synapses in the amygdala, leaving a physical trace of threat from the outside world.

LEARNING TO FORGET

IF OUR BRAINS WERE WIRED ONLY TO ACQUIRE FEAR, WE'D BE IN big trouble. We all experience lots of bad things in the course of a lifetime, or even just a day. If we held on to fear memories every time something bad happened, we'd be paralyzed by fears and unable to function. Fortunately, that doesn't happen. Natural selection has also given us a mechanism for letting go of our fears. It's called "fear extinction." Thankfully, when our worst fears are not realized, they often lose their grip—that is, they can be extinguished. In the jargon of conditioning, when we repeatedly find that a conditioned stimulus is not followed by a real threat, the stimulus gradually loses its power. Extinction allows us to drop fears that are no longer relevant. Chances are, after a few more turbulence-free flights, the sight of the seat belt light and the sound of the pilot's voice would go back to being just another part of the background of your life.

Surprisingly, overcoming learned fears doesn't involve forgetting so much as learning—in other words, laying down new memories that separate the safe present from the dangerous past. Fear extinction involves writing a new story that's more compelling and reassuring than the old one. That process requires a new round of learning—only this time we learn that the conditioned stimulus predicts safety rather than harm. Though the fear memory still exists, extinction works by giving priority to the "safe" memory. But if we're exposed to the same frightening situation, the fear can easily come back. So rather than disappearing or fading away, our fears are submerged by a new extinction memory.

Fear extinction involves many of the same emotional brain regions that we use to learn fear in the first place: the basolateral amygdala, the medial prefrontal cortex, and the hippocampus.[13, 14] Like acquiring fears, extinguishing them involves synaptic plasticity in the basolateral amygdala, where glutamate binds to NMDA receptors, but this time the amygdala learns that the conditioned stimulus is associated with safety not harm. Meanwhile, the amygdala and the hippocampus tell the prefrontal cortex that there is an absence of danger signals. In turn, the prefrontal cortex sends inhibitory signals back to the basolateral amygdala, telling it there's nothing to fear.[15] The basolateral amygdala and prefrontal cortex convey these signals to inhibitory neurons (intercalated cells) within the amygdala, which then suppress firing of neurons in the amygdala's central nucleus.[16] As a consequence, the central nucleus no longer triggers fear responses and the circuit consolidates a memory of safety.

THE ANXIOUS BRAIN

PEOPLE OFTEN USE THE WORDS *FEAR* AND *ANXIETY* AS THOUGH they were synonyms. But psychologists and neuroscientists would tell you they are not quite the same. The distinction has to do with the nature of the threat. *Fear* refers to the emotional, behavioral, and physiological reaction to an immediate threat of harm—a clear and present danger. Imagine stepping off the curb and seeing a car barreling toward you. Suddenly, you experience a sense of terror, you startle, you may freeze or try to jump back onto the sidewalk, your body is flooded with stress hormones, your heart pounds, your breathing quickens, and you break into a sweat. Your fear system sensed the threat and mobilized your defenses (stress responses and fight-or-flight reactions). That's fear.

Anxiety, on the other hand, is about the anticipation of threat, feelings that are often described as apprehension, vigilance, and hyperarousal. The threat may be distant, vague, or even undefined. And the feeling is often more chronic than sudden. The element of anticipation and apprehension adds a more cognitive quality to human anxiety. We don't just panic—we worry and even dread. These are experiences that are uniquely human, that we alone bear. Perhaps they are the price we pay for being able to project ourselves beyond the present moment and imagine a future. In that sense, they are like another uniquely human experience, one that may be our compensation for the burden of anxiety: hope.

Although fear and anxiety are not quite the same, the brain systems that generate them are quite similar. Anxiety, like fear, involves coordinated activity of the amygdala, prefrontal cortex and hippocampus, and other regions of the emotional brain.[17] In both fear and anxiety, stress hormones like cortisol, epinephrine, and norepinephrine are released; the sympathetic nervous system triggers fight-or-flight responses; and regions of the cortex generate fearful thoughts.

Anxiety, like fear, is an inescapable and universal part of the human condition. And, like fear, anxiety can be good for us. It drives us to do our best but prepare for the worst. The anxiety you felt before taking a test or giving a speech probably helped you focus your thoughts and motivated you to be prepared. But fear and anxiety can also take a toll. When they are intense and persistent, they can damage body and mind. In rare cases, they can even be deadly.

DEATH BY FEAR

IN 1915 THE GREAT HARVARD PHYSIOLOGIST WALTER B. CANNON developed the idea of the fight-or-flight response: the notion that, when we're threatened, strong emotional reactions like fear and rage create physical symptoms by triggering the sympathetic nervous system and the release of adrenaline (epinephrine) from the adrenal glands. These emotional reactions, Cannon claimed, evolved to protect an animal from danger by preparing the body for "flight or conflict" (p. 277).[18]

In 1942 Cannon described a series of cases of "primitive people in widely scattered parts of the world" who had literally been scared to death: "When subjected to spells or sorcery or the use of 'black magic,' men may be brought to death," he wrote. In the typical case, Cannon reported, some poor soul is condemned for breaking a tribal taboo or cursed by an enemy. In a gesture pregnant with drama, the enemy points a bone at the victim who now knows his fate has been sealed. Cannon cites a description of the terror that follows:

> The man who discovers that he is being boned by an enemy is, indeed, a pitiable sight. He stands aghast, and with his hands lifted as though to ward off the lethal medium, which he imagines is pouring into his body. His cheeks blanch, and his eyes become glassy and the expression of his face become horribly distorted. . . . He attempts to shriek but usually the sound chokes in his throat, and all

*that one might see is froth at his mouth. His body begins
to tremble, and the muscles twist involuntarily. He sways
backwards and falls to the ground, and after a short time
appears to be in a swoon; but soon after he writhes as if
in mortal agony, and, covering his face with his hands,
begins to moan. Unless help is forthcoming in the shape
of a countercharm administered by the hands of the Nan-
garri, or medicine-man, his death is only a matter of a
comparatively short time. (p. 184)*[19]

Voodoo death, Cannon concluded, was the "fatal power of the
imagination working through unmitigated terror" (p. 183), and he
felt certain that "the rapidly fatal result is due to a persistent excessive activity of the sympathico-adrenal system" (p. 187). In other
words, fear itself had driven the fight-or-flight system over a cliff.

At the time, Cannon based his conjecture on little evidence.
In the years that followed, however, a number of detailed physiological studies supported many of his assumptions. Now cases of
sudden death related to excessive fear or stress are recognized to be
more than a curiosity of Cannon's "primitive" cultures. As Cannon
surmised, the cause of death seems to be a sympathetic storm, or
rather, a paroxysm of neural activity that grips the heart. Neurons
from the sympathetic nervous system tell the adrenal glands to release epinephrine into the bloodstream while simultaneously releasing norepinephrine into the cardiac muscle. Overstimulated muscle
cells in the heart contract in a crushing squeeze, ultimately flooding
them with calcium that poisons them and leaves them to die, locked
in a state of contraction.[20] Less dramatic episodes of anxiety may
also affect our physical health.[21, 22] For example, my colleagues and I
asked a group of 3,369 older women if they had recently experienced
a panic attack—a sudden attack of fear or anxiety accompanied by
physical symptoms and fearful thoughts.[23] We found that, over the
course of about the next five years, the women who had reported at

least one panic attack had a threefold increased risk of heart attack or stroke and were nearly twice as likely to die from any cause.

But when fear and anxiety go awry, they're more likely to take a toll on our mental health. For most people, fear and anxiety won't kill you, but for some, they can create a living hell.

THE AGE OF ANXIETY

THE PROBLEM OF DRAWING LINES BETWEEN NORMAL AND PATHO-logic fear and anxiety is not at all straightforward. More than perhaps any other psychiatric symptoms, fear and anxiety are common to both mental health and mental illness.

The idea that there can be disorders of fear and anxiety has only been around since the mid-nineteenth century.[24] Between 1860 and 1900, European physicians began to group symptoms of anxiety into medical syndromes with names like *neurasthenia, agoraphobia*, and *soldier's heart*. At the close of the nineteenth century, Sigmund Freud coined the term *anxiety neurosis* to describe a syndrome that included general irritability, anxious expectation and worry, sudden anxiety attacks, phobias, and a variety of physical symptoms.[25]

The cause was—as you might guess with Freud—sexual, although he also claimed there was usually a hereditary predisposition at work. In particular, Freud believed anxiety neurosis was the result of excess sexual tension arising from sexual abstinence, coitus interruptus, or any other situation in which sexual gratification was frustrated.[26] He later came to see anxiety as a more general response to danger—but the danger, according to Freud, often originated within the mind itself.[27] Anxiety, he claimed, was a signal that the ego was in danger of being overwhelmed by unacceptable impulses and unconscious wishes. To contain this threat, Freud believed that the ego applies defense mechanisms like repression and displacement.

For the much of the twentieth century, the psychoanalytic theories of Freud and his intellectual descendants prevailed in psychiatry. As a result, psychiatrists and psychologists placed little emphasis on dissecting different syndromes of anxiety, in part because there was one treatment—psychoanalysis—that worked for everything.

But in the 1950s and 1960s, behavioral psychologists, influenced by John Watson and B. F. Skinner, were developing therapies designed to treat phobias, and psychiatrists began experimenting with drugs to treat agoraphobia and anxiety neurosis. Suddenly, it mattered what kind of anxiety you had. So when American psychiatrists unveiled the third edition of the DSM (DSM-III) in 1980, they introduced a whole portfolio of anxiety disorders. The most recent edition of the diagnostic manual, DSM-IV, includes seven major anxiety disorders: panic disorder, agoraphobia without a history of panic, generalized anxiety disorder, specific phobias, social phobia, posttraumatic stress disorder, and obsessive-compulsive disorder. Each of the syndromes has its own flavor of pathologic anxiety, but they all represent variations on normal fear and anxiety mechanisms that evolved to keep us safe in a dangerous world.

One of the most surprising facts about the anxiety disorders is just how common they are. The largest survey of U.S. adults found that about 29 percent—more than a quarter of the population!—will meet criteria for an anxiety disorder at some time in their lives.[28] As a group, anxiety disorders are the most common form of psychiatric disorder, edging out mood disorders such as depression and bipolar disorder, and substance use disorders such as alcohol and drug addiction. And anxiety disorders are not benign—in addition to incurring an enormous burden of personal suffering and economic costs (tens of billions of dollars per year in the United States alone), they are also among the most disabling chronic illnesses in all of medicine.

A recent study looked at a number of mental and physical illnesses in terms of how many days per month each illness knocked people out of work or their usual activities. The researchers found that anxiety disorders ranked second only to musculoskeletal disorders and far outpaced cancer, chronic obstructive pulmonary disease, digestive diseases, and even heart disease.[29]

So can the biology of fear and emotional memory help us understand what's going on in anxiety disorders? For the most common anxiety disorders, there is good reason to believe that a problem with the functioning of fear circuitry is at work. For example, compared to other people, those with anxiety disorders acquire conditioned fears more easily and have more difficulty extinguishing them.[30] People who suffer from panic disorder have a tendency to overgeneralize conditioned fear—that is, their fear responses are leaky and can attach to a broad range of stimuli.[31] A large body of neuroimaging studies have shown that people with anxiety disorders have abnormalities in the structure and function of fear circuitry regions that include the amygdala, insula, prefrontal cortex, and hippocampus.[14, 32–35]

As John Watson himself conjectured, phobias provide the most obvious example of fear conditioning gone awry. A phobia is an exaggerated fear and avoidance of some object, situation, or experience. Clinicians recognize three major flavors of phobic disorders, depending on what it is that you fear. For example, someone with a *specific phobia* has an extreme fear of a specific object or situation, like snakes, enclosed spaces, or thunderstorms. When the fear is of other people—that is, embarrassment and anxiety in social situations—it's called *social phobia*; and when the fear is focused on the experience of paniclike sensations, we call it *agoraphobia*.

It's important to remember that all of these fears can be normal, and some are universal. But normal fear becomes a phobia when it is excessive, persistent, and causes suffering or impairment. People with phobic disorders usually avoid whatever it is they fear—and that can lead to a very constricted life. If you have a specific phobia of enclosed spaces (claustrophobia), you may not be able to ride an elevator, travel by airplane, or have an MRI without panicking. If you fear social situations, you may avoid dating, eating in restaurants, going on job interviews, or participating in meetings. And if you're afraid of having panicky feelings, you may avoid any situation where those feelings have occurred or where it might be hard to escape if they do—going to the mall, traveling over bridges, or even just leaving home.

The popular press has a fondness for coming up with names of esoteric phobias. Stick the word *phobia* onto any Greek or Latin root and presto—you've got a disorder: homichlophobia (fear of fog); chronophobia (fear of time); socerophobia (fear of in-laws); triskaidekaphobia (fear of the number thirteen); and so on. But as someone who's been treating patients with anxiety disorders for nearly twenty years, I can tell you that these are not major public health problems. I've never seen a case of metrophobia (fear of poetry), let alone hippopotomonstrosesquippedaliophobia (fear of long words).

In reality, the most common fears and phobias have to do with dangers that we have been prepared to fear because they've been real threats throughout our evolutionary history: snakes, insects, and other dangerous animals, storms, heights, the sight of blood or injury, enclosed spaces, and social threats.

We can think of these kinds of threats as unconditioned fear stimuli—things we readily learn to fear because fearing them kept our ancestors safe.[36] Some of these—like the fear of heights and the fear of being separated from a caregiver—appear to be innate and ready to go the first time we ever encounter the threat. For others, we seem to be biologically prepared to learn them by fear conditioning.

CROCODILE FEARS

IN A SENSE, OUR TENDENCY TO PICK UP THESE FEARS IS ANOTHER example of experience-expectant learning: we're innately tuned to respond to signs of danger, but we need some instruction about what's really harmful and what isn't.

In a series of experiments, psychologists Susan Mineka, Arne Öhman, and their colleagues have shown that primates are selectively biased to learn to fear certain things, and often do so by watching the reactions of others. Rhesus monkeys rapidly acquire fears of natural threats by watching their fellow monkeys—a process known as vicarious conditioning.[36] Monkeys in the wild have a fear of snakes, but lab-reared monkeys don't, suggesting that some kind of learning is involved. If you show a lab monkey a video of another monkey reacting with fear to a toy snake or toy crocodile, the lab monkey quickly develops a fear of those animals.[37, 38] But when the video is doctored to make it look like the monkey was afraid of something that's not naturally dangerous—flowers or a toy rabbit— the lab monkey doesn't develop any fear. It's as though the monkey's fear-learning system is biased to pick up fears of natural threats but blind to phony ones.

The same kind of biased fear learning happens in human babies. If you present a snake to an infant, he or she is likely to be curious and unafraid—just like Little Albert was when he first encountered a rat. But infants as young as nine months will rapidly associate snakes with fear responses if they're paired with the sound of a frightened voice.[39] And numerous studies have found that people can be conditioned to fear primal dangers (snakes, spiders, angry faces) much more easily and intensely than neutral stimuli (flowers, mushrooms, shapes) or even modern dangers (pointed guns, knives, damaged electrical outlets).[36]

Phobias take hold when our normal fear-conditioning mechanisms—the same fear circuits that we all use to learn about danger—lock onto a perceived threat and don't let go. For some people, a combination of genes and life experiences make that fear circuitry more sticky—they are particularly susceptible to picking up and holding on to fears that many of us would let pass. And they may not even recall an event that created their phobia of spiders or snakes.

Prisoners of War

There is one anxiety disorder where the fear-conditioning event is unforgettable. In fact, it's one of the only disorders in all of psychiatry where a life experience is part of its very definition. I'm talking, of course, about posttraumatic stress disorder (PTSD).

The fact that trauma or extreme stress can cause debilitating anxiety and impairment became clear thanks to the oldest man-made disaster: war. After the American Civil War, physicians began recognizing long-lasting problems in soldiers who'd seen a lot of combat. In 1871 an army surgeon names Jacob Mendez Da Costa described a syndrome he called *irritable heart* (later known as Da Costa's syndrome) among Civil War vets who seemed physically well but had bouts of chest pain, palpitations, shortness of breath, and exhaustion. One of Da Costa's cases was an eighteen-year-old soldier who had "served with his regiment for one year, doing much marching and being much exposed. For a couple of months before he left it, he was frequently attacked at night with smothering or suffocating sensations, and with palpitations; and even prior to this had found it difficult to do his duty, and had signs of cardiac distress" (p. 21).[40]

After World War I, tens of thousands of soldiers were crippled by what came to be known as *shell shock*—an incapacitating syndrome of panic, intrusive memories, nightmares, insomnia, and dissociation, which almost entirely captures the modern definition of PTSD. A typical case, reported in 1918 in the medical journal *The Lancet* hints at the torment these soldiers lived with. The victim was an officer who had been buried alive by the explosion of a shell, and later "collapsed altogether after a very trying experience, in which he had gone out to seek a fellow officer and had found his body blown into pieces, with head and limbs lying separated from the trunk":

> *From that time he had been haunted at night by the vision*
> *of his dead and mutilated friend. When he slept he had*

nightmares in which his friend appeared, sometimes as he had seen him mangled on the field, sometimes in the still more terrifying aspect of one whose limbs and features had been eaten away by leprosy. The mutilated or leprous officer of the dream would come nearer and nearer until the patient suddenly awoke pouring with sweat and in a state of the utmost terror. He dreaded to go to sleep and spent each day looking forward in painful anticipation of the night. He had been advised to keep all thoughts of war from his mind, but the experience which recurred so often at night was so insistent that he could not keep it wholly from his thoughts, much as he tried to do so. (p. 174)[41]

Devastated by the brutal trench warfare of the Great War, as many as two hundred thousand soldiers were reportedly exempted from further duty as a result of shell shock.[42] With each successive war, the story has been the same: thousands of young people carrying mental scars that would not heal.

As awareness of the syndrome grew, clinicians began recognizing it among victims of disasters, motor vehicle accidents, and violent crime. In the 1970s and 1980s two streams of cultural uproar brought the phenomenon of traumatic stress and its sequelae to center stage. The first was an apparent epidemic of psychological disability among Vietnam-era veterans (post-Vietnam syndrome), and the second was a growing conviction that childhood sexual and physical abuse were much more common and destructive than anyone had previously imagined. Advocates, families, and clinicians clamored for greater recognition of the scourge of psychic trauma. And so in 1980, when the DSM-III unveiled a new diagnostic system for psychiatry, the diagnosis of PTSD was formally born.

In addition to requiring that a person has suffered a terrifying traumatic event, the diagnosis involves three clusters of impairing symptoms that must persist for at least a month. The first of these clusters is known as *reexperiencing symptoms*. These are essentially trauma-related memories that come back and haunt the victim in the form of flashbacks, nightmares, unwanted recollections, or fear reactions triggered by cues that evoke the traumatic event. The second cluster of symptoms, known as *avoidance and numbing symptoms*, include efforts to avoid thoughts, activities, people, and feelings that remind the victim of the trauma. Victims may also feel emotionally numb or detached and have difficulty connecting with other people or imagining a normal future. And finally, there are *hyperarousal symptoms*, which resemble a state of persistent fear and alarm: insomnia, anger outbursts, trouble concentrating, hypervigilance, and exaggerated startle responses.

Despite widespread agreement that trauma can cause debilitating psychological symptoms, PTSD has become one of the most controversial and politically charged diagnoses in modern psychiatry. Even within the profession, some have criticized the diagnostic criteria for including symptoms that are common to many mood and anxiety disorder and not specific to traumatic causes,[43] or, on the other hand, may simply be normal reactions to major stress and not a disorder at all.[44]

What's more, the criteria for what constitutes a traumatic event have expanded over time. In the DSM's current definition, a traumatic exposure requires that "the person experienced, witnessed, or was confronted with an event or events that involved actual or threatened death or serious injury, or a threat to the physical integrity of self or others," and which evoked "intense fear, helplessness, or horror."[45] The words *confronted with* had not appeared in prior editions of the diagnostic manual and opened the door to a much broader range of potentially traumatic experiences. For example, one can now qualify for a diagnosis of PTSD after having merely heard or read about a terrible event that happened to other people. The Harvard psychologist Richard J. McNally notes, "such secondhand exposure seems qualitatively distinct from being subjected to artillery bombardment for days on end while huddled in a muddy trench" (p. 231).[46] McNally believes that diluting the definition of trauma might make the PTSD diagnosis into such a heterogeneous mixture that it will be impossible to get a firm handle on what is going on at a psychobiological level: "it's like nailing Jell-O to a wall," he said.

But few question that major trauma can create conditioned fear reactions that are debilitating. There is increasing evidence that PTSD is not so much a disorder of fear learning, but one of fear extinction.[47] Virtually everyone can develop fears after a life-threatening trauma. What distinguishes PTSD is that the fears don't go away.

UNFORGETTABLE IN EVERY WAY

AMY RAPPELL CAME TO MY OFFICE SEEKING HELP FOR ANXIETY. She was dressed in a business suit and spoke with a formality that I took as an indication of her discomfort at meeting with a psychiatrist. She had been having panic attacks that seemed to come on mostly at night. Her marriage of five years was falling apart, and she thought the first attack may have happened one night after her husband came home intoxicated and tried to pressure her to have sex.

Since then, the attacks occurred several nights a week. She and her husband were arguing frequently and had recently decided to separate. She denied having any symptoms of depression, except difficulty sleeping, and reported no prior history of anxiety problems or substance abuse. After an extensive review of her history and symptoms, she said that her only concern was getting relief from the panic attacks. Her busy schedule, she said, would make it hard for her to commit to psychotherapy, and she wondered if there was any medication that might help her. We tried an SSRI antidepressant, but after six weeks, she reported that the medicine only seemed to make her more jittery. We then began a trial of clonazepam, an antianxiety medication, and within two weeks, she reported that the panic had improved. Unfortunately, her recovery was short-lived: over the next month, she became increasingly depressed and again began experiencing daily anxiety attacks. But before we could address the problem, she stopped coming to her appointments.

Then one day a few months later, I received a call from a local hospital. "Dr. Smoller, this is Dr. Benham. We have your patient Amy Rappell here in our Emergency Room—she came in complaining of chest pain, but she's ruled out for an MI [heart attack], and it looks like it was a panic attack. But she also tells us she's been drinking daily for the past month, and she's agreed to come in for detox."

At our next meeting, Ms. Rappell told me the full story she had been too ashamed to confide before. The night her husband pressured her to have sex had reawakened terrifying memories of abuse she had suffered as a child at the hands of a family friend. She began having awful nightmares about the abuse, and almost every night she was flooded with unwanted memories that she couldn't suppress.

The terror she felt was as intense as what she had experienced as a little girl, and she took to medicating her panic with alcohol. This would temporarily relieve her anxiety, but the horrifying memories would return and overwhelm her. She soon began drinking more and more, first to quell the anxiety, and eventually to achieve a nearly unconscious state that would finally blot out the fear memories.

She found that the fear and intrusive images could be triggered by anything that remotely reminded her of her childhood trauma—a crime drama on TV, an angry word from her husband, a dark room. She and her husband separated, and she found herself unable to sleep with the lights off. As her fear escalated, she installed a dead bolt lock on her bedroom and lay awake watching the door, her mind in the grip of memories so powerful that she now lived in a constant state of alarm, unable to distinguish the past from the present.

Thankfully, most of us will never experience the conspiracy of fear and memory that tormented Amy Rappell. And yet the suffering caused by PTSD and other anxiety disorders involves a subversion of the same emotional memory circuits our brains routinely use to remember which situations are safe and which we should fear.

In the case of PTSD, neuroimaging studies have found abnormalities in the activity of the ventromedial prefrontal cortex (vmPFC)—a key area in fear extinction learning. Research by my colleague Mohammed Milad at Massachusetts General Hospital has shown that people who are less able to overcome learned fears have thinner brain tissue and less activity in the vmPFC.[33, 48, 49] The vmPFC has a crucial role in teaching the amygdala that what we once learned to fear is actually safe. The hippocampus, in turn, helps retrieve these safety memories when we encounter reminders of the dangerous context. And so, just as we might expect, compared to healthy individuals, victims of PTSD have reduced activity in the vmPFC, smaller and less active hippocampi, and more highly reactive amygdalae (presumably because of less inhibition by the vmPFC and hippocampus).[14, 32, 34] It's as though the entire extinction circuit is dysfunctional: fear and startle reactions are more intense, and fears that were learned in one context now generalize to others that bear even vague connections to the trauma.

THE NATURE AND NURTURE OF ANXIETY

IF YOU THINK ABOUT IT, THOUGH, FEAR CONDITIONING CAN'T BE the whole story behind anxiety disorders like phobias and PTSD. We're all exposed to heights, closed spaces, and spiders, but most of us don't develop phobias. More than 60 percent of Americans are exposed to trauma at some point in their lives, including physical assault, life-threatening accidents, and the sudden loss of a loved one.[50] Trauma is even more common in areas of the world that have been ravaged by war, ethnic conflict, and natural disasters, such as the Middle East, Asia, and Africa. And yet most people who experience trauma don't develop PTSD. For example, the lifetime rate of PTSD in the United States is about 7 percent, roughly one-tenth the proportion of people who experience traumatic events.[28]

So being exposed to something frightening or even traumatic is not enough to cause an anxiety disorder. The obvious question then is what makes some people vulnerable to the transformation of normal fear and anxiety into disorder?

The answer is not entirely clear, but the evidence brings us back to the intersection of genes and experience. Studies have shown over and over that all the anxiety disorders run in families. For example, first-degree relatives (siblings and children) of people with phobias are about five times more likely to develop a phobia compared to relatives of people without the disorder. The same is true of panic disorder, agoraphobia, social phobia, generalized anxiety disorder, PTSD, and obsessive-compulsive disorder.[51] Twin studies find that the heritability of all of these anxiety disorders is about 25 to 45 percent.[52] At the same time, because the heritability is not 100 percent, environmental factors, including life experiences, must explain an even larger share of the vulnerability.

The leading theory of how these genetic and environmental influences make some people vulnerable and others resilient is often called the "stress-diathesis" model.[53] The word *diathesis* refers to an underlying predisposition. The idea is that there's broad range of normal variation when it comes to fear and anxiety, but some of us, by virtue of our particular genetic endowment and life histories, are more sensitive to threat and prone to experiencing anxiety than others. When we encounter threatening experiences, those of us on the higher side of anxiety-proneness are susceptible to more intense and persistent anxiety symptoms. In other words, the combination of an anxiety-prone diathesis and stressful life events results in a tendency to go from normal anxiety to pathologic anxiety.

Anxiety-prone people often have a cognitive bias that makes them see danger where others might not.[54] Their attention is more easily drawn to any sign of threat, and they have a heightened sensitivity of brain regions involved in fear conditioning and emotion processing—especially the amygdala, insula, and prefrontal cortex.[55–62] And that seems to be part of the story of how variations in our genes cause variations in our emotional response to life—including how anxious and fearful we tend to be.

In Chapter 2, I told you the story of how my colleagues and I found that a gene called *RGS2* is associated with anxiety proneness in children as well as hyperreactivity of the amygdala and insula when people look at fearful and angry faces.[63] A large number of studies have found similar effects for a variant of the serotonin transporter gene (*SLC6A4*) that seems to affect anxious temperament in animals from mice to monkeys to humans.[64–67] As I explained in Chapter 2, the protein made by this gene is the target of SSRI drugs that are widely used to treat anxiety disorders and depression.

There is a common genetic variant in the promoter of the gene that affects how actively the gene is expressed. The short allele version is missing forty-four base pairs of DNA compared to the long allele. These missing base pairs makes the serotonin transporter gene less active. Brain-imaging studies have shown that people who carry the short allele have a greater amygdala reaction when they look at emotional faces—especially fearful faces.[68] Moreover, the short allele has been associated with weaker connectivity between the amygdala and a region of the prefrontal cortex important in inhibiting the amygdala and extinguishing fear.[69] In other words, those of us who carry this specific genetic variant may have more difficulty controlling fear and anxiety reactions.

A similar story is emerging for a gene called *BDNF*, which makes brain-derived neurotrophic factor, a protein that promotes the growth and health of neurons. When animals are exposed to stress, BDNF levels increase in the hippocampus, where it has stress-buffering and antidepressant effects.* BDNF also plays a critical role in fear learning, extinction, and memory in the amygdala, prefrontal cortex, and hippocampus by promoting synaptic plasticity,[71, 72, 73] and BDNF injections into the medial prefrontal

---

* On the other hand, BDNF has pro-depressant effects in reward centers of the brain.[70]

cortex have been shown to extinguish conditioned fears in rats.[74] Several studies have now shown that people carrying a single variation in the DNA sequence of this gene have higher levels of anxiety proneness, amygdala reactivity, and impairments in memory, including fear extinction memory. The DNA variation causes one amino acid, methionine, to be substituted for valine in the BDNF protein and seems to be common among people of European-American ancestry.

In 2006 a team led by scientists at Cornell Medical College reported a remarkable experiment that directly linked the *BDNF* valine/methionine variation to anxiety proneness. Using genetic engineering, they inserted the human methionine allele (Met-allele) into a line of mice, creating animals that carry this single human DNA change.[75] Sure enough, the *met*-allele* mice had increased stress-related anxiety behavior and impairments in memory—the same thing that had been reported in humans carrying this genetic variant.

Then, in 2010, the researchers performed fear-conditioning experiments with both the BDNF-*met* mice and with a group of human volunteers.[76] Both the mice and humans carrying the *met* allele had impaired fear extinction. The researchers then put the human subjects in an fMRI scanner during extinction learning, and discovered that the Met carriers had reduced activity in the ventromedial prefrontal cortex (vmPFC) and increased activity in the amygdala—exactly what we would expect because we know that fear extinction involves inhibition of the amygdala by the vmPFC. So here is a specific genetic variation that seems to increase anxiety proneness and interfere with letting go of fear by weakening the brakes on our fear circuitry.

The search for genes that affect anxiety proneness is an ongoing challenge. So far the few genes that have been found explain only a tiny fraction of the heritability estimated from twin studies. And that's what we'd expect given that anxiety proneness and anxi-

---

* A word on notation: the methionine allele is abbreviated "Met" in humans and "*met*" in mice.

ety disorders are very complex and probably involve many (perhaps thousands) of genes that have small effects on their own. But these early results suggest that genes operate by subtly biasing the sensitivity of brain circuits involved in perceiving and processing fear and other emotions—adjusting the lens of salience through which we experience the world around us.

As we saw in Chapter 3, however, life experience can have the same effect. It's now clear that experiences in childhood also shape how our brains process emotions and stressful life events. Child abuse, neglect, and other forms of adversity can have lasting effects on how anxious and fearful we become, especially if they occur during sensitive periods of brain development when thresholds for detecting threat and regulating our emotions are being set in the emotional circuits of our brains. Recall that early adversity can modify the chemistry of our chromosomes, causing epigenetic changes that may program anxiety and stress hormone responses in long-lasting ways that sensitize fear circuits. Epigenetic effects on *BDNF* gene expression, for example, can affect how tightly an animal holds on to fear memories.[77]

For some of us, our genes may contribute to anxiety problems by actually increasing the chance that we'll be exposed to dangerous situations. One twin study of Vietnam veterans found a heritability of 35 to 47 percent for exposure to combat-related trauma—that is, the likelihood that a soldier would find himself in harm's way was itself influenced by genes.[78] Another study of civilians found that exposure to assaultive trauma, such as being the victim of a mugging, or sexual assault, had a heritability of 20 percent.[79]

How could a person's genes affect whether she would be in the wrong place at the wrong time? The most likely answer is again related to temperament and personality. Genes that contribute to risk-taking, seeking out new experiences, and even antisocial traits may increase the chance that you'll find yourself in situations where trouble happens.[80]

Boundary Violations

There's an interesting implication to the idea that vulnerability to anxiety disorders begins with a general exaggeration of normal fear and anxiety systems. If that idea is true, we ought to see a lot of overlap in the biology of different anxiety disorders. The numerous anxiety disorders should share features with one another and with normal temperamental variation in anxiety proneness. And that's exactly what the evidence shows.

It turns out that the genes that influence anxious personality and a broad range of anxiety disorders overlap substantially.[81-83] And there's strong overlap at the level of brain function as well. Recall that neuroimaging studies find that healthy but anxiety-prone people have heightened reactivity of the amygdala when they're presented with emotional stimuli—especially fearful faces.[56, 84] The same is true of people with a broad range of anxiety disorders: specific phobias, social anxiety disorder, agoraphobia, panic disorder, generalized anxiety disorder, and PTSD.[14] Perhaps it's not surprising then that the major treatments for anxiety disorders—antidepressants and cognitive behavioral therapy—tend to work for all anxiety disorders.

That's not to say that all of these different anxiety disorders are really just different names for the same thing. They have different constellations of symptoms and a variety of different risk factors. But it does suggest that a clear boundary between normal and abnormal anxiety and even between the various anxiety disorders may be impossible to draw.

Take Back the Fright

Understanding the biology of fear and anxiety may help us understand how anxiety disorders develop. But for people afflicted with these disorders and those of us who try to help them, the real hope is that we can use this knowledge to improve treatment. In the last several years, psychologists, psychiatrists, and neuroscientists have come together to make good on that hope. Using insights from the molecular neuroscience of emotional memory, researchers and clinicians are developing methods to prevent, suppress, and perhaps even erase traumatic fears and painful memories. The results, though preliminary, are startling.

As we'll see, each of these strategies exploits the biology of emotional memory to try to hack into the system and use it to our advantage. The first tries to use medications to block traumatic memories from taking hold to begin with. The second offers a drug that can boost the effects of psychotherapy to help extinguish fears that already exist. And the third hijacks the molecular basis of memory to rewrite a painful past.

FIRST RESPONDERS

WHAT IF WE COULD HAVE PREVENTED AMY RAPPELL'S TRAUMA from gelling into an indelible memory? Imagine the pain and suffering she would have been spared.

After a traumatic event, there's a brief window of time before the fear becomes entrenched as a memory, and that creates an opportunity to intervene. One way to do that might be to quell the storm of stress hormones that prime the amygdala to generate fear memories.

The emotional arousal that accompanies stress and trauma involves a surge of epinephrine (adrenaline, from the adrenal glands) and its chemical cousin norepinephrine that act on the amygdala and help fuel the cascade of cellular events that consolidate emotional memories.[85] Drugs that trigger or mimic norepinephrine release in the brain enhance fear conditioning in rats and bias the amygdala toward threat in humans,[86, 87, 88] while drugs that block norepinephrine interfere with remembering emotionally charged events.[89]

That kind of evidence suggested a bold idea to Harvard psychiatrist Roger Pitman and his colleagues—what if we could block the stress hormones that help etch traumatic memories into our synapses?

The drug propranolol has been used for decades to treat high blood pressure by blocking receptors for epinephrine and norepinephrine. These receptors are known as beta-adrenergic receptors, and so drugs like propanolol are referred to as beta-blockers. Propranolol is also well known to musicians and actors as a pill that can relieve stage fright—mainly by quieting the pounding heart, shaky voice, and tremulous hands that let people know you're nervous.

Over the course of ten days, Pitman and his colleagues gave either propanolol or a placebo to patients who came to the emergency room at Massachusetts General Hospital after experiencing car accidents and other traumatic events.[90] Three months later, when the researchers reexamined the patients for symptoms of PTSD and physiological responses while recalling the traumatic event, those who had received propranolol had significantly lower stress responses. The implication was powerful: by interfering with the brain's normal tendency to create powerful memories of stressful events, a commonly used blood pressure drug could spare trauma victims the pain of unforgettable fears. The jury is still out on just how effective propranolol really is, but the possibility of protecting the emotional brain from psychic trauma is an exciting one.

One reason that it may prove difficult to actually prevent the formation of traumatic memories has to do with timing. Our brains are very good at learning to fear things that are life-threatening. In the face of overwhelming stress—like a physical or sexual assault—the window for blocking memory consolidation may close quickly. Interfering with that process might require intervening within minutes of exposure to the danger. By the time a victim gets to an emergency room, it's probably too late. Even if propranolol proves to be protective, it may just be unfeasible to get it into the brain before the window of conditioning closes for most people.

The importance of haste in blocking memory consolidation was clear in a study of nearly seven hundred Iraq War military personnel who had been severely injured on the battlefield by IEDs, gunshots, mortar, or rocket-propelled grenades.[91] It's common for injured soldiers to be given IV morphine to treat their pain early in the course of their emergency medical care. Pain is known to trigger an outpouring of stress hormones like epinephrine and norepinephrine, and there's evidence that morphine, like propranolol, can interfere with memory consolidation by a beta-blocking effect.

The authors of the military personnel study wondered if the morphine might have a protective effect against developing PTSD. When they compared those who got morphine to those who didn't, the effect was dramatic. Within two years of their injury, the incidence of PTSD was nearly cut in half among those who got morphine. And, importantly, more than 70 percent of those who were treated with IV morphine got it within an hour of their trauma. We don't know for sure what would have happened if the morphine had been delayed, but it's likely that immediate treatment made a difference.

THE PETER PRINCIPLE

OF COURSE, FOR PEOPLE WHO ALREADY HAVE PTSD, PREVENTION is not an option. Fortunately, our memories are not written in stone but in synapses, and synapses are plastic. Just as we can learn to fear, we can also learn not to fear—that's what fear extinction is all about.

One of the oldest and most effective treatments for anxiety disorders is all about fear extinction. With the case of Little Albert, behaviorists established that fears and phobias can be learned. Four years after Watson's "success" in terrorizing Little Albert, another American psychologist, Mary Cover Jones, reported the first successful use of conditioning to eliminate fear. Her first subject was a three-year-old boy named Peter who, like Albert, had developed a fear of a white rabbits. Jones and her team repeatedly exposed Peter to the rabbit while at the same time giving the boy food and candy. After a series of these sessions, Peter's fear was gone.[92]

Beginning in the 1950s, psychiatrists and psychologists began developing more formal therapies that moved behaviorism from the laboratory to the clinic. If fear reactions can be learned by so-called classical (Pavlovian) and operant (Skinnerian) conditioning, these new therapies showed that they can be conquered the same way. By gradually exposing a phobic patient to his worst fear in a safe context, the fear loses its power.

In the 1970s a psychiatrist named Aaron Beck at the University of Pennsylvania developed a variant of behavior therapy that he called cognitive therapy. The idea was to teach patients to recognize and challenge the automatic thoughts and faulty assumptions that were causing and perpetuating their problems. In the 1980s a merger of sorts occurred and cognitive-behavior therapy (CBT)— combining the active ingredients of both cognitive therapy and behavioral approaches—became increasingly popular.

From the perspective of modern neuroscience, we can think of behavior therapies and CBT as techniques that help extinguish painful emotional memories. By systematically desensitizing the patient to what he fears and reframing his catastrophic thoughts, CBT allows him to form safety memories where there were only fear memories. Gradually exposing someone to a feared (conditioned) stimulus in a safe setting extinguishes fear. When this treatment is successful, it's also been shown to normalize amygdala reactivity when people are later shown the thing they most feared.[93]

Over the past twenty-five years, a mountain of studies has established that behavior therapies, including CBT, are effective treatments for a broad range of anxiety disorders as well as depression, eating disorders, addiction, and even psychotic disorders. In fact, behavior therapies are the only treatment shown to provide lasting benefit for all the anxiety disorders, with success rates on the order of 50 to 80 percent.

In the months after Amy Rappell revealed the secret she had been keeping—the trauma that had hijacked her mind and revisited her in terrifying waves of panic—we struggled together to find a way to quiet her fears. When I recommended a course of CBT, she balked, saying she had tried therapy before and it had only made her feel worse. She was afraid that any kind of therapy would mean she would have to relive her trauma and that would be too frightening to bear. "I just need to forget," she told me with a desperate look. We tried a series of medications—antidepressants, anticonvulsants, even antipsychotics—but any relief she found was short-lived. Finally, exhausted and nearly hopeless, she agreed to see the cognitive-behavioral therapist I'd referred her to.

We met again after she had completed her first month of CBT. Those first weeks had been difficult, and she confessed that she'd been on the verge of quitting before every session. I told her I admired her strength in persisting despite her impulses to quit and tried to reassure her that the feelings she had were not uncommon. We needed to help her stay with the treatment long enough for it to have a chance of working.

After a second month of treatment, she seemed to be making progress. She was sleeping through the night. Her bedroom door was unbolted. And then, when we met two months later, something had clearly changed. She looked different somehow. As I talked with her, I realized that I was seeing her face without the mask of tense vigilance that had covered it like a taut film for so long. She told me she was finally free of panic attacks and was able to focus on her work again. She'd even spent the weekend away, visiting friends for the first time since she and her husband had separated.

After five months, Amy completed the CBT, confident that she had been able to put the past in its place. In the years that have followed, that confidence has sometimes been shaken. Once, after she unexpectedly ran into the son of the "family friend" who'd been the source of her pain, the panic came back in full force, along with a relapse of drinking. She resumed CBT and was able to regain her balance. Today, she is doing well, thankful for the truce between her past and present but mindful that it may be provisional.

TURBO-CHARGED THERAPY

WHAT IF THERE WERE A WAY TO BOOST THE EFFECT OF BEHAVIOR therapy by exploiting the biology of emotional memory? Imagine a drug that could help extinguish pathologic fears.

Recall that fear extinction involves new learning and that, at a molecular level, this learning depends on synaptic changes that occur when the neurotransmitter glutamate activates NMDA receptors in the amygdala and prefrontal cortex. If you want a drug that could boost fear extinction, glutamate (or something like it) would be an obvious choice. The problem is chemicals that mimic glutamate itself can be toxic to the brain.

Fortunately, Emory University neuroscientist Michael Davis and his colleagues found another way. Davis, who had been one of the pioneers in defining the biology of fear extinction in rats, knew that glutamate needs a partner to do its job. Glycine, another neurotransmitter (or its cousin d-serine), has to bind to the NMDA receptor at the same time as the glutamate. In 2002 the Davis lab reported that directly infusing d-cycloserine (DCS), a drug that mimics glycine and d-serine, into the amygdala can enhance extinction learning in rats; that is, it helps them to let go of fear.[94] Obviously that strategy won't work in humans because no one would agree to have amygdala injections (if it were even possible). But their results were exciting because DCS pills had been safely used in humans since the 1950s. Through an entirely different mechanism, DCS acts as an antibiotic and had long been used as a fallback drug for the treatment of tuberculosis.

Kerry Ressler, a psychiatry resident working in the lab at the time, had a hunch. If behavior therapy works for anxiety disorders by extinguishing fears, maybe DCS could make behavior therapy even more effective. Ressler, now a professor at Emory, and his colleague Barbara Rothbaum treated twenty-seven volunteers who had a phobia of heights (acrophobia) with virtual reality behavior therapy.[95] Before the treatment started, the subjects were randomized to receive either DCS or a placebo and told to take the pill about two to four hours before their therapy sessions. During the therapy sessions, subjects wore a virtual reality helmet that simulated the experience of being in a moving glass elevator. From inside the elevator, they could look down over a virtual railing. The therapy consisted of gradual exposure and desensitization to (virtual) heights. Each subject had two therapy sessions two weeks apart—a lot less than the usual two-month course needed to treat phobias.

At the end of the treatment, the results were clear. The DCS group had extinguished their fear of heights with only two sessions while the placebo group were still quaking in their virtual boots. Three months later, when the researchers had the subjects return and repeat the virtual elevator experience, the DCS group remained cured of their phobias. And the benefit wasn't just in the virtual reality world—the DCS group had significantly decreased their avoidance of heights in the real world.

As word about the success of DCS-enhanced behavior therapy got around, other researchers put it to the test. Clinical trials have now shown that DCS can improve outcomes in a broad range of anxiety disorders that are traditionally treated with behavior therapy and CBT including obsessive-compulsive disorder, panic disorder, social anxiety disorder, public speaking phobia, spider phobia, and even dental phobia.[96–102]

Consider for a moment what we're talking about here: a drug that you could take before a therapy session to make the therapy work faster and better. There is one caveat about using DCS to turbo-charge psychotherapy—it has to be given intermittently. It turns out that the brain adapts quickly to DCS, so daily dosing won't work. But that fits well with its use to augment psychotherapy, which is typically conducted on a weekly basis—a long enough interval to prevent the development of tolerance to the drug.

There's still work to be done to better understand how and when DCS should be used, and so far, it's not FDA-approved for the treatment of any psychiatric or emotional conditions. Still, if the early studies hold up, the DCS story will stand as a remarkable example of how understanding the biology of fear and memory can translate into therapeutic discoveries.

THE SPOTLESS MIND

RECENTLY, NEUROSCIENTISTS HAVE BEGUN TO PUSH THE ENVElope of conquering fear and anxiety even further by trying to delete memories that cause fear and anxiety. Behavior therapy can extinguish fears but it doesn't erase them because fear extinction just creates a new memory of safety that inhibits the old memory of threat. But the original memory hasn't gone away. Our deepest fears can still lurk in our synapses, even after we've subdued them. Under the right circumstances, when the original stimulus and context align or when you're under stress, they can return.

But what if we could free ourselves of painful emotional memories—not just suppress them, but get rid of them? That was the premise of the 2004 film *Eternal Sunshine of the Spotless Mind*. The title was taken from Alexander Pope's poignant poem *Eloisa to Abelard* about the epic, but illicit, romance between Heloise and the great scholar Pierre Abelard. After her vengeful uncle arranges for Abelard to be castrated, Heloise flees to a convent, where she is tormented by her separation from her lover and teacher. In Pope's poem, Heloise is left only with her memories of their love and prays that she can find relief by forgetting.

> *Of all affliction taught a lover yet,*
> *'Tis sure the hardest science to forget!*

If only their illicit love had never been, she would have been spared her endless longing and could have had the peace of mind that innocent souls enjoy:

> *How happy is the blameless Vestal's lot!*
> *The world forgetting, by the world forgot:*
> *Eternal sunshine of the spotless mind!*

In the 2004 film, scientists at a commercial outfit called Lacuna, Inc. have perfected a technique for "the focused erasure of troubling memories." The introverted and emotionally subdued Joel (Jim Carrey) signs up after he's wounded by the news that his vibrant but impulsive ex-girlfriend Clementine (Kate Winslet) has already erased him from her mind. The procedure begins with a detailed interview at the Lacuna offices during which Joel is asked to bring in mementos of the relationship. He's asked to recount the details of their love affair, which Lacuna uses to create a detailed brain map of his memories. Later, technicians will hunt down and zap away the memories using some kind of MRI technology. The film doesn't bother much with scientific details—it's a fantasy after all.

At least so far. In the few years since the movie's release, scientists have begun to explore the possibility of rewriting our emotional past. The groundwork that made this possible was a surprising set of discoveries about how we remember. The common view of how memories are formed is analogous to how a computer stores files: like saving a document to a hard drive, our brains store a memory so that it can be retrieved when we need it. Calling up the memory is like opening and reading the saved file.

But that's not what really happens. In fact, remembering is a dynamic process, in which a memory is re-created each time we retrieve it. As Harvard psychiatrist Roger Pitman put it, "when you remember something, you don't remember what originally happened; what you remember is what you remembered the last time you remembered it. Each time you destabilize a memory by remembering it, you are updating it and changing the memory. So theoretically it's impossible for us to ever remember something that originally happened to us. All we're remembering is what we remembered the last time, which in turn is what we remembered the time before that, and so on. Memory is continually being sculpted."

So when we retrieve a memory, it becomes unstable. To use the computer analogy, it's as though we've called up the file and begun working on it again. The memory has briefly entered a vulnerable or labile state—like a computer file that's in temporary (RAM) memory. If we don't save ("reconsolidate") it again, it can be lost.*

The fragility of our memory seems like a problem. Why would natural selection have allowed for our minds to change or erase information that we presumably needed to remember? Perhaps because it also allows us to update our memories—to incorporate new information about changes in the world around us. Whatever the reason, this fragile window of memory formation may work to our advantage when the past is too much with us. For the victim of PTSD, there are some memories that may be better lost. The idea of erasing a traumatic memory would be like getting the memory into RAM and then corrupting the file before it can be saved again. But is such a thing really possible?

It turns out that the synaptic changes involved in reconsolidating memories require that neurons make new proteins. And it's during this process that memories are in the vulnerable state. Once the

---

* It would be as though you started writing a book chapter and worked on it for hours and then your computer crashed and you realized you hadn't saved it and you started pounding the table and yelling "What the hell is the matter with this !@#*ing thing!" and everyone in Starbucks turned and looked at you like you were out of your mind. It can happen.

protein synthesis is completed—within a few hours—the memory is reconsolidated (saved to the hard drive, if you will).

In a set of experiments published in 2000, NYU neuroscientists Karim Nader, Glenn Schafe, and Joseph LeDoux reported that they could disrupt memory reconsolidation of conditioned fears by injecting rats at just the right moment with a drug that interferes with protein synthesis. They had effectively erased the fear memory by hacking into the amygdala during the reconsolidation window. That was an exciting discovery, but it involved infusing a potentially dangerous drug directly into the amygdala. Not exactly an option for clinical use. In the years that followed, other researchers reported success in erasing fear memories in rats by equally dramatic measures, including poisoning amygdala neurons that had encoded the fear.[103]

Then in 2009 LeDoux and his colleagues hit on a much safer alternative. Their idea was to evoke the fear memory, bringing it into the reconsolidation window, and then overwrite it by extinction learning.[104] Using a classic paradigm, they conditioned rats to fear the sound of a tone that had been paired with a shock. Then they divided the rats into three groups. For the first (early retrieval) group, they played the tone once, waited for up to an hour, and then performed the usual extinction procedure by presenting a series of the tones without the shock. The idea was that the isolated tone would evoke the memory, bringing it into the fragile state, and then they could overwrite it with extinction learning before the reconsolidation window had closed. The second (late retrieval) group got the same treatment as the first but instead of waiting an hour between the isolated tone and extinction, they waited at least six hours (by which time the reconsolidation window would have closed). And the third (control) group were simply put back in the conditioning cage without any tones or shock.

The results were fairly astounding. When they tested the rats a day later, the fear had been extinguished in all three groups. But when they retested the rats after a month, the fear had spontaneously returned for the late retrieval group and the control group. Not so for the early retrieval group—their fear was gone. In other words, by extinguishing the fear during the reconsolidation window, the researchers had rewritten the fear memory as a safe memory. They had, in effect, erased the fear.

The next question was whether something like this could work in humans. In 2010 LeDoux, Elizabeth Phelps, and their colleagues showed that it could.[105] Just as they'd done with rats, the scientists created a conditioned fear response in human volunteers and then eliminated the fear memory by performing extinction training within minutes of retrieving the memory. A comparison group who underwent extinction six hours after the memory was evoked (after the reconsolidation window had closed) had their fear return. And the difference was still there a year later.

Another team of researchers has now achieved the same effect—effectively erasing fear—by giving subjects propranolol during the reconsolidation window.[106, 107] It turns out that propranolol may act by indirectly inhibiting the protein synthesis required for memory reconsolidation.

In actuality, these techniques don't erase the memory entirely. They preserve what's called "declarative memory"—that is, the simple ability to recall that an event occurred. But they decouple the event from the emotion that was attached to it. Thinking back to the turbulent airplane ride example, you would presumably recall that you experienced a rough ride, but the fear you felt would be gone. It's analogous to what happens when you remember a stomachache—you might recall that you had a pain in your belly, but the visceral experience of pain itself is hard to re-create in your mind.

Still, this is heady stuff. Rather than simply suppressing fears, these studies suggest that we can use behavioral treatments or drugs to actually delete them. Much work remains to be done before something like these techniques could be offered to people suffering from PTSD. For one thing, these studies used simple conditioning methods and mild fear stimuli in a laboratory setting. It's not at all clear that the kind of intense, real-world trauma that leads to PTSD could be overwritten in the same way. But the stage has been set to find out.

Beyond the remarkable scientific achievement, these studies raise a more intriguing question about who we are and how far we want to go in altering our memories. If someone offered you the chance to erase an emotional memory of your past, would you take it? Victims of PTSD, for whom the past has become a prison, might welcome that chance. But in principle, the same techniques that might erase the pain of trauma could be used to scrub other life experiences of their emotional residue. How much of ourselves would we lose in the process? Many of our most meaningful experiences draw their beauty from indivisible shadings of shadow and light. Love entails loss, redemption can only follow a fall.

After they have been estranged by amnesia, Joel and Clementine find each other again and learn that they had both erased the hurt of their shared past. And yet, fully aware of the pain it may bring, they fall in love again. It's worth it.

A REMEMBRANCE

IN THE CENTURY SINCE LITTLE ALBERT FIRST LEARNED TO FEAR rats and rabbits, the science of emotional memory has taken us from behavior to the molecular biology of the cell. By unraveling the mysteries of normal fear and anxiety, we are beginning to understand how anxiety can overwhelm the mind and how it can be quieted.

But as the science marched on, another mystery hung in the air: whatever happened to Little Albert, the infant who started it all? Watson's case study of Little Albert was the last research paper he published, creating what has been called "one of the greatest mysteries in the history of psychology" (p. 605).[108] Did Watson ever try to decondition Albert's fear or did he live out his days with a phobic fear of furry animals? Did he even know about the contribution he'd made to modern psychology?

In 2009 psychologist Hall Beck and his colleagues revealed that they had solved the mystery. After several years of painstaking detective work, they concluded that Albert's real name was Douglas Merritte. He was the son of an unwed young woman who was the wet nurse at the Harriet Lane Home mentioned by Watson. Sadly, the record shows that little Douglas's life was painful and brief. In 1922, at the age of three, he developed hydrocephalus and died three years later. As Beck wrote, "Our search of seven years was longer than the little boy's life."

But as we've seen, Albert's legacy lives on. His fate is both poignant and fitting for a little boy whose story was so important to the study of emotion and memory: gone, but not forgotten.

Chapter Eight

The New Normal

WHEN IT COMES TO THE HUMAN MIND, WE'VE LONG had an uneasy relationship with the concept of normal. For one thing, it's hard to define. And more than that, it has connotations that can imply something derogatory about people who struggle with life's challenges. For some, normal is the way things ought to be. You're either normal or you're not, and abnormal is something alien and possibly inferior.

But I've argued for a very different view—another side of normal. Normal is not the ideal, the average, or even the state of being healthy. It is more like a landscape of human possibility whose contours have been shaped by the design features of the mind and brain. As I suggested in Chapter 1, like the statistician's concept of the normal distribution, variation is part of its very definition.

A central premise of this book has been the deceptively simple claim that there is a biology of normal—a complex but fathomable basis for how the mind and the brain operate.

We are still filling in the details, but the broad outline is taking shape from a convergence of evolutionary biology, psychology, neuroscience, and genetics. To illustrate that premise, I've focused on what we're learning about some of the most fundamental challenges that our brains and minds were designed to tackle. Of course, the examples I've given cover only a tiny fraction of our mental lives, but they reveal some important themes about the biology of normal. Here are a few.

1. THE RHYME BEHIND REASON

OUR MINDS ARE ORGANIZED AROUND SOLVING PROBLEMS THAT mattered to the reproductive fitness of our ancestors: avoiding harm, understanding the thoughts and feelings of other people, forming attachments, selecting mates, and learning from the past, to name a few. But, for the most part, natural selection is a sketch artist, providing the lines and shading that help us make sense of the world, but leaving us to fill in the details. It has given us mental rules of thumb, honed over millennia of biological competition, that we use to navigate our lives. We inherit the instructions for building neural circuitry and mental algorithms that tune our brains to process the salient signals out of the infinite noise of life. But it would be impossible to build a brain that anticipated all of the important contingencies of human life. And so the most important tool that evolution has given us is the capacity to adjust to life, to learn from experience.

2. GREAT EXPECTATIONS: THE POWER OF SENSITIVE PERIODS

SOME EXPERIENCES MATTER MORE THAN OTHERS. ONE OF THE most remarkable features of human development is that the brain uses the world to wire itself. Many of the universal functions of the mind are programmed early in life by experiences that evolution has prepared our brains to expect. Our brains lie in wait for the inputs we need to wire foundational systems like vision, language, attachment, and social cognition. This process of experience-expectant plasticity involves sensitive periods during which the brain is supersticky— acutely responsive to instructions from the world around it. But these windows of opportunity create a double-edged sword. On the one hand, sensitive periods ensure that we can extract what we need from the environment. Children will acquire language as long as they are exposed to some speakers and they will form an attachment system as long as there is some sort of caregiver around. On the other, sensitive periods create windows of vulnerability. If the experiences we expect are corrupted or absent, the damage may be hard to undo. A child who suffers severe neglect early in life may have attachment problems that last a lifetime.

Our brains expect a good enough environment. We need certain inputs—exposure to visual information, language, caregiving—to wire key systems in the brain. Disrupting those inputs during sensitive periods can have profound effects on the downside, but trying to supercharge the environment is not likely to get you a supernormal child. It's true that when the environment is less than good enough, enrichment can make a big difference. And specific skills and talents—like musical abilities—can be enhanced by training. But the fundamentals of brain development don't need perfection. That may come as a disappointment to parents who are frantically trying to optimize every detail of their two-year-old's environment in the hope of supersizing their abilities. But it should also be a relief: barring catastrophe, most children will develop just fine.

3. BUFFERING AND BUFFETING

THERE'S NO QUESTION THAT EARLY EXPERIENCE MATTERS, BUT our fate isn't sealed by the age of five. The unique trajectory each of us travels from cradle to grave is continually shaped by the buffeting and buffering effects of genes and experience. And through it all, the brain continues to be capable of change—the result of experience-dependent plasticity. Even children who endure early neglect or disrupted attachments may do well if they later find a nurturing home. Cultural and social influences can shape our desires. Psychotherapy can extinguish long-standing fears. Love can mend a broken heart. As a psychiatrist I am continually humbled by the remarkable resilience of the human mind. Even in the face of tremendous adversity, we can find ways to adapt and carry on.

4. THE LITTLE THINGS

SOMETIMES THE EFFECTS OF NATURE AND NURTURE SEEM PLAIN and simple. The distinctive personality profile of Williams syndrome can be related to a chunk of missing DNA on chromosome 7. Exposure to alcohol or toxins in the womb can cause lifelong cognitive impairments. Major abuse and neglect can have devastating effects on emotional development. But most of the broad spectrum of individual differences we see—the variance within the normal distribution—has much more complex roots.

For example, consider the impact of a small change in how we perceive other people's emotions. We've seen that specific variants of genes involved in emotion processing can cause slight differences in amygdala sensitivity, influencing temperament and adjusting the emotional lens through which we look at the world. Depending on the combination of genetic variants you carry, you may be slightly more (or less) sensitive to detecting fear or anger in the faces of other people. And we've seen that life experience can have similar effects: a child raised in a hostile environment may also be more attuned to seeing anger and aggression. Again and again we find that genetic variations and the vagaries of experience produce small differences in how our minds/brains are tuned to the world around us. They calibrate and recalibrate our brain circuits in subtle ways that are largely invisible to us, but over time they shape who we are and what we care about.

5. THE UNITY OF NATURE AND NURTURE

IN HIS BOOK *NATURE VIA NURTURE*, MATT RIDLEY TAKES ON THE age-old but misguided nature vs. nurture debate. Ridley explains that genes and environment are always interacting and it is meaningless to try to apportion their effects. Genes are only expressed in the context of the environments they inhabit. As he puts it, "Nature can only act via nurture. It can only act by nudging people to seek out the environmental influences that will satisfy their appetites" (pp. 92–93).[1] In other words, genes affect our behavior only with the complicity of the environment.

In recent years, however, our understanding of gene expression has taken this insight even further—down to a molecular level— hammering perhaps the final nail in the coffin of the nature-nurture dichotomy. We now know that in addition to the genome, there is an epigenome—a parallel code through which the environment can turn genes on and off. We've only begun to unravel this enormously complex system, but we've already seen examples of how experience can modify the epigenome. In animal studies, variation in maternal care can cause long-lasting abnormalities in the stress response by attaching an off switch to genes involved in the stress response. Some of the same changes have been found in human suicide victims who had been maltreated as children. In other words, nurture acts in part by modifying the chemistry of our chromosomes. In the nuclear core of our cells, nurture *is* nature.

The science of epigenetics has uncapped an entirely new source of normal variation. Over the course of our lives, epigenetic changes, or marks, accumulate and fluctuate, creating physical traces of the experiences that make us unique. That helps explain why identical twins—who are genetic clones—are not identical in their behavior, personality, or risk of diseases. Beginning at fertilization, each twin begins to accumulate differences in their epigenomes that alter the expression of their genes and nudge their development in different directions. Recent research has shown that some epigenetic changes can even be transmitted across generations, raising the possibility that we may inherit not only our ancestors' genes but the effect of their environments.[2] In other words, it's conceivable that your genes are being affected by experiences your grandmother had. If that's the case, we're talking about inheriting nurture.

Epigenetic research has also revealed another force that shapes the trajectory of our lives—and it's a little disturbing. It turns out that some, and maybe even most, of the epigenetic marks that regulate the expression of our genes are the result of random chance. At a molecular level, random (or "stochastic") variation may turn genes on or off, with a cascading influence—like the proverbial "butterfly effect"—creating new twists and turns in the trajectory of development.[2, 3] And so to natural selection, genes, and experience, we must add chance as a force in creating the distribution of normal.

PATHOLOGIZING NORMAL?

THERE'S ANOTHER THEME THAT EMERGES FROM STUDYING THE DEvelopment and functioning of the mind. Mapping the biology and psychology of normal will not only demystify what makes us tick, it can tell us something about how things go awry. The more we learn about the architecture of the mind, the more we see that conditions we recognize as disorders are variations of the same biological and psychological systems that operate in all of us.

In Chapter 1, I argued that normal and abnormal are like day and night: we recognize them as different, but there is no sharp line between them. And that creates a dilemma. If the science doesn't support a clear boundary between normal and abnormal, doesn't that undermine the whole idea of defining psychiatric disorders?

In April 2010 the American Psychiatric Association released its provisional plans for revising the most recent version of the DSM. Once again, a group of leading experts was charged with reevaluating and improving what has become the standard classification of mental illness. That news reenergized a chorus of vocal critiques within the media, blogosphere, and the mental health professions. Columnist George Will warned that "childhood eccentricities, sometimes inextricable from creativity, might be labeled 'disorders' to be 'cured.' If seven-year-old Mozart tried composing his concertos today, he might be diagnosed with attention-deficit hyperactivity disorder and medicated into barren normality."[4] Writing in the *Wall Street Journal*, historian Edward Shorter argued that psychiatry was pursuing a misguided program of "reshuffling" symptoms rather than identifying real diseases: "With DSM-V, American psychiatry is headed in exactly the opposite direction: defining ever-widening circles of the population as mentally ill with vague and undifferentiated diagnoses and treating them with powerful drugs."[5]

The debate over how to define *abnormal* is far from simply "inside baseball" for mental health clinicians and scientists. Where we draw the line between mental health and mental illness has far-reaching implications. Insurance companies typically require a psychiatric diagnosis for reimbursement of mental health care. Government agencies use these categories to determine disability benefits. Pharmaceutical companies use the DSM categories to get approval for psychiatric drugs. National studies find that a little more than half of the U.S. population will meet the DSM's criteria for at least one mental disorder in their lifetimes. And the evidence suggests the rates of several psychiatric disorders, including autism, ADHD, and depression, have been climbing over the past several decades.[6-8] With more people seeking mental health care, the use of psychiatric medication has also increased substantially in recent years. Between the 1990s and 2000s, there was a relative 400 percent increase in the prescription of antidepressants to adults.[9] And in less tangible ways, our views of normal and abnormal affect how we judge ourselves and one another.

BIBLE STORIES

IN RECENT YEARS, PSYCHIATRY'S SYSTEM OF CLASSIFYING MENTAL disorders—the DSM—has become a popular whipping boy. The field has been accused of "disease mongering"—basically creating, expanding, and hyping new definitions of disease that could apply to anyone.

The shortcomings of the DSM are self-evident. The fact is that all psychiatric disorders are currently defined by checklists of symptoms that are based on a consensus of experts. Many of the diagnostic criteria seem arbitrary. For example, a diagnosis of panic disorder requires recurrent unexpected panic attacks that are followed by a month or more of worry about additional attacks or a change in behavior as a result of the attacks. Why a month? Why not two? Panic attacks are defined by the presence of at least four out of thirteen anxiety symptoms (why four out of thirteen?) that have to reach their peak intensity within ten minutes (as though ten minutes is meaningfully different than fifteen). And in some cases, categories and criteria are based more on accidents of history than standards of evidence.

But before we simply dismiss the current diagnostic system out of hand, it's worth appreciating some of the challenges the field has faced in trying to diagnose and treat people who are seeking help for symptoms that are often disabling.

So try this thought experiment. If you were asked to develop a better way of diagnosing mental illness, what would you recommend?

It's not so easy. Imagine you are a psychiatrist. Your job is to help people suffering from mental distress. A woman comes to your office and begins to sob as she tells you that her life has become unbearable. She's been crying daily for no reason. She hasn't been able to sleep well in months. She spends most of her day lying in bed, ruminating with guilt about wasting her life. She hasn't been able to work for the past two years. Nothing seems to matter anymore and even the things she used to enjoy seem meaningless. She's begun to believe that her family wants her dead, and now she's convinced they would be better off without her. She's caused them nothing but pain. Last week she almost took an overdose of Tylenol but stopped herself because she was afraid she'd go to hell. "What is wrong with me?" she asks. "How do I stop feeling this way?"

What do you tell her? Does she have a disorder? Are there treatments you can offer her? Without a system of diagnosis, it's hard to answer those questions.

Before 1980 psychiatric diagnosis was a little like the Wild West. There were no clear criteria that practicing psychiatrists and psychologists agreed on for deciding whether someone had a disorder or what kind of disorder they had. If there is no common language for diagnosing, say, depression, and if the definitions people use are idiosyncratic, how can we learn anything about it? If you want to begin developing answers to questions that are important to people seeking help ("How long will I feel this way?" "Is there a treatment that can help?" and even "Are my children likely to develop this problem?") you need to start with a definition of the problem.

The arrival of the DSM-III in 1980 meant that, for the first time, mental health clinicians had a standard set of criteria for making diagnoses and treatment decisions. And researchers had a common starting point for testing the validity of these diagnoses and evaluating the effectiveness of treatments.

It's clear that many of the disorders defined by the DSM capture important syndromes that affect many people. In 2001 the World Health Organization (WHO) catalogued the leading causes of chronic disability worldwide for young adults (ages fifteen to forty-four), including everything from heart disease and infectious disease to accidents and malnutrition. Remarkably, four of the top five slots were occupied by psychiatric disorders: depression (#1), alcoholism (#2), schizophrenia (#3), and bipolar disorder (#5).[10] Our modern categories of mental illness have also been used to make important discoveries. In the past several years, by using advanced DNA chip technology and ever-larger studies, psychiatric genetic researchers have been able to identify specific DNA risk factors for schizophrenia, autism, and bipolar disorder.[11-14] By pinpointing these genetic variations, we have opened new windows onto biological pathways that contribute to these disorders.

The DSM approach has clearly been useful; the problem is that it was only a starting point. Nevertheless, the appeal of some kind of formal criteria rapidly established the DSM as psychiatry's bible. As it infiltrated clinical practice, the DSM went from being *a* standard to *the* standard for American (and, later, global) psychiatry. Scientists often focus their research on the existing categories of mental illness as though their validity is a given. For many people, both within and outside of psychiatry, the DSM categories have taken on the status of settled law.

Steven Hyman, a psychiatric neuroscientist and the former director of the National Institute of Mental Health (NIMH), has argued that this reification of the DSM has become an obstacle to improving the validity of psychiatric diagnosis and classification. Without a deep understanding of the causes of psychiatric disorders, there was no real alternative to starting with a simply descriptive system of classification. But the inevitable mismatch between rigid lists of specific diagnostic criteria and the real-world diversity of clinical presentations has become obvious. Hyman points out the irony that a classification system designed to advance research and clinical practice is now in danger of stifling them.

So what can be done to improve this state of affairs?

AN IMMODEST PROPOSAL

FOR THE REASONS I'VE ALREADY DISCUSSED, HAVING A SYSTEM for making diagnoses and treating those who suffer is important, and that requires drawing some boundaries between disorder and health. We can accept that drawing boundaries sometimes has pragmatic benefits, even though we may recognize that such boundaries inevitably involve imperfect judgments. And because our state of knowledge is evolving, the categories we construct today may not turn out to be the most scientifically sound categories possible. The challenge is to improve their validity in the most thoughtful, ethical, and conceptually coherent way that we can. That means recognizing that the lines we draw are provisional and being open to revising them as new evidence accumulates and practical priorities evolve. As the psychiatrist Kenneth Kendler and philosopher Peter Zachar have proposed,[15–17] one way to do this is to iteratively refine our categories by testing how well they capture a coherent set of causal mechanisms and how well they serve the clinical purposes of diagnosis: predicting prognosis, optimizing treatment, maximizing distinctions from other diagnoses, and minimizing stigma.

Since the birth of modern psychiatry and psychology, efforts to understand mental function and mental illness have followed the notion attributed to William James nearly a century ago: "the best way to understand the normal is to study the abnormal." And with good reason. As James's colleague E. E. Southard put it, "Normality is baffling."[18]

With only the most fragmentary picture of how the brain works, the focus in psychiatry was necessarily on the extremes—the qualitative differences that might shine a light into the black box. Lesions that knock out territories of the brain and functions of the mind provided crucial clues about neural circuitry and the architecture of mental function. Dramatic symptoms—psychosis, mania, compulsions, panic attacks, and self-induced starvation—provided a basis for constructing psychiatric syndromes.

Unfortunately, this approach has also constrained our understanding. Focusing on the abnormal led to a system of classification and diagnosis—the DSM—based on constructing categories from constellations of symptoms. Without a map of how these symptoms connect to the functional organization of the mind and brain, it's hard to evaluate their validity.

So, as I proposed at the outset of this book, one hundred years after William James proposed his agenda for abnormal psychology, the time has come to turn his formula on its head: to encourage a different project for twenty-first-century psychiatry and psychology, guided by the principle that the best way to understand the abnormal is to study the normal. Rather than simply starting at the edges and working our way back, our goal should be to illuminate the full and vast distribution of normal. As we fill out the center, we can see its connections to the extremes—how and where the functions of the mind can be perturbed or disrupted. We're hardly there yet, but as I've suggested in this book, the work is well under way.

To get there, we'll need to begin moving beyond drawing boundaries around disorders by consensus definitions of the abnormal in favor of developing a basic understanding of how the mind and the brain develop and function—the biology and psychology of normal.

The first step is to have a conceptual framework—a way of organizing our understanding of the biology and psychology of normal. There are many ways to approach this,[17] but I find Jerome Wakefield's model of harmful dysfunction (introduced in Chapter 1) particularly helpful. You'll recall that Wakefield defines disorders as harmful "failure of some internal mechanism to perform a function for which it was biologically designed. . . ."[19]

But understanding dysfunction has to start with an understanding of function. And that's the second, more difficult step, for which the project of illuminating the biology and psychology of normal becomes essential. Of course, we can infer dysfunction even without knowing the details of function—before we knew the causes of delirium, it still would have been clear that something was wrong with brain function. But appreciating what the dysfunction is about—how it might be treated or prevented—requires a more basic understanding. That kind of understanding can dramatically improve how well we draw lines between normal and abnormal and how we interpret symptoms and disorders.

We can see the perils of ignoring these insights when we look at some of psychiatry's ideas about mental illness before the modern era of evidence-based psychiatry. In 1953 the psychoanalyst John Rosen expressed the prevailing view of the cause of schizophrenia when he wrote, "A schizophrenic is always one who is reared by a woman who suffers from a perversion of the maternal instinct. Schizophrenia . . . is caused by the mother's inability to love her child." Sadly, it took too long for these kinds of theories to yield to the evidence that discredited them.

A "bottom-up" approach to the brain and the mind—that is, one that is based on how the mind works and is not constrained by our current diagnostic categories—may force us to reconsider some cherished dogma. Boundaries among diagnoses may need to be re-drawn. Findings from our research group and others' are revealing overlapping genetic influences on syndromes that have traditionally been considered quite different, including schizophrenia, autism, ADHD, bipolar disorder, depression, and anxiety disorders.[20-24] We also now know that single variations in stretches of DNA called copy number variants can lead to a range of neurodevelopmental disorders, including autism, ADHD, epilepsy, schizophrenia, and intellectual disability.[25-29]

And some of the syndromes that the DSM has treated as qualitative categories—anxiety disorders or personality disorders, for example—may be more accurately and usefully treated as dimensions of normal. Genetic, psychological, and developmental studies of neuropsychiatric and behavioral conditions increasingly point to the conclusion that many (though not all) are quantitative extremes of a normal distribution.[30-32]

There may be other profound implications of grounding our approach to psychopathology in the biology of normal. For example, there's at least one fundamental domain of the mind that's been nearly invisible to psychiatry's classification of mental dysfunction: anger and aggression. Like fear and anxiety, these are universal, evolved, and hardwired functions. They allow us to defend ourselves and others from harm and exploitation. Just as anxiety disorders are dysfunctions of a fear system and mood disorders are (at least partly) a dysfunction of reward systems, shouldn't we expect dysfunctions of an aggression system? Psychiatry recognizes multiple disorders of anxiety and mood, but there is no category of anger or aggression disorders.

While I was writing this book, Thomas Insel, director of the National Institute of Mental Health, announced the launch of the Research Domain Criteria Project (RDoC). The project aims to develop new ways of classifying psychopathology by studying "basic dimensions of functioning . . . across multiple levels of analysis, from genes to neural circuits to behaviors, cutting across disorders as traditionally defined."[33, 34] This is an important effort that may well begin the transition to a bottom-up understanding of function and dysfunction.

And, finally, one of the great hopes of pursuing the biology and psychology of normal is that it will lead to more effective strategies for preventing and relieving suffering. The stark reality is that all of the widely used medications and psychotherapies for treating mental disorders—depression, bipolar disorder, schizophrenia, anxiety, obsessive-compulsive disorder—are based on a handful of discoveries that date to the 1970s or before. Medications have been helpful for many and lifesaving for some, but too often they fall short or have intolerable side effects. And for many other conditions—autism, intellectual disability, dementias—the options are even more limited. But we've seen how an understanding of how the brain and the mind work can open new doors. For instance, insights into the biology of social cognition led to the finding that oxytocin may enhance social functioning in autism spectrum disorders, and insights into the science of emotional memory led to the discovery of treatments that can extinguish and perhaps eliminate traumatic fears.

Illuminating the biology of normal means enlarging our view to understand not only our limitations but our talents, not only vulnerability but resilience. And, in the end, an appreciation of the full breadth of how the human mind and brain adapt to life will allow us to see ourselves and one another with compassion and wonder.

ACKNOWLEDGMENTS

I AM PROFOUNDLY GRATEFUL TO THE MANY PEOPLE WHO MADE it possible for me to write this book. First and foremost, to my amazing wife, Alexis, who was supernormal in her support, love, and patience as this project evolved from an idea to an obsession, and finally a reality. Her impeccable eye for style and simplicity also saved me many times from straying into arcane technicalities and tedious prose. The book was born when my dearest friends, Jed Rubenfeld and Amy Chua, insisted that I stop talking and start writing about the science by which I have been so captivated over the past several years. Throughout the writing of this book they have been extraordinarily generous with their comments, suggestions, and enthusiasm for the project. Without their exceptional insight and unwavering encouragement, this book would never have come to be.

The greatest challenge in writing a book about something as complex as the workings of the mind and brain is to make it readable and engaging without oversimplifying the science. If this book has managed to strike that balance, it's only because of the brilliant guidance of my agent, Suzanne Gluck at William Morris Endeavor Entertainment, and the extraordinary talents of my editor, Henry Ferris, and his team at William Morrow.

A remarkable group of scientists and scholars generously shared their time, insights, and findings with me in the course of my research for this book. Many of them also read, provided detailed comments, and helped me fact-check drafts of the chapters. My special thanks go to Coren Apicella, Lisa Feldman Barrett, Anne Becker, R. James Blair, Ray Blanchard, Hans Breiter, Randy Buckner, Sue Carter, Dante Cicchetti, Leda Cosmides, Stacy Drury, Alice

Flaherty, John Gunderson, Takao Hensch, Steven Hyman, Martin Kafka, Jerome Kagan, Kenneth Kendler, Karlen Lyons-Ruth, Richard McNally, Mohammed Milad, Ben Neale, Charles Nelson, Roger Pitman, Barbara Pober, Harrison G. Pope Jr., Scott Rauch, Kerry Ressler, Joshua Roffman, Rebecca Saxe, Carl Schwartz, Alexandra Shields, Jack Shonkoff, Regina Sullivan, Helen Tager-Flusberg, John Tooby, Danielle Truxaw, Jerome Wakefield, Larry Young, and Leslie Zebrowitz. I am especially grateful to my friends Leda Cosmides and John Tooby, who have inspired my excitement about the human mind ever since I met them more than thirty years ago. They also allowed me to spend time with them and their colleagues at the UCSB Center for Evolutionary Psychology, which was crucial in developing and alpha-testing ideas for this book. I am grateful to Tracy and Eric Novack for generously lending me their home in the woods of Vermont where chunks of this book were written. James Hong and Jim Young graciously shared background information on the story of their Am I Hot or Not? website. I am indebted to Mike Bornstein for allowing me to share his remarkable story, and to Walter Austerer, Mary Carmichael, Sophia Chua-Rubenfeld, Anne Dailey, Stephen Gilman, Linda Kraft, Dave Mendenhall, Alisha Pollastri, Lidia Rosenbaum, and Malorie Snider for invaluable comments and feedback on earlier drafts of the book. I am tremendously thankful to Leslie Gaffney for her help with illustrations, and Stefanie Block and Patience Gallagher provided essential assistance with library research. Any misstatements, oversimplifications, or failed attempts at humor are entirely my own.

I have been privileged to work with people who make every day an adventure in learning and discovery (well, almost every day). My particular thanks to Jerry Rosenbaum and Maurizio Fava for their leadership, friendship, and unflagging support and for creating an environment of remarkable collegiality in the MGH Department of Psychiatry; and to Jim Gusella for his invaluable guidance, wisdom, and leadership of the MGH Center for Human Genetic Research. I am also deeply grateful to my fellow scientists, trainees, and research staff in the Psychiatric and Neurodevelopmental Genetics Unit. Their enthusiasm, dedication, and intellectual energy are a constant inspiration to me, and I can't imagine a community of people I'd rather spend my working hours with.

And, finally, my eternal gratitude and admiration go to Sylvia Wassertheil-Smoller—my role model for resilience, compassion, and intellectual curiosity. Thank you for that and for everything else.

SOURCES

PROLOGUE

1. I. Hacking, *The Taming of Chance* (Cambridge, UK: Cambridge University Press, 1990).
2. J. Tooby and L. Cosmides, "Conceptual Foundations of Evolutionary Psychology," in *The Handbook of Evolutionary Psychology*, ed. D. M. Buss (Hoboken, NJ: John Wiley & Sons, 2005), 5–67.
3. F. Galton, "Biometry," *Biometrika* 1 (1901): 7–10.
4. F. Galton, *English Men of Science: Their Nature and Nurture* (New York: D. Appleton and Company, 1875).

CHAPTER 1: "WE'RE ALL MAD HERE"

1. R. C. Kessler, P. Berglund, O. Demler, R. Jin, K. R. Merikangas, and E. E. Walters, "Lifetime Prevalence and Age-of-Onset Distributions of DSM-IV Disorders in the National Comorbidity Survey Replication," *Arch Gen Psychiatry* 62, no. 6 (2005): 593–602.
2. M. Stobbe, "Autism Rate in Children Higher Than Estimated, CDC Reports," *Boston Globe*, February 9, 2007.
3. R. Weiss, "1 in 150 Children in U.S. Has Autism, New Survey Finds," *Washington Post*, February 9, 2007.
4. C. Moreno, G. Laje, C. Blanco, H. Jiang, A. B. Schmidt, and M. Olfson, "National Trends in the Outpatient Diagnosis and Treatment of Bipolar Disorder in Youth," *Arch Gen Psychiatry* 64, no. 9 (2007): 1032–39.
5. G. Santayana, *The Middle Span*, vol. 2 (New York: Charles Scribner's Sons, 1945).
6. D. L. Rosenhan, "On Being Sane in Insane Places," *Science* 179 (January 19, 1973): 250–58.
7. J. Zubin and B. J. Gurland, "The United States–United Kingdom Project on Diagnosis of the Mental Disorders," *Ann NY Acad Sci* 285 (March 18, 1977): 676–86.
8. American Psychiatric Association, *Diagnostic and Statistical Manual of Mental Disorders*, 4th ed., rev. ed. (Washington, DC: American Psychiatric Association, 2000).
9. R. L. Spitzer, J. B. Williams, and A. E. Skodol, "DSM-III: The Major Achievements and an Overview," *Am J Psychiatry* 137, no. 2 (1980): 151–64.
10. P. R. McHugh, "Witches, Multiple Personalities, and Other Psychiatric Artifacts," *Nature Med* 1, no. 2 (1995): 110–14.
11. H. G. Pope Jr., M. B. Poliakoff, M. P. Parker, M. Boynes, and J. I. Hudson, "Is Dissociative Amnesia a Culture-Bound Syndrome? Findings from a Survey of Historical Literature," *Psychol Med* 37, no 2 (February 2007): 225–33.
12. E. T. Carlson, "Multiple Personality and Hypnosis: The First One Hundred Years," *J Hist Behav Sci* 25, no. 4 (October 1989): 315–22.
13. S. W. Mitchell, "Mary Reynolds: A Case of Double Consciousness," *Transactions of the College of Physicians of Philadelphia*, 1889.
14. I. Hacking, *Mad Travelers: Reflections on the Reality of Transient Mental Illnesses* (Cambridge, MA: Harvard University Press, 1998).
15. A. Jablensky, "Epidemiology of Schizophrenia: The Global Burden of Disease and Disability," *Eur Arch Psychiatry Clin Neurosci* 250, no. 6 (2000): 274–85.

16. H. Rin, "A Study of the Aetiology of Koro in Respect to the Chinese Concept of Illness," *Int J Soc Psychiatry* 11 (1965): 7–13.

17. B. Y. Ng, "History of Koro in Singapore," *Singapore Med J* 38, no. 8 (August 1997): 356–57.

18. J. J. Mattelaer and W. Jilek, "Koro—the Psychological Disappearance of the Penis," *J Sex Med* 4, no. 5 (September 2007): 1509–15.

19. T. M. Chong, "Epidemic Koro in Singapore," *Br Med J* 1, no. 5592 (March 9, 1968): 640–41.

20. C. Buckle, Y. M. Chuah, C. S. Fones, and A. H. Wong, "A Conceptual History of Koro," *Transcult Psychiatry* 44, no. 1 (March 2007): 27–43.

21. The Middle East Media Research Institute, "Panic in Khartoum: Foreigners Shake Hands, Make Penises Disappear," October 22, 2003, Special Dispatch No. 593, http://www.memri.org/report/en/0/0/0/0/0/0/976.htm.

22. V. A. Dzokoto and G. Adams, "Understanding Genital-Shrinking Epidemics in West Africa: Koro, Juju, or Mass Psychogenic Illness?" *Cult Med Psychiatry* 29, no. 1 (March 2005): 53–78.

23. A. Sumathipala, S. H. Siribaddana, and D. Bhugra, "Culture-Bound Syndromes: The Story of Dhat Syndrome," *Br J Psychiatry* 184 (March 2004): 200–209.

24. J. C. Wakefield, "Disorder as Harmful Dysfunction: A Conceptual Critique of DSM-III-R's Definition of Mental Disorder," *Psychol Rev* 99, no. 2 (April 1992): 232–47.

25. J. C. Wakefield, "The Concept of Mental Disorder: Diagnostic Implications of the Harmful Dysfunction Analysis," *World Psychiatry* 6, no. 3 (October 2007): 149–56.

26. J. C. Wakefield, M. F. Schmitz, M. B. First, and A. V. Horwitz, "Extending the Bereavement Exclusion for Major Depression to Other Losses: Evidence from the National Comorbidity Survey," *Arch Gen Psychiatry* 64, no. 4 (April 2007): 433–40.

27. D. Mataix-Cols, "Deconstructing Obsessive-Compulsive Disorder: A Multidimensional Perspective," *Curr Opin Psychiatry* 19, no. 1 (January 2006): 84–89.

28. D. L. Feygin, J. E. Swain, and J. F. Leckman, "The Normalcy of Neurosis: Evolutionary Origins of Obsessive-Compulsive Disorder and Related Behaviors," *Prog Neuropsychopharmacol Biol Psychiatry* 30, no. 5 (July 2006): 854–64.

29. J. F. Leckman, L. C. Mayes, R. Feldman, D. W. Evans, R. A. King, and D. J. Cohen, "Early Parental Preoccupations and Behaviors and Their Possible Relationship to the Symptoms of Obsessive-Compulsive Disorder," *Acta Psychiatr Scand Suppl* 396 (1999): 1–26.

30. Associated Press, "U.S. Businesses Hype Hand Sanitizers," CBC News, January 4, 2007, http://www.cbc.ca/news/story/2007/01/04/hand-sanitizer.html.

31. D. Mapes, "Pass the Purell: It's Hip to Be Germ-Free," msnbc.com, December 10, 2007.

32. M. Oaten, R. J. Stevenson, and T. I. Case, "Disgust as a Disease-Avoidance Mechanism," *Psychol Bull* 135, no. 2 (March 2009): 303–21.

33. P. Rozin, L. Millman, and C. Nemeroff, "Operation of the Laws of Sympthetic Magic in Disgust and Other Domains," *J Pers Soc Psychol* 50 (1986): 703–12.

34. D. Mataix-Cols, S. Wooderson, N. Lawrence, M. J. Brammer, A. Speckens, and M. L. Phillips, "Distinct Neural Correlates of Washing, Checking, and Hoarding Symptom Dimensions in Obsessive-Compulsive Disorder," *Arch Gen Psychiatry* 61, no. 6 (June 2004): 564–76.

35. D. Mataix-Cols, S. Cullen, K. Lange, F. Zelaya, C. Andrew, E. Amaro, M. J. Brammer, S. C. Williams, A. Speckens, and M. L. Phillips, "Neural Correlates of Anxiety Associated with Obsessive-Compulsive Symptom Dimensions in Normal Volunteers," *Biol Psychiatry* 53, no. 6 (March 15, 2003): 482–93.

36. D. S. Husted, N. A. Shapira, and W. K. Goodman, "The Neurocircuitry of Obsessive-Compulsive Disorder and Disgust," *Prog Neuropsychopharmacol Biol Psychiatry* 30, no. 3 (May 2006): 389–99.

37. M. P. Paulus and M. B. Stein, "An Insular View of Anxiety," *Biol Psychiatry* 60, no. 4 (August 15, 2006): 383–87.

38. S. A. Simon, I. E. de Araujo, R. Gutierrez, and M. A. Nicolelis, "The Neural Mechanisms of Gustation: A Distributed Processing Code," *Nat Rev Neurosci* 7, no. 11 (November 2006): 890–901.

39. S. Yaxley, E. T. Rolls, and Z. J. Sienkiewicz, "Gustatory Responses of Single Neurons in the Insula of the Macaque Monkey," *J Neurophysiol* 63, no. 4 (April 1990): 689–700.

40. W. Penfield and M. E. Faulk Jr., "The Insula: Further Observations on Its Function," *Brain* 78, no. 4 (1955): 445–70.

41. P. Wright, G. He, N. A. Shapira, W. K. Goodman, and Y. Liu, "Disgust and the Insula: fMRI Responses to Pictures of Mutilation and Contamination," *Neuroreport* 15, no. 15 (October 25, 2004): 2347–51.

42. P. Dawson, I. Han, M. Cox, C. Black, and L. Simmons, "Residence Time and Food Contact Time Effects on Transfer of Salmonella Typhimurium from Tile, Wood and Carpet: Testing the Five-Second Rule," *J Appl Microbiol* 102, no. 4 (April 2007): 945–53.

43. P. Rozin, L. Hammer, H. Oster, T. Horowitz, and V. Marmora, "The Child's Conception of Food: Differentiation of Categories of Rejected Substances in the 16 Months to 5 Year Age Range," *Appetite* 7, no. 2 (June 1986): 141–51.

44. P. Rozin, J. Haidt, and C. R. McCauley, "Disgust," in *Handbook of Emotions,* 3rd ed., ed. M. Lewis, J. M. Haviland-Jones, and L. M. Feldman Barrett (New York: Guilford Press, 2008), 757–76.

45. R. J. Stevenson, M. J. Oaten, T. I. Case, B. M. Repacholi, and P. Wagland, "Children's Response to Adult Disgust Elicitors: Development and Acquisition," *Dev Psychol* 46, no. 1 (January 2010): 165–77.

46. B. Wicker, C. Keysers, J. Plailly, J. P. Royet, V. Gallese, and G. Rizzolatti, "Both of Us Disgusted in My Insula: The Common Neural Basis of Seeing and Feeling Disgust," *Neuron* 40, no. 3 (October 30, 2003): 655–64.

CHAPTER 2: HOW GENES TUNE THE BRAIN

1. J. Kagan, *Galen's Prophecy* (New York: BasicBooks, 1994).

2. S. W. Jackson, "A History of Melancholia and Depression," in *History of Psychiatry and Medical Psychology,* ed. E. G. Wallace and J. Gach (New York: Springer, 2008), 443–60.

3. P. F. Merenda, "Toward a Four-Factor Theory of Temperament and/or Personality," *J Pers Assess* 51, no. 3 (Fall 1987): 367–74.

4. A. Thomas and S. Chess, "Genesis and Evolution of Behavioral Disorders: From Infancy to Early Adult Life," *Am J Psychiatry* 141, no. 1 (January 1984): 1–9.

5. S. J. Haggbloom, R. Warnick, J. E. Warnick, V. K. Jones, G. L. Yarbrough, T. M. Russell, C. M. Borecky, R. McGahhey, J. L. Powell, J. Beavers, and E. Monte, "The 100 Most Eminent Psychologists of the 20th Century," *Rev Gen Psychol* 6 (2002): 139–52.

6. J. Kagan and N. Snidman, *The Long Shadow of Temperament* (Cambridge, MA: Belknap Press, 2004).

7. J. Kagan, S. Reznick, and N. Snidman, "Biological Bases of Childhood Shyness," *Science* 240 (1988): 167–71.

8. J. Kagan, N. Snidman, V. Kahn, and S. Towsley, "The Preservation of Two Infant Temperaments into Adolescence," *Monogr Soc Res Child Dev* 72, no. 2 (2007): 1–75; vii; discussion 76–91.

9. C. Schwartz, N. Snidman, and J. Kagan, "Adolescent Social Anxiety as an Outcome of Inhibited Temperament in Childhood," *J Am Acad Child Adolesc Psychiatry* 38 (1999): 1008–15.

10. C. Darwin, *The Expression of the Emotions in Man and Animals* (New York: D. Appleton and Co., 1913).

11. H. C. Breiter, N. L. Etcoff, P. J. Whalen, W. A. Kennedy, S. L. Rauch, R. L. Buckner, M. M. Strauss, S. E. Hyman, and B. R. Rosen, "Response and Habituation of the Human Amygdala During Visual Processing of Facial Expression," *Neuron* 17, no. 5 (November 1996): 875–87.

12. J. S. Morris, C. D. Frith, D. I. Perrett, D. Rowland, A. W. Young, A. J. Calder, and R. J. Dolan, "A Differential Neural Response in the Human Amygdala to Fearful and Happy Facial Expressions," *Nature* 383, no. 6603 (October 31, 1996): 812–15.
13. C. E. Schwartz, C. I. Wright, L. M. Shin, J. Kagan, P. J. Whalen, K. G. McMullin, and S. L. Rauch, "Differential Amygdalar Response to Novel Versus Newly Familiar Neutral Faces: A Functional MRI Probe Developed for Studying Inhibited Temperament," *Biol Psychiatry* 53, no. 10 (May 15, 2003): 854–62.
14. C. E. Schwartz, C. I. Wright, L. M. Shin, J. Kagan, S. L. Rauch, "Inhibited and Uninhibited Infants 'Grown Up': Adult Amygdalar Response to Novelty," *Science* 300, no. 5627 (June 20, 2003): 1952–53.
15. K. Perez-Edgar, R. Roberson-Nay, M. G. Hardin, K. Poeth, A. E. Guyer, E. E. Nelson, E. B. McClure, H. A. Henderson, N. A. Fox, D. S. Pine, and M. Ernst, "Attention Alters Neural Responses to Evocative Faces in Behaviorally Inhibited Adolescents," *Neuroimage* 35, no. 4 (May 1, 2007): 1538–46.
16. C. E. Schwartz, P. S. Kunwar, D. N. Greve, L. R. Moran, J. C. Viner, J. M. Covino, J. Kagan, S. E. Stewart, N. C. Snidman, M. G. Vangel, and S. R. Wallace, "Structural Differences in Adult Orbital and Ventromedial Prefrontal Cortex Predicted by Infant Temperament at 4 Months of Age," *Arch Gen Psychiatry* 67, no. 1 (January 2010): 78–84.
17. C. E. Schwartz, P. S. Kunwar, D. N. Greve, J. Kagan, N. C. Snidman, and R. B. Bloch, "A Phenotype of Early Infancy Predicts Reactivity of the Amygdala in Male Adults," *Mol Psychiatry* (September 6, 2011).
18. A. S. Fox, S. E. Shelton, T. R. Oakes, R. J. Davidson, and N. H. Kalin, "Trait-like Brain Activity During Adolescence Predicts Anxious Temperament in Primates," *PLoS ONE* 3, no. 7 (2008): e2570.
19. N. H. Kalin, C. Larson, S. E. Shelton, and R. J. Davidson, "Asymmetric Frontal Brain Activity, Cortisol, and Behavior Associated with Fearful Temperament in Rhesus Monkeys," *Behav Neurosci* 112, no. 2 (1998): 286–92.
20. A. Caspi, "The Child Is Father of the Man: Personality Continuities from Childhood to Adulthood," *J Pers Soc Psychol* 78, no. 1 (January 2000): 158–72.
21. J. F. Rosenbaum, J. Biederman, E. A. Bolduc-Murphy, S. V. Faraone, J. Chaloff, D. Hirshfeld, and J. Kagan, "Behavioral Inhibition in Childhood: A Risk Factor for Anxiety Disorders," *Harvard Rev Psychiatry* 1 (1993): 2–16.
22. J. Biederman, D. R. Hirshfeld-Becker, J. F. Rosenbaum, C. Herot, D. Friedman, N. Snidman, J. Kagan, and S. V. Faraone, "Further Evidence of Association Between Behavioral Inhibition and Social Anxiety in Children," *Am J Psychiatry* 158, no. 10 (2001): 1673–79.
23 J. F. Rosenbaum, J. Biederman, D. R. Hirshfeld-Becker, J. Kagan, N. Snidman, D. Friedman, A. Nineberg, D. J. Gallery, and S. V. Faraone, "A Controlled Study of Behavioral Inhibition in Children of Parents with Panic Disorder and Depression," *Am J Psychiatry* 157, no. 12 (2000): 2002–10.
24. J. P. Lorberbaum, S. Kose, M. R. Johnson, G. W. Arana, L. K. Sullivan, M. B. Hamner, J. C. Ballenger, R. B. Lydiard, P. S. Brodrick, D. E. Bohning, and M. S. George, "Neural Correlates of Speech Anticipatory Anxiety in Generalized Social Phobia," *Neuroreport* 15, no. 18 (December 24, 2004): 2701–2705.
25. K. S. Blair, M. Geraci, N. Hollon, M. Otero, J. DeVido, C. Majestic, M. Jacobs, R. J. Blair, and D. S. Pine, "Social Norm Processing in Adult Social Phobia: Atypically Increased Ventromedial Frontal Cortex Responsiveness to Unintentional (Embarrassing) Transgressions," *Am J Psychiatry* 167, no. 12 (December 2010): 1526–32.
26. M. B. Stein, P. R. Goldin, J. Sareen, L. T. Zorrilla, and G. G. Brown, "Increased Amygdala Activation to Angry and Contemptuous Faces in Generalized Social Phobia," *Arch Gen Psychiatry* 59, no. 11 (November 2002): 1027–34.
27. K. L. Phan, D. A. Fitzgerald, P. J. Nathan, and M. E. Tancer, "Association Between Amygdala Hyperactivity to Harsh Faces and Severity of Social Anxiety in Generalized Social Phobia," *Biol Psychiatry* 59, no. 5 (March 1, 2006): 424–29.

28. A. Caspi, D. Begg, N. Dickson, H. Harrington, J. Langley, T.E. Moffitt, and P. A. Silva, "Personality Differences Predict Health-Risk Behaviors in Young Adulthood: Evidence from a Longitudinal Study," *J Pers Soc Psychol* 73, no. 5 (November 1997): 1052–63.

29. D. R. Hirshfeld-Becker, J. Biederman, S. Calltharp, E. D. Rosenbaum, S. V. Faraone, and J. F. Rosenbaum, "Behavioral Inhibition and Disinhibition as Hypothesized Precursors to Psychopathology: Implications for Pediatric Bipolar Disorder," *Biol Psychiatry* 53, no. 11 (June 1, 2003): 985–99.

30. D. R. Hirshfeld-Becker, J. Biederman, A. Henin, S. V. Faraone, J. A. Micco, A. van Grondelle, B. Henry, and J. F. Rosenbaum, "Clinical Outcomes of Laboratory-Observed Preschool Behavioral Disinhibition at Five-Year Follow-Up," *Biol Psychiatry* 62, no. 6 (September 15, 2007): 565–72.

31. K. H. Rubin, K. B. Burgess, and P. D. Hastings, "Stability and Social-Behavioral Consequences of Toddlers' Inhibited Temperament and Parenting Behaviors," *Child Dev* 73, no. 2 (March–April 2002): 483–95.

32. N. A. Fox, H. A. Henderson, K. H. Rubin, S. D. Calkins, and L. A. Schmidt, "Continuity and Discontinuity of Behavioral Inhibition and Exuberance: Psychophysiological and Behavioral Influences Across the First Four Years of Life," *Child Dev* 72, no. 1 (January–February 2001): 1–21.

33 G. L. Gladstone, G. B. Parker, and G. S. Malhi, "Do Bullied Children Become Anxious and Depressed Adults?: A Cross-Sectional Investigation of the Correlates of Bullying and Anxious Depression," *J Nerv Ment Dis* 194, no. 3 (March 2006): 201–208.

34. G. W. Allport and H. S. Odbert, "Trait Names: A Psycho-Lexical Study," *Psychol Monogr* 47, no. 211 (1936).

35. R. R. McCrae and P. T. Costa Jr., "Personality Trait Structure as a Human Universal," *Am Psychol* 52, no. 5 (May 1997): 509–16.

36. P. J. Rentfrow, S. D. Gosling, and J. Potter, "A Theory of the Emergence, Persistence, and Expression of Geographic Variation in Psychological Characteristics," *Perspect Psychol Sci* 3 (2008): 339–69.

37. R. R. McCrae and A. Terracciano, "Personality Profiles of Cultures: Aggregate Personality Traits," *J Pers Soc Psychol* 89, no. 3 (September 2005): 407–25.

38. J. M. French, "Assessment of Donkey Temperament and the Influence of Home Environment," *Applied Anim Behav Science* 36 (1993): 249–57.

39. http://www.akc.org/breeds/labrador_retriever/. Accessed September 1, 2008.

40. L. J. Eaves and H. Eysenck, "The Nature of Extraversion: A Genetical Analysis," *J Pers Soc Psychol* 1 (1975): 102–12.

41. B. Floderus-Myrhed, N. Pederscn, and I. Rasmuson, "Assessment of Heritability for Personality, Based on a Short-Form of the Eysenck Personality Inventory: A Study of 12,898 Twin Pairs," *Behav Genet* 10 (1980): 153–62.

42. J. Horn, R. Plomin, and R. Rosenman, "Heritability of Personality Traits in Adult Male Twins," *Behav Genet* 6 (1976): 17–30.

43. K. L. Jang, W. J. Livesley, and P. A. Vernon, "Heritability of the Big Five Personality Dimensions and Their Facets: A Twin Study," *J Pers* 64, no. 3 (September 1996): 577–91.

44. R. Riemann, A. Angleitner, and J. Strelau, "Genetic and Environmental Influences on Personality: A Study of Twins Reared Together Using the Self- and Peer Report NEO-FFI Scales," *J Pers* 65 (1997): 449–75.

45. R. Rose, M. Koskenvuo, J. Kaprio, S. Sarna, and H. Langinvainio, "Shared Genes, Shared Experiences, and Similarity of Personality: Data from 14,288 Adult Finnish Co-Twins," *J Pers Soc Psychol* 54 (1988): 161–71.

46. M. B. Stein, K. L. Jang, and W. J. Livesley, "Heritability of Social Anxiety-Related Concerns and Personality Characteristics: A Twin Study," *J Nerv Ment Dis* 190, no. 4 (2002): 219–24.

47. S. E. Young, M. C. Stallings, R. P. Corley, K. S. Krauter, and J. K. Hewitt, "Genetic and Environmental Influences on Behavioral Disinhibition," *Am J Med Genet* 96, no. 5 (2000): 684–95.

48. M. Tassabehji, "Williams-Beuren Syndrome: A Challenge for Genotype-Phenotype Correlations," *Hum Mol Genet* 12, spec. no. 2 (October 15, 2003): R229–37.
49. M. A. Martens, S. J. Wilson, and D. C. Reutens, "Research Review: Williams Syndrome: A Critical Review of the Cognitive, Behavioral, and Neuroanatomical Phenotype," *J Child Psychol Psychiatry* 49, no. 6 (June 2008): 576–608.
50. A. Meyer-Lindenberg, A. R. Hariri, K. E. Munoz, C. B. Mervis, V. S. Mattay, C. A. Morris, and K. F. Berman, "Neural Correlates of Genetically Abnormal Social Cognition in Williams Syndrome," *Nat Neurosci* 8, no. 8 (August 2005): 991–93.
51. B. W. Haas, D. Mills, A. Yam, F. Hoeft, U. Bellugi, and A. Reiss, "Genetic Influences on Sociability: Heightened Amygdala Reactivity and Event-Related Responses to Positive Social Stimuli in Williams Syndrome," *J Neurosci* 29, no. 4 (January 28, 2009): 1132–39.
52. H. F. Dodd and M. A. Porter, "I See Happy People: Attention Bias Towards Happy But Not Angry Facial Expressions in Williams Syndrome," *Cogn Neuropsychiatry* 15, no. 6 (November 2010): 549–67.
53. S. Ripke, A. R. Sanders, K. S. Kendler, D. F. Levinson, P. Sklar, P. A. Holmans, D. Y. Lin, J. Duan, et al., "Genome-Wide Association Study Identifies Five New Schizophrenia Loci," *Nature Genetics* 43, no. 10 (2011): 969–76.
54. P. Sklar, S. Ripke, L. J. Scott, O. A. Andreassen, S. Cichon, N. Craddock, H. J. Edenberg, J. I. Nurnberger Jr., M. Rietschel, et al., "Large-Scale Genome-Wide Association Analysis of Bipolar Disorder Identifies a New Susceptibility Locus near ODZ4," *Nature Genetics* 43, no. 10 (2011): 977–83.
55. K.-P. Lesch, D. Bengel, A. Heils, S. Sabol, B. Greenberg, S. Petri, J. Benjamin, C. Muller, D. Hamer, and D. Murphy, "Association of Anxiety-Related Traits with a Polymorphism in the Serotonin Transporter Gene Regulatory Region," *Science* 274 (1996): 1527–31.
56. M. R. Munafo, T. Clark, and J. Flint, "Does Measurement Instrument Moderate the Association Between the Serotonin Transporter Gene and Anxiety-Related Personality Traits? A Meta-Analysis," *Mol Psychiatry* 10, no. 4 (April 2005): 415–19.
57. J. A. Schinka, R. M. Busch, and N. Robichaux-Keene, "A Meta-Analysis of the Association Between the Serotonin Transporter Gene Polymorphism (5-HTTLPR) and Trait Anxiety," *Mol Psychiatry* 9, no. 2 (February 2004): 197–202.
58. S. Sen, M. Burmeister, and D. Ghosh, "Meta-Analysis of the Association Between a Serotonin Transporter Promoter Polymorphism (5-HTTLPR) and Anxiety-Related Personality Traits," *Am J Med Genet B Neuropsychiatr Genet* 127, no. 1 (May 15, 2004): 85–89.
59. A. R. Hariri, V. S. Mattay, A. Tessitore, B. Kolachana, F. Fera, D. Goldman, M. F. Egan, and D. R. Weinberger, "Serotonin Transporter Genetic Variation and the Response of the Human Amygdala," *Science* 297, no. 5580 (July 19, 2002): 400–403.
60. M. R. Munafo, S. M. Brown, and A. R. Hariri, "Serotonin Transporter (5-HTTLPR) Genotype and Amygdala Activation: A Meta-Analysis," *Biol Psychiatry* 63, no. 9 (May 1, 2008): 852–57.
61. L. Pezawas, A. Meyer-Lindenberg, E. M. Drabant, B. A. Verchinski, K. E. Munoz, B. S. Kolachana, M. F. Egan, V. S. Mattay, A. R. Hariri, and D. R. Weinberger, "5-HTTLPR Polymorphism Impacts Human Cingulate-Amygdala Interactions: A Genetic Susceptibility Mechanism for Depression," *Nat Neurosci* 8, no. 6 (June 2005): 828–34.
62. N. A. Fox, K. E. Nichols, H. A. Henderson, K. Rubin, L. Schmidt, D. Hamer, M. Ernst, and D. S. Pine, "Evidence for a Gene-Environment Interaction in Predicting Behavioral Inhibition in Middle Childhood," *Psychol Sci* 16, no. 12 (December 2005): 921–26.
63. T. Canli, M. Qiu, K. Omura, E. Congdon, B. W. Haas, A. Amin, M. J. Herrmann, R. T. Constable, and K. P. Lesch, "Neural Correlates of Epigenesis," *Proc Natl Acad Sci USA* 103, no. 43 (October 24, 2006): 16033–38.
64. R. H. Waterston, K. Lindblad-Toh, E. Birney, J. Rogers, J. F. Abril, P. Agarwal, R. Agarwala, R. Ainscough, M. Alexandersson, P. An, S. E. Antonarakis, J. Attwood, R.

Baertsch, et al., "Initial Sequencing and Comparative Analysis of the Mouse Genome," *Nature* 420, no. 6915 (December 5, 2002): 520–62.

65. J. Flint, R. Corley, J. C. DeFries, D. W. Fulker, J. A. Gray, S. Miller, and A. C. Collins, "A Simple Genetic Basis for a Complex Psychological Trait in Laboratory Mice," *Science* 269 (1995): 1432–35.

66. B. Yalcin, S. A. Willis-Owen, J. Fullerton, A. Meesaq, R. M. Deacon, J. N. Rawlins, R. R. Copley, A. P. Morris, J. Flint, and R. Mott, "Genetic Dissection of a Behavioral Quantitative Trait Locus Shows That Rgs2 Modulates Anxiety in Mice," *Nat Genet* 36, no. 11 (November 2004): 1197–1202.

67. A. J. Oliveira-Dos-Santos, G. Matsumoto, B. E. Snow, D. Bai, F. P. Houston, I. Q. Whishaw, S. Mariathasan, T. Sasaki, A. Wakeham, P. S. Ohashi, J. C. Roder, C. A. Barnes, D. P. Siderovski, and J. M. Penninger, "Regulation of T cell Activation, Anxiety, and Male Aggression by RGS2," *Proc Natl Acad Sci USA* 97, no. 22 (October 24, 2000): 12272–77.

68. V. Gross, J. Tank, M. Obst, R. Plehm, K. J. Blumer, A. Diedrich, J. Jordan, and F. C. Luft, "Autonomic Nervous System and Blood Pressure Regulation in RGS2-Deficient Mice," *Am J Physiol Regul Integr Comp Physiol* 288, no. 5 (May 2005): R1134–42.

69. T. Ingi and Y. Aoki, "Expression of RGS2, RGS4 and RGS7 in the Developing Postnatal Brain," *Eur J Neurosci* 15, no. 5 (March 2002): 929–36.

70. T. Ingi, A. M. Krumins, P. Chidiac, G. M. Brothers, S. Chung, B. E. Snow, C. A. Barnes, A. A. Lanahan, D. P. Siderovski, E. M. Ross, A. G. Gilman, and P. F. Worley, "Dynamic Regulation of RGS2 Suggests a Novel Mechanism in G-Protein Signaling and Neuronal Plasticity," *J Neurosci* 18, no. 18 (September 15, 1998): 7178–88.

71. R. R. Neubig and D. P. Siderovski, "Regulators of G-Protein Signalling as New Central Nervous System Drug Targets," *Nat Rev Drug Discov* 1, no. 3 (March 2002): 187–97.

72. J. W. Smoller, M. P. Paulus, J. A. Fagerness, S. Purcell, L. Yamaki, D. Hirshfeld-Becker, J. Biederman, J. F. Rosenbaum, J. Gelernter, and M. B. Stein, "Influence of *RGS2* on Anxiety-Related Temperament, Personality, and Brain Function," *Arch Gen Psychiatry* 65 (2008): 298–308.

73. K. P. Lesch, J. Meyer, K. Glatz, B. Flugge, A. Hinney, J. Hebebrand, S. M. Klauck, A. Poustka, F. Poustka, D. Bengel, R. Mossner, P. Riederer, and A. Heils, "The 5-HT Transporter Gene-Linked Polymorphic Region (5-HTTLPR) in Evolutionary Perspective: Alternative Biallelic Variation in Rhesus Monkeys. Rapid Communication," *J Neural Transm* 104, no. 11–12 (1997): 1259–66.

74. D. Nettle, "The Evolution of Personality Variation in Humans and Other Animals," *Am Psychol* 61, no. 6 (September 2006): 622–31.

75. L. J. Matthews and P. M. Butler, "Novelty-Seeking DRD4 Polymorphisms Are Associated with Human Migration Distance Out-of-Africa After Controlling for Neutral Population Gene Structure," *Am J Phys Anthropol* 145, no. 3 (July 2011): 382–89.

76. M. R. Munafo, B. Yalcin, S. A. Willis-Owen, and J. Flint, "Association of the Dopamine D4 Receptor (DRD4) Gene and Approach-Related Personality Traits: Meta-Analysis and New Data," *Biol Psychiatry* 63, no. 2 (January 2008): 197–206.

77. R. P. Ebstein, "The Molecular Genetic Architecture of Human Personality: Beyond Self-Report Questionnaires," *Mol Psychiatry* 11, no. 5 (May 2006): 427–45.

78. K. Hejjas, J. Vas, E. Kubinyi, M. Sasvari-Szekely, A. Miklosi, and Z. Ronai, "Novel Repeat Polymorphisms of the Dopaminergic Neurotransmitter Genes Among Dogs and Wolves," *Mamm Genome* 18, no. 12 (December 2007): 871–79.

79. K. Hejjas, J. Vas, J. Topal, E. Szantai, Z. Ronai, A. Szekely, E. Kubinyi, Z. Horvath, M. Sasvari-Szekely, and A. Miklosi, "Association of Polymorphisms in the Dopamine D4 Receptor Gene and the Activity-Impulsivity Endophenotype in Dogs," *Anim Genet* 38, no. 6 (December 2007): 629–33.

80. Y. Momozawa, Y. Takeuchi, R. Kusunose, T. Kikusui, and Y. Mori, "Association Between Equine Temperament and Polymorphisms in Dopamine D4 Receptor Gene," *Mamm Genome* 16, no. 7 (July 2005): 538–44.

81. P. Korsten, J. C. Mueller, C. Hermannstadter, K. M. Bouwman, N. J. Dingemanse, P. J. Drent, M. Liedvogel, E. Matthysen, K. van Oers, T. van Overveld, S. C. Patrick, J. L. Quinn, B. C. Sheldon, et al., "Association Between DRD4 Gene Polymorphism and Personality Variation in Great Tits: A Test Across Four Wild Populations," *Mol Ecol* 19, no. 4 (February 2010): 832–43.

82. D. Li, P. C. Sham, M. J. Owen, and L. He, "Meta-Analysis Shows Significant Association Between Dopamine System Genes and Attention Deficit Hyperactivity Disorder (ADHD)," *Hum Mol Genet* 15, no. 14 (July 15, 2006): 2276–84.

83. E. Wang, Y. C. Ding, P. Flodman, J. R. Kidd, K. K. Kidd, D. L. Grady, O. A. Ryder, M. A. Spence, J. M. Swanson, and R. K. Moyzis, "The Genetic Architecture of Selection at the Human Dopamine Receptor D4 (DRD4) Gene Locus," *Am J Hum Genet* 74, no. 5 (May 2004): 931–44.

84. F. M. Chang, J. R. Kidd, K. J. Livak, A. J. Pakstis, and K. K. Kidd, "The World-Wide Distribution of Allele Frequencies at the Human Dopamine D4 Receptor Locus," *Hum Genet* 98, no. 1 (July 1996): 91–101.

85. M. H. de Moor, P. T. Costa, A. Terracciano, R. F. Krueger, E. J. de Geus, T. Toshiko, B. W. Penninx, T. Esko, P. A. Madden, J. Derringer, N. Amin, G. Willemsen, J. J. Hottenga, et al., "Meta-Analysis of Genome-Wide Association Studies for Personality," *Mol Psychiatry* (December 21, 2010).

86. A. Terracciano, T. Esko, A. Sutin, M. de Moor, O. Meirelles, G. Zhu, I. Giegling, T. Nutile, A. Realo, J. Allik, N. Hansell, M. Wright, et al., "Meta-Analysis of Genome-Wide Association Studies Identifies Common Variants in CTNNA2 Associated with Excitement-Seeking," *Translational Psychiatry* (October 18, 2011): e49.

87. E. K. Speliotes, C. J. Willer, S. I. Berndt, K. L. Monda, G. Thorleifsson, A. U. Jackson, H. L. Allen, C. M. Lindgren, J. Luan, R. Magi, J. C. Randall, S. Vedantam, T. W. Winkler, et al., "Association Analyses of 249,796 Individuals Reveal 18 New Loci Associated with Body Mass Index," *Nat Genet* 42, no. 11 (November 2010): 937–48.

88. J. W. Smoller, "Genetic Boundary Violations: Phobic Disorders and Personality," *Am J Psychiatry* 164, no. 11 (November 2007): 1631–33.

89. J. M. Hettema, M. C. Neale, J. M. Myers, C. A. Prescott, and K. S. Kendler, "A Population-Based Twin Study of the Relationship Between Neuroticism and Internalizing Disorders," *Am J Psychiatry* 163, no. 5 (May 2006): 857–64.

90. C. J. Harmer, G. M. Goodwin, and P. J. Cowen, "Why Do Antidepressants Take So Long to Work? A Cognitive Neuropsychological Model of Antidepressant Drug Action." *Brit J Psychiatry* 195, no. 2 (August 2009): 102–108.

91. C. J. Harmer, C. E. Mackay, C. B. Reid, P. J. Cowen, and G. M. Goodwin, "Antidepressant Drug Treatment Modifies the Neural Processing of Nonconscious Threat Cues," *Biol Psychiatry* 59, no. 9 (May 1, 2006): 816–20.

92. S. E. Murphy, R. Norbury, U. O'Sullivan, P. J. Cowen, and C. J. Harmer, "Effect of a Single Dose of Citalopram on Amygdala Response to Emotional Faces," *Br J Psychiatry* 194, no. 6 (June 2009): 535–40.

93. C. Windischberger, R. Lanzenberger, A. Holik, C. Spindelegger, P. Stein, U. Moser, F. Gerstl, M. Fink, E. Moser, and S. Kasper, "Area-Specific Modulation of Neural Activation Comparing Escitalopram and Citalopram Revealed by Pharmaco-fMRI: A Randomized Cross-Over Study," *Neuroimage* 49, no. 2 (January 15, 2010): 1161–70.

94. R. Plomin, C. M. Haworth, and O. S. Davis, "Common Disorders Are Quantitative Traits," *Nat Rev Genet* 10, no. 12 (December 2009): 872–78.

CHAPTER 3: BLIND CATS AND BABY EINSTEINS

1. F. H. Rauscher, G. L. Shaw, and K. N. Ky, "Music and Spatial Task Performance," *Nature* 365, no. 6447 (October 14, 1993): 611.

2. R. L. Hotz, "Study Finds That Mozart Music Makes You Smarter," *Los Angeles Times*, October 14, 1993, 1.

3. D. Campbell, *The Mozart Effect for Children* (New York: Quill, 2002).

4. CBS, The Early Show, "Going Beyond Baby Einstein" (transcript), 2005.

5. S. Mead, *Million Dollar Babies: Why Infants Can't Be Hardwired for Success* (Washington, DC: Education Sector, 2007).

6. K. Sack, "Georgia's Governor Seeks Musical Start for Babies," *New York Times,* January 15, 1998.

7. C. F. Chabris, "Prelude or Requiem for the 'Mozart Effect'?" *Nature* 400, no. 6747 (August 26, 1999): 826–27; author reply 827–28.

8. E. G. Schellenberg and S. Hallam, "Music Listening and Cognitive Abilities In 10- and 11-Year-Olds: The Blur Effect," *Ann NY Acad Sci* 1060 (December 2005): 202–209.

9. W. F. Thompson, E. G. Schellenberg, and G. Husain, "Arousal, Mood, and the Mozart Effect," *Psychol Sci* 12, no. 3 (May 2001): 248–51.

10. B. E. Rideout, S. Dougherty, and L. Wernert, "Effect of Music on Spatial Performance: A Test of Generality," *Percept Mot Skills* 86, no. 2 (April 1998): 512–14.

11. R. Jones, "Mozart's Nice But Doesn't Increase IQs," *WebMD,* August 25, 1999.

12. F. H. Rauscher, "Author Reply: Prelude or Requiem for the 'Mozart Effect'?" *Nature* 400 (1999): 827–28.

13. Henry J. Kaiser Family Foundation, *A Teacher in the Living Room? Educational Media for Babies, Toddlers, and Preschoolers* (Menlo Park, CA: December 14, 2005). www .kff.org/entmedia/7427.cfm.

14. F. J. Zimmerman, D. A. Christakis, and A. N. Meltzoff, "Television and DVD/Video Viewing in Children Younger Than 2 years," *Arch Pediatr Adolesc Med* 161, no. 5 (May 2007): 473–79.

15. "Media Education. American Academy of Pediatrics. Committee on Public Education," *Pediatrics* 104, no. 2, part 1 (August 1999): 341–43.

16. "Media Use by Children Younger Than 2 Years," *Pediatrics,* October 17, 2011.

17. S. Tomopoulos, B. P. Dreyer, S. Berkule, A. H. Fierman, C. Brockmeyer, and A. L. Mendelsohn, "Infant Media Exposure and Toddler Development," *Arch Pediatr Adolesc Med* 164, no. 12 (December 2010): 1105–11.

18. F. J. Zimmerman, D. A. Christakis, and A. N. Meltzoff, "Associations Between Media Viewing and Language Development in Children Under Age 2 years," *J Pediatr* 151, no. 4 (October 2007): 364–68.

19. D. A. Christakis, F. J. Zimmerman, D. L. DiGiuseppe, and C. A. McCarty, "Early Television Exposure and Subsequent Attentional Problems in Children," *Pediatrics* 113, no. 4 (April 2004): 708–13.

20. F. J. Zimmerman and D. A. Christakis, "Associations Between Content Types of Early Media Exposure and Subsequent Attentional Problems," *Pediatrics* 120, no. 5 (November 2007): 986–92.

21. D. Bavelier, C. S. Green, and M. W. Dye, "Children, Wired: For Better and for Worse," *Neuron* 67, no. 5 (September 9, 2010): 692–701.

22. M. B. Robb, R. A. Richert, and E. A. Wartella, "Just a Talking Book? Word Learning from Watching Baby Videos," *Br J Dev Psychol* 27, part 1 (March 2009): 27–45.

23. M. E. Schmidt, M. Rich, S. L. Rifas-Shiman, E. Oken, and E. M. Taveras, "Television Viewing in Infancy and Child Cognition at 3 Years of Age in a US Cohort," *Pediatrics* 123, no. 3 (March 2009): e370–75.

24. J. S. DeLoache, C. Chiong, K. Sherman, N. Islam, M. Vanderborght, G. L. Troseth, G. A. Strouse, and K. O'Doherty, "Do Babies Learn from Baby Media?" *Psychol Sci* 21, no. 11 (November 1, 2010): 1570–74.

25. S. McLain, "Baby Einstein Sets the Record Straight on Refund," www.babyeinstein .com/refund/. Accessed June 26, 2010.

26. T. Lewin, "'Baby Einstein' Founder Goes to Court," *New York Times,* January 13, 2010.

27. T. N. Wiesel, "Postnatal Development of the Visual Cortex and the Influence of Environment," *Nature* 299, no. 5884 (October 14, 1982): 583–91.

28. G. J. Fisher, "Does Maternal Mental Influence Have Any Constructive or Destructive

Power in the Production of Malformations or Monstrosities at Any Stage of Embryonic Development?" *Am J Insanity* 26 (1870): 241–95.

29. M. Newton, *Savage Girls and Wild Boys: A History of Feral Children* (New York: Picador, 2002).

30. A. S. Benzaquen, "Kamala of Midnapore and Arnold Gesell's Wolf Child and Human Child: Reconciling the Extraordinary and the Normal," *Hist Psychol* 4 (2001): 59–78.

31. W. T. Greenough, J. E. Black, and C. S. Wallace, "Experience and Brain Development," *Child Dev* 58, no. 3 (June 1987): 539–59.

32. S. B. Hofer, T. D. Mrsic-Flogel, T. Bonhoeffer, and M. Hubener, "Experience Leaves a Lasting Structural Trace in Cortical Circuits," *Nature* 457, no. 7227 (January 15, 2009): 313–17.

33. P. K. Kuhl, "Early Language Acquisition: Cracking the Speech Code," *Nat Rev Neurosci* 5, no. 11 (November 2004): 831–43.

34. P. Kuhl and M. Rivera-Gaxiola, "Neural Substrates of Language Acquisition," *Annu Rev Neurosci* 31 (2008): 511–34.

35. P. K. Kuhl, B. T. Conboy, S. Coffey-Corina, D. Padden, M. Rivera-Gaxiola, and T. Nelson, "Phonetic Learning as a Pathway to Language: New Data and Native Language Magnet Theory Expanded (NLM-e)," *Philos Trans R Soc Lond B Biol Sci* 363, no. 1493 (March 12, 2008): 979–1000.

36. C. L. Darwin, "The Expression of the Emotions in Man and Animals," in *From So Simple a Beginning: The Four Great Books of Charles Darwin,* ed. E. O. Wilson (New York: W.W. Norton, 2006), 1255–1477.

37. O. Pascalis, M. de Haan, and C. A. Nelson, "Is Face Processing Species-Specific During the First Year of Life?" *Science* 296, no. 5571 (May 17, 2002): 1321–23.

38. J. M. Leppanen and C. A. Nelson, "Tuning the Developing Brain to Social Signals of Emotions," *Nat Rev Neurosci* 10, no. 1 (January 2009): 37–47.

39. J. M. Leppanen, M. C. Moulson, V. K. Vogel-Farley, and C. A. Nelson, "An ERP Study of Emotional Face Processing in the Adult and Infant Brain," *Child Dev* 78, no. 1 (January–February 2007): 232–45.

40. J. F. Sorce, R. N. Emde, J. Campos, and M. D. Klinnert, "Maternal Emotional Signaling: Its Effect on the Visual Cliff Behavior of 1-year-Olds," *Dev Psychol* 21 (1985): 195–200.

41. D. Cicchetti and S. L. Toth, "Child Maltreatment," *Annu Rev Clin Psychol* 1 (2005): 409–38.

42. S. D. Pollak and P. Sinha, "Effects of Early Experience on Children's Recognition of Facial Displays of Emotion," *Dev Psychol* 38, no. 5 (September 2002): 784–91.

43. S. D. Pollak, M. Messner, D. J. Kistler, and J. F. Cohn, "Development of Perceptual Expertise in Emotion Recognition," *Cognition* 110, no. 2 (February 2009): 242–47.

44. S. D. Pollak and D. J. Kistler, "Early Experience Is Associated with the Development of Categorical Representations for Facial Expressions of Emotion," *Proc Natl Acad Sci USA* 99, no. 13 (June 25, 2002): 9072–76.

45. K. Breslau, "Overplanned Parenthood: Ceausescu's Cruel Law," *Newsweek,* January 22, 1990, 35.

46. W. Moskoff, "Pronatalist Policies in Romania," *Econ Dev Cultur Change* 28 (1980): 597–614.

47. H. P. David and A. Baban, "Women's Health and Reproductive Rights: Romanian Experience," *Patient Educ Couns* 28, no. 3 (August 1996): 235–45.

48. T. J. Keil and V. Andreescu, "Fertility Policy in Ceausescu's Romania," *J Fam Hist* 24, no. 4 (October 1999): 478–92.

49. P. Stephenson, M. Wagner, M. Badea, and F. Serbanescu, "Commentary: The Public Health Consequences of Restricted Induced Abortion—Lessons from Romania," *Am J Public Health* 82, no. 10 (October 1992): 1328–31.

50. P. Gloviczki, "Ceausescu's Children: The Process of Democratization and the Plight of Romania's Orphans," *Critique: J Socialist Theory* (Spring 2004): 116–25.

51. C. H. Zeanah, C. A. Nelson, N. A. Fox, A. T. Smyke, P. Marshall, S. W. Parker, and S. Koga, "Designing Research to Study the Effects of Institutionalization on Brain and Behavioral Development: The Bucharest Early Intervention Project," *Dev Psychopathol* 15, no. 4 (Fall 2003): 885–907.

52. C. A. Nelson III, C. H. Zeanah, N. A. Fox, P. J. Marshall, A. T. Smyke, and D. Guthrie, "Cognitive Recovery in Socially Deprived Young Children: The Bucharest Early Intervention Project," *Science* 318, no. 5858 (December 21, 2007): 1937–40.

53. N. A. Fox, A. N. Almas, K. A. Degnan, C. A. Nelson, and C. H. Zeanah, "The Effects of Severe Psychosocial Deprivation and Foster Care Intervention on Cognitive Development at 8 Years of Age: Findings from the Bucharest Early Intervention Project," *J Child Psychol Psychiatry* 52, no. 9 (September 2011): 919–928.

54. M. M. Ghera, P. J. Marshall, N. A. Fox, C. H. Zeanah, C. A. Nelson, A. T. Smyke, and D. Guthrie, "The Effects of Foster Care Intervention on Socially Deprived Institutionalized Children's Attention and Positive Affect: Results from the BEIP Study," *J Child Psychol Psychiatry* 50, no. 3 (March 2009): 246–53.

55. K. Bos, C. H. Zeanah, N. A. Fox, S. S. Drury, K. A. McLaughlin, C. A. Nelson, "Psychiatric Outcomes in Young Children with a History of Institutionalization." *Harv Rev Psychiatry* 19, no. 1 (January–February 2011): 15–24.

56. C. H. Zeanah, H. L. Egger, A. T. Smyke, C. A. Nelson, N. A. Fox, P. J. Marshall, and D. Guthrie, "Institutional Rearing and Psychiatric Disorders in Romanian Preschool Children," *Am J Psychiatry* 166, no. 7 (July 2009): 777–85.

57. J. van Vliet, N. A. Oates, and E. Whitelaw, "Epigenetic Mechanisms in the Context of Complex Diseases," *Cell Mol Life Sci* 64, no. 12 (June 2007): 1531–38.

58. C. C. Wong, A. Caspi, B. Williams, I. W. Craig, R. Houts, A. Ambler, T. E. Moffitt, and J. Mill, "A Longitudinal Study of Epigenetic Variation in Twins," *Epigenet* 5, no. 6 (August 4, 2010).

59. B. P. Rutten and J. Mill, "Epigenetic Mediation of Environmental Influences in Major Psychotic Disorders," *Schizophr Bull* 35, no. 6 (November 2009): 1045–56.

60. M. J. Meaney, "Maternal Care, Gene Expression, and the Transmission of Individual Differences in Stress Reactivity Across Generations," *Annu Rev Neurosci* 24 (2001): 1161–92.

61. A. A. Hane and N. A. Fox, "Ordinary Variations in Maternal Caregiving Influence Human Infants' Stress Reactivity," *Psychol Sci* 17, no. 6 (June 2006): 550–56.

62. P. O. McGowan, A. Sasaki, A. C. D'Alessio, S. Dymov, B. Labonte, M. Szyf, G. Turecki, and M. J. Meaney, "Epigenetic Regulation of the Glucocorticoid Receptor in Human Brain Associates with Childhood Abuse," *Nat Neurosci* 12, no. 3 (March 2009): 342–48.

63. C. Murgatroyd, A. V. Patchev, Y. Wu, V. Micale, Y. Bockmuhl, D. Fischer, F. Holsboer, C. T. Wotjak, O. F. Almeida, and D. Spengler, "Dynamic DNA Methylation Programs Persistent Adverse Effects of Early-Life Stress," *Nat Neurosci* 12, no. 12 (December 2009): 1559–66.

64. S. Uchida, K. Hara, A. Kobayashi, K. Otsuki, H. Yamagata, T. Hobara, T. Suzuki, N. Miyata, and Y. Watanabe, "Epigenetic Status of Gdnf in the Ventral Striatum Determines Susceptibility and Adaptation to Daily Stressful Events," *Neuron* 69, no. 2 (January 27, 2011): 359–72.

65. T. Y. Zhang, I. C. Hellstrom, R. C. Bagot, X. Wen, J. Diorio, and M. J. Meaney, "Maternal Care and DNA Methylation of a Glutamic Acid Decarboxylase 1 Promoter in Rat Hippocampus," *J Neurosci* 30, no. 39 (September 29, 2010): 13130–37.

66. "Why Orbit," www.orbitbaby.com/why/rotate.php. Accessed March 7, 2009.

67. M. D. Seery, E. A. Holman, and R. C. Silver, "Whatever Does Not Kill Us: Cumulative Lifetime Adversity, Vulnerability, and Resilience," *J Pers Soc Psychol* 99, no. 6 (December 2010): 1025–41.

68. H. E. Stevens, J. F. Leckman, J. D. Coplan, and S. J. Suomi, "Risk and Resilience: Early Manipulation of Macaque Social Experience and Persistent Behavioral and Neu-

rophysiological Outcomes," *J Am Acad Child Adolesc Psychiatry* 48, no. 2 (February 2009): 114–27.

69. X. Ge, M. N. Natsuaki, J. M. Neiderhiser, and D. Reiss, "The Longitudinal Effects of Stressful Life Events on Adolescent Depression Are Buffered by Parent-Child Closeness," *Dev Psychopathol* 21, no. 2 (Spring 2009): 621–35.

70. J. Kaufman, B. Z. Yang, H. Douglas-Palumberi, S. Houshyar, D. Lipschitz, J. H. Krystal, and J. Gelernter, "Social Supports and Serotonin Transporter Gene Moderate Depression in Maltreated Children," *Proc Natl Acad Sci USA* (November 24, 2004).

71. A. Feder, E. J. Nestler, and D. S. Charney, "Psychobiology and Molecular Genetics of Resilience," *Nat Rev Neurosci* 10, no. 6 (June 2009): 446–57.

72. G. L. Ming and H. Song, "Adult Neurogenesis in the Mammalian Brain: Significant Answers and Significant Questions," *Neuron* 70, no. 4 (May 26, 2011): 687–702.

73. J. S. Snyder, A. Soumier, M. Brewer, J. Pickel, and H. A. Cameron, "Adult Hippocampal Neurogenesis Buffers Stress Responses and Depressive Behaviour," *Nature* 476, no. 7361 (August 25, 2011): 458–61.

74. L. Santarelli, M. Saxe, C. Gross, A. Surget, F. Battaglia, S. Dulawa, N. Weisstaub, J. Lee, R. Duman, O. Arancio, C. Belzung, and R. Hen, "Requirement of Hippocampal Neurogenesis for the Behavioral Effects of Antidepressants," *Science* 301, no. 5634 (August 8, 2003): 805–809.

75. S. Brene, A. Bjornebekk, E. Aberg, A. A. Mathe, L. Olson, and M. Werme, "Running Is Rewarding and Antidepressive," *Physiol Behav* 92, no. 1–2 (September 10, 2007): 136–40.

76. O. Berton, C. A. McClung, R. J. Dileone, V. Krishnan, W. Renthal, S. J. Russo, D. Graham, N. M. Tsankova, C. A. Bolanos, M. Rios, L. M. Monteggia, D. W. Self, and E. J. Nestler, "Essential Role of BDNF in the Mesolimbic Dopamine Pathway in Social Defeat Stress," *Science* 311, no. 5762 (February 10, 2006): 864–68.

77. V. Krishnan, M. H. Han, D. L. Graham, O. Berton, W. Renthal, S. J. Russo, Q. Laplant, A. Graham, M. Lutter, D. C. Lagace, S. Ghose, R. Reister, P. Tannous, et al., "Molecular Adaptations Underlying Susceptibility and Resistance to Social Defeat in Brain Reward Regions," *Cell* 131, no. 2 (October 19, 2007): 391–404.

78. V. Vialou, A. J. Robison, Q. C. Laplant, H. E. Covington III, D. M. Dietz, Y. N. Ohnishi, E. Mouzon, A. J. Rush III, E. L. Watts, D. L. Wallace, S. D. Iniguez, Y. H. Ohnishi, M. A. Steiner, et al., "DeltaFosB in Brain Reward Circuits Mediates Resilience to Stress and Antidepressant Responses," *Nat Neurosci* 13, no. 6 (June 2010): 746–752.

79. J. Harper, "To Air Is Human at Oxygen Bars," *Washington Times,* January 11, 1999, A.

80. A. Depalma, "Just When You Thought Air Was Free," *New York Times,* June 22, 1997.

81. J. T. Bruer, *The Myth of the First Three Years* (New York: The Free Press, 1999).

82. M. Grunwald, "Reiner Puts Tots on National Stage," *Boston Globe,* February 5, 1997, Living section.

83. J. Nithianantharajah and A. J. Hannan, "Enriched Environments, Experience-Dependent Plasticity and Disorders of the Nervous System," *Nat Rev Neurosci* 7, no. 9 (September 2006): 697–709.

84. W. S. Barnett, "Effectiveness of Early Educational Intervention," *Science* 333, no. 6045 (August 19, 2011): 975–78.

85. K. Woollett, H. J. Spiers, and E. A. Maguire, "Talent in the Taxi: A Model System for Exploring Expertise," *Philos Trans R Soc Lond B Biol Sci* 364, no. 1522 (May 27, 2009): 1407–16.

86. I. S. Park, K. J. Lee, J. W. Han, N. J. Lee, W. T. Lee, K. A. Park, I. J. Rhyu, "Basketball Training Increases Striatum Volume," *Human Movement Sci* 30, no. 1 (February 2011): 56–62.

87. L. Bezzola, S. Merillat, C. Gaser, and L. Jancke, "Training-Induced Neural Plasticity in Golf Novices," *J Neurosci* 31, no. 35 (August 31, 2011): 12444–48.

88. M. Stein, A. Federspiel, T. Koenig, M. Wirth, W. Strik, R. Wiest, D. Brandeis, and T. Dierks, "Structural Plasticity in the Language System Related to Increased Second Language Proficiency," *Cortex* (October 28, 2010).

89. J. Hanggi, S. Koeneke, L. Bezzola, and L. Jancke. "Structural Neuroplasticity in the Sensorimotor Network of Professional Female Ballet Dancers," *Human Brain Mapping* 31, no. 8 (August 2010): 1196–1206.

90. K. L. Hyde, J. Lerch, A. Norton, M. Forgeard, E. Winner, A. C. Evans, G. Schlaug, "Musical Training Shapes Structural Brain Development," *J Neurosci* 29, no. 10 (March 11, 2009): 3019–25.

CHAPTER 4: DOGS, POKER, AND AUTISM

1. M. Bar, M. Neta, and H. Linz, "Very First Impressions," *Emotion* 6, no. 2 (May 2006): 269–78.

2. B. Duchaine and K. Nakayama, "Dissociations of Face and Object Recognition in Developmental Prosopagnosia," *J Cogn Neurosci* 17, no. 2 (February 2005): 249–61.

3. B. C. Duchaine and K. Nakayama, "Developmental Prosopagnosia: A Window to Content-Specific Face Processing," *Curr Opin Neurobiol* 16, no. 2 (April 2006): 166–73.

4. R. Russell, B. Duchaine, and K. Nakayama, "Super-Recognizers: People with Extraordinary Face Recognition Ability," *Psychon Bull Rev* 16, no. 2 (April 2009): 252–57.

5. J. B. Wilmer, L. Germine, C. F. Chabris, G. Chatterjee, M. Williams, E. Loken, K. Nakayama, and B. Duchaine, "Human Face Recognition Ability Is Specific and Highly Heritable," *Proc Natl Acad Sci USA* 107, no. 11 (March 16, 2010): 5238–41.

6. N. Kanwisher, J. McDermott, and M. M. Chun, "The Fusiform Face Area: A Module in Human Extrastriate Cortex Specialized for Face Perception," *J Neurosci* 17, no. 11 (June 1, 1997): 4302–11.

7. N. Kanwisher, "Functional Specificity in the Human Brain: A Window into the functional Architecture of the Mind," *Proc Natl Acad Sci USA* 107, no. 25 (June 22, 2010): 11163–70.

8. K. S. Scherf, M. Behrmann, K. Humphreys, and B. Luna, "Visual Category-Selectivity for Faces, Places and Objects Emerges Along Different Developmental Trajectories," *Dev Sci* 10, no. 4 (July 2007): F15–30.

9. M. de Haan and A. Matheson, "The Development and Neural Bases of Processing Emotion in Faces and Voices," in *Handbook of Developmental Social Neuroscience,* ed. M. de Haan and M. R. Gunnar (New York: Guilford Press, 2009), 107–21.

10. A. Senju and G. Csibra, "Gaze Following in Human Infants Depends on Communicative Signals," *Curr Biol* 18, no. 9 (May 6, 2008): 668–71.

11. M. Tomasello, M. Carpenter, J. Call, T. Behne, and H. Moll, "Understanding and Sharing Intentions: The Origins of Cultural Cognition," *Behav Brain Sci* 28, no. 5 (October 2005): 675–91; discussion 691–735.

12. M. Tomasello, M. Carpenter, and U. Liszkowski, "A New Look at Infant Pointing," *Child Dev* 78, no. 3 (May–June 2007): 705–22.

13. T. Grossmann and M. H. Johnson, "The Development of the Social Brain in Human Infancy," *Eur J Neurosci* 25, no. 4 (February 2007): 909–19.

14. A. Whiten, "The Second Inheritance System of Chimpanzees and Humans," *Nature* 437, no. 7055 (September 1, 2005): 52–55.

15. A. Whiten, V. Horner, and F. B. de Waal, "Conformity to Cultural Norms of Tool Use in Chimpanzees," *Nature* 437, no. 7059 (September 29, 2005): 737–40.

16. G. Csibra and G. Gergely, "Natural Pedagogy," *Trends Cogn Sci* 13, no. 4 (April 2009): 148–53.

17. E. Herrmann, J. Call, M. V. Hernandez-Lloreda, B. Hare, and M. Tomasello, "Humans Have Evolved Specialized Skills of Social Cognition: The Cultural Intelligence Hypothesis," *Science* 317, no. 5843 (September 7, 2007): 1360–66.

18. A. Lillard, "Pretend Play and Cognitive Development," in *Blackwell Handbook of Childhood Cognitive Development,* ed. U. Goswami (Oxford: Blackwell Publishers, 2004), 188–205.

19. O. Friedman and A. M. Leslie, "The Conceptual Underpinnings of Pretense: Pretending Is Not 'Behaving-as-If,'" *Cognition* 105, no. 1 (October 2007): 103–24.

20. A. M. Leslie, "Pretense and Representation: The Origins of 'Theory of Mind,'" *Psychol Rev* 94 (1987): 412–26.

21. R. Baillargeon, R. M. Scott, and Z. He, "False-Belief Understanding in Infants," *Trends Cogn Sci* 14, no. 3 (March 2010): 110–18.

22. F. Heider and M. Simmel, "An Experimental Study of Apparent Behavior," *Am J Psychol* 57 (1944): 243–59.

23. D. Premack and G. Woodruff, "Does the Chimpanzee Have a Theory of Mind?" *Behav Brain Sci* 4 (1978): 515–26.

24. J. Call and M. Tomasello, "Does the Chimpanzee Have a Theory of Mind? 30 Years Later," *Trends Cogn Sci* 12, no. 5 (May 2008): 187–92.

25. D. Buttelmann, J. Call, and M. Tomasello, "Do Great Apes Use Emotional Expressions to Infer Desires?" *Dev Sci* 12, no. 5 (September 2009): 688–98.

26. D. Buttelmann, M. Carpenter, J. Call, and M. Tomasello, "Enculturated Chimpanzees Imitate Rationally," *Dev Sci* 10, no. 4 (July 2007): F31–38.

27. B. Hare, J. Call, and M. Tomasello, "Chimpanzees Deceive a Human Competitor by Hiding," *Cognition* 101, no. 3 (October 2006): 495–514.

28. D. Premack, "Human and Animal Cognition: Continuity and Discontinuity," *Proc Natl Acad Sci USA* 104, no. 35 (August 28, 2007): 13861–67.

29. J. M. Dally, N. J. Emery, and N. S. Clayton, "Food-Caching Western Scrub-Jays Keep Track of Who Was Watching When," *Science* 312, no. 5780 (June 16, 2006): 1662–65.

30. B. Hare, M. Brown, C. Williamson, and M. Tomasello, "The Domestication of Social Cognition in Dogs," *Science* 298, no. 5598 (November 22, 2002): 1634–36.

31. B. Hare and M. Tomasello, "Human-like Social Skills in Dogs?" *Trends Cogn Sci* 9, no. 9 (September 2005): 439–44.

32. A. Miklosi, *Dog Behavior, Evolution, and Cognition* (New York: Oxford University Press, 2007).

33. E. K. Karlsson and K. Lindblad-Toh, "Leader of the Pack: Gene Mapping in Dogs and Other Model Organisms," *Nat Rev Genet* 9, no. 9 (September 2008): 713–25.

34. A. Miklosi, *Dog Behavior, Evolution, and Cognition* (New York: Oxford University Press, 2009).

35. M. A. Udell, N. R. Dorey, and C. D. Wynne, "Can Your Dog Read Your Mind? Understanding the Causes of Canine Perspective Taking," *Learn Behav* (June 4, 2011).

36. L. N. Trut, "Early Canid Domestication: The Farm-Fox Experiment," *Am Sci* 87 (1999): 160–69.

37. T. C. Spady and E. A. Ostrander, "Canine Behavioral Genetics: Pointing Out the Phenotypes and Herding Up the Genes," *Am J Hum Genet* 82, no. 1 (January 2008): 10–18.

38. D. K. Belyaev, "The Wilhelmine E. Key 1978 Invitational Lecture. Destabilizing Selection as a Factor in Domestication," *J Hered* 70, no. 5 (September–October 1979): 301–308.

39. B. Hare, I. Plyusnina, N. Ignacio, O. Schepina, A. Stepika, R. Wrangham, and L. Trut, "Social Cognitive Evolution in Captive Foxes Is a Correlated By-Product of Experimental Domestication," *Curr Biol* 15, no. 3 (February 8, 2005): 226–30.

40. D. C. Penn and D. J. Povinelli, "On the Lack of Evidence That Non-Human Animals Possess Anything Remotely Resembling a 'Theory of Mind,'" *Philos Trans R Soc Lond B Biol Sci* 362, no. 1480 (April 29, 2007): 731–44.

41. D. C. Penn and D. J. Povinelli, "Causal Cognition in Human and Nonhuman Animals: A Comparative, Critical Review," *Annu Rev Psychol* 58 (2007): 97–118.

42. R. Saxe and N. Kanwisher, "People Thinking About Thinking People. The Role of the Temporo-Parietal Junction in 'Theory of Mind,'" *Neuroimage* 19, no. 4 (August 2003): 1835–42.

43. M. D. Bauman and D. G. Amaral, "Neurodevelopment of Social Cognition," in *Handbook of Developmental Cognitive Neuroscience*, 2nd ed., ed. C. A. Nelson and M. Luciana (Cambridge, MA: MIT Press, 2008), 161–86.

44. R. Saxe, "Uniquely Human Social Cognition," *Curr Opin Neurobiol* 16, no. 2 (April 2006): 235–39.

45. M. A. Sabbagh, L. C. Bowman, L. E. Evraire, and J. M. Ito, "Neurodevelopmental Correlates of Theory of Mind in Preschool Children," *Child Dev* 80, no. 4 (July–August 2009): 1147–62.

46. C. Hughes, S. R. Jaffee, F. Happe, A. Taylor, A. Caspi, and T. E. Moffitt, "Origins of Individual Differences in Theory of Mind: From Nature to Nurture?" *Child Dev* 76, no. 2 (March–April 2005): 356–70.

47. A. Ronald, E. Viding, F. Happe, and R. Plomin, "Individual Differences in Theory of Mind Ability in Middle Childhood and Links with Verbal Ability and Autistic Traits: A Twin Study," *Soc Neurosci* 1, no. 3–4 (2006): 412–25.

48. D. F. Polit and T. Falbo, "The Intellectual Achievement of Only Children," *J Biosoc Sci* 20, no. 3 (July 1988): 275–85.

49. N. Howe, H. Petrakos, C. M. Rinaldi, and R. LeFebvre, " 'This Is a Bad Dog, You Know . . .': Constructing Shared Meanings During Sibling Pretend Play," *Child Dev* 76, no. 4 (July–August 2005): 783–94.

50. L. M. Youngblade and J. Dunn, "Individual Differences in Young Children's Pretend Play with Mother and Sibling: Links to Relationships and Understanding of Other People's Feelings and Beliefs," *Child Dev* 66, no. 5 (October 1995): 1472–92.

51. C. C. Peterson, "Kindred Spirits: Influence of Siblings' Perspectives on Theory of Mind," *Cogn Dev* 15 (2000): 435–55.

52. S. G. Shamay-Tsoory, R. Tomer, and J. Aharon-Peretz, "The Neuroanatomical Basis of Understanding Sarcasm and Its Relationship to Social Cognition," *Neuropsychol* 19, no. 3 (May 2005): 288–300.

53. L. Kanner, "Autistic Disturbances of Affective Contact," *Nerv Child* 2 (1943): 217–50.

54. CDC, "Prevalence of Autism Spectrum Disorders: Autism and Developmental Disabilities Monitoring Network, United States, 2006," *MMWR* 58, SS–10 (2009).

55. Y. S. Kim, B. L. Leventhal, Y. J. Koh, E. Fombonne, E. Laska, E. C. Lim, K. A. Cheon, S. J. Kim, Y. K. Kim, H. Lee, D. H. Song, and R. R. Grinker, "Prevalence of Autism Spectrum Disorders in a Total Population Sample," *Am J Psychiatry* 168, no. 9 (September 2011): 904–12.

56. E. Fombonne, "Epidemiology of Pervasive Developmental Disorders," *Pediatr Res* 65, no. 6 (June 2009): 591–98.

57. F. R. Volkmar, M. State, and A. Klin, "Autism and Autism Spectrum Disorders: Diagnostic Issues for the Coming Decade," *J Child Psychol Psychiatry* 50, no. 1–2 (January 2009): 108–15.

58. S. Baron-Cohen, A. M. Leslie, and U. Frith, "Does the Autistic Child Have a 'Theory of Mind'?" *Cognition* 21, no. 1 (October 1985): 37–46.

59. S. Baron-Cohen, *Mindblindness: An Essay on Autism and Theory of Mind* (Cambridge, MA: The MIT Press, 1995).

60. R. Redon, S. Ishikawa, K. R. Fitch, L. Feuk, G. H. Perry, T. D. Andrews, H. Fiegler, M. H. Shapero, A. R. Carson, W. Chen, E. K. Cho, S. Dallaire, J. L. Freeman, et al., "Global Variation in Copy Number in the Human Genome," *Nature* 444, no. 7118 (November 23, 2006): 444–54.

61. J. T. Glessner, K. Wang, G. Cai, O. Korvatska, C. E. Kim, S. Wood, H. Zhang, A. Estes, C. W. Brune, J. P. Bradfield, M. Imielinski, E. C. Frackelton, J. Reichert, et al., "Autism Genome-Wide Copy Number Variation Reveals Ubiquitin and Neuronal Genes," *Nature* 459, no. 7246 (May 28, 2009): 569–78.

62. J. Sebat, B. Lakshmi, D. Malhotra, J. Troge, C. Lese-Martin, T. Walsh, B. Yamrom, S. Yoon, A. Krasnitz, J. Kendall, A. Leotta, D. Pai, R. Zhang, et al., "Strong Association of De Novo Copy Number Mutations with Autism," *Science* 316, no. 5823 (April 20, 2007): 445–49.

63. L. A. Weiss, Y. Shen, J. M. Korn, D. E. Arking, D. T. Miller, R. Fossdal, E. Saemundsen, H. Stefansson, M. A. Ferreira, T. Green, O. S. Platt, D. M. Ruderfer, C. A.

Walsh, et al., "Association Between Microdeletion and Microduplication at 16p11.2 and Autism," *N Engl J Med* 358, no. 7 (February 14, 2008): 667–75.

64. D. Pinto, A. T. Pagnamenta, L. Klei, R. Anney, D. Merico, R. Regan, J. Conroy, T. R. Magalhaes, C. Correia, B. S. Abrahams, J. Almeida, E. Bacchelli, G. D. Bader, et al., "Functional Impact of Global Rare Copy Number Variation in Autism Spectrum Disorders," *Nature* 466, no. 7304 (July 15, 2010): 368–372.

65. A. Guilmatre, C. Dubourg, A. L. Mosca, S. Legallic, A. Goldenberg, V. Drouin-Garraud, V. Layet, A. Rosier, S. Briault, F. Bonnet-Brilhault, F. Laumonnier, S. Odent, G. Le Vacon, et al., "Recurrent Rearrangements in Synaptic and Neurodevelopmental Genes and Shared Biologic Pathways in Schizophrenia, Autism, and Mental Retardation," *Arch Gen Psychiatry* 66, no. 9 (September 2009): 947–56.

66. J. Sebat, D. L. Levy, and S. E. McCarthy, "Rare Structural Variants in Schizophrenia: One Disorder, Multiple Mutations; One Mutation, Multiple Disorders," *Trends Genet* 25, no. 12 (December 2009): 528–35.

67. M. D. King, C. Fountain, D. Dakhlallah, and P. S. Bearman, "Estimated Autism Risk and Older Reproductive Age," *Am J Public Health* 99, no. 9 (September 2009): 1673–79.

68. D. Malaspina, S. Harlap, S. Fennig, D. Heiman, D. Nahon, D. Feldman, and E. S. Susser, "Advancing Paternal Age and the Risk of Schizophrenia," *Arch Gen Psychiatry* 58, no. 4 (April 2001): 361–67.

69. F. Zhang, W. Gu, M. E. Hurles, and J. R. Lupski, "Copy Number Variation in Human Health, Disease, and Evolution," *Annu Rev Genomics Hum Genet* 10 (2009): 451–81.

70. J. F. Shelton, D. J. Tancredi, and I. Hertz-Picciotto, "Independent and Dependent Contributions of Advanced Maternal and Paternal Ages to Autism Risk," *Autism Res* 3, no. 1 (February 2010): 30–39.

71. A. Di Martino, K. Ross, L. Q. Uddin, A. B. Sklar, F. X. Castellanos, and M. P. Milham, "Functional Brain Correlates of Social and Nonsocial Processes in Autism Spectrum Disorders: An Activation Likelihood Estimation Meta-analysis," *Biol Psychiatry* 65, no. 1 (January 1, 2009): 63–74.

72. M. V. Lombardo, B. Chakrabarti, E. T. Bullmore, and S. Baron-Cohen, "Specialization of Right Temporo-Parietal Junction for Mentalizing and Its Relation to Social Impairments in Autism," *Neuroimage* 56, no. 3 (June 1, 2011): 1832–38.

73. J. N. Constantino, "The Quantitative Nature of Autistic Social Impairment," *Pediatr Res* 69, no. 5 Pt 2 (May 2011): #55R–62R.

74. R. S. Hurley, M. Losh, M. Parlier, J. S. Reznick, and J. Piven, "The Broad Autism Phenotype Questionnaire," *J Autism Dev Disord* 37, no. 9 (October 2007): 1679–90.

75. M. Losh, R. Adolphs, M. D. Poe, S. Couture, D. Penn, G. T. Baranek, and J. Piven, "Neuropsychological Profile of Autism and the Broad Autism Phenotype," *Arch Gen Psychiatry* 66, no. 5 (May 2009): 518–26.

76. M. Losh, D. Childress, K. Lam, and J. Piven, "Defining Key Features of the Broad Autism Phenotype: A Comparison Across Parents of Multiple- and Single-Incidence Autism Families," *Am J Med Genet B Neuropsychiatr Genet* 147B, no. 4 (June 5, 2008): 424–33.

77. M. Losh and J. Piven, "Social-Cognition and the Broad Autism Phenotype: Identifying Genetically Meaningful Phenotypes," *J Child Psychol Psychiatry* 48, no. 1 (January 2007): 105–12.

78. S. Baron-Cohen, S. Wheelwright, R. Skinner, J. Martin, and E. Clubley, "The Autism-Spectrum Quotient (AQ): Evidence from Asperger Syndrome/High-Functioning Autism, Males and Females, Scientists and Mathematicians," *J Autism Dev Disord* 31, no. 1 (February 2001): 5–17.

79. E. B. Robinson, K. C. Koenen, M. C. McCormick, K. Munir, V. Hallett, F. Happe, R. Plomin, A. Ronald, "Evidence That Autistic Traits Show the Same Etiology in the General Population and at the Quantitative Extremes (5%, 2.5%, and 1%)," *Arch Gen Psychiatry* 68, no. 11 (November 2011): 1113–21.

80. S. Baron-Cohen, E. Ashwin, C. Ashwin, T. Tavassoli, and B. Chakrabarti, "Talent in Autism: Hyper-Systemizing, Hyper-Attention to Detail and Sensory Hypersensitivity," *Philos Trans R Soc Lond B Biol Sci* 364, no. 1522 (May 27, 2009): 1377–83.

81. *Temple Grandin: The World Needs All Kinds of Minds,* TED Conferences, LLC, 2010.

82. A. Fenton and T. Krahn, "Autism, Neurodiversity and Equality Beyond the 'Normal,'" *J Ethics Ment Health* 2 (2007): 1–6.

83. C. Brownlow, "Re-presenting Autism: The Construction of 'NT Syndrome,'" *J Med Humanit* 31 (2010): 243–55.

84. S. Bellini and J. K. Peters, "Social Skills Training for Youth with Autism Spectrum Disorders," *Child Adolesc Psychiatr Clin N Am* 17, no. 4 (October 2008): 857–73, x.

85. P. A. Rao, D. C. Beidel, and M. J. Murray, "Social Skills Interventions for Children with Asperger's Syndrome or High-Functioning Autism: A Review and Recommendations," *J Autism Dev Disord* 38, no. 2 (February 2008): 353–61.

86. E. A. Laugeson, F. Frankel, A. Gantman, A. R. Dillon, and C. Mogil, "Evidence-Based Social Skills Training for Adolescents with Autism Spectrum Disorders: The UCLA PEERS Program," *J Autism Dev Disord* (August 20, 2011).

87. V. Talwar, H. M. Gordon, and K. Lee, "Lying in the Elementary School Years: Verbal Deception and Its Relation to Second-Order Belief Understanding," *Dev Psychol* 43, no. 3 (May 2007): 804–10.

88. C. F. Bond and B. M. Depaulo, "Individual Differences in Judging Deception: Accuracy and Bias," *Psychol Bull* 134, no. 4 (July 2008): 477–92.

89. P. Ekman and M. O'Sullivan, "Who Can Catch a Liar?" *Am Psychol* 46, no. 9 (September 1991): 913–20.

90. P. Ekman, M. O'Sullivan, and M. G. Frank, "A Few Can Catch a Liar," *Psychol Sci* 10 (1999): 263–66.

91. M. O'Sullivan, "Home Runs and Humbugs: Comment on Bond and DePaulo (2008)," *Psychol Bull* 134, no. 4 (July 2008): 493–97; discussion 501–493.

92. J. Navarro, *Phil Hellmuth Presents Read 'Em and Reap* (New York: Collins, 2006).

93. D. Sklansky, *The Theory of Poker,* 4th ed. (Henderson, NV: Two Plus Two Publishing, 1999).

94. E. J. Schlicht, S. Shimojo, C. F. Camerer, P. Battaglia, and K. Nakayama, "Human Wagering Behavior Depends on Opponents' Faces," *PLoS One* 5, no. 7 (2010): e11663.

95. D. Matsumoto and B. Willingham, "Spontaneous Facial Expressions of Emotion of Congenitally and Noncongenitally Blind Individuals," *J Pers Soc Psychol* 96, no. 1 (January 2009): 1–10.

96. J. L. Tracy and D. Matsumoto, "The Spontaneous Expression of Pride and Shame: Evidence for Biologically Innate Nonverbal Displays," *Proc Natl Acad Sci USA* 105, no. 33 (August 19, 2008): 11655–60.

97. D. Matsumoto, D. Keltner, M. N. Shiota, M. O'Sullivan, and M. Frank,. "Facial Expressions of Emotion," in *Handbook of Emotions,* 3rd ed., ed. M. Lewis, J. M. Haviland-Jones, and L. Feldman Barrett (New York: Guilford Press, 2009).

98. J. M. Susskind and A. K. Anderson, "Facial Expression Form and Function," *Commun Integr Biol* 1, no. 2 (2008): 148–49.

99. A. Smith, *The Theory of Moral Sentiments,* new ed. (London: Henry G. Bohn, 1853).

100. T. L. Chartrand and J. A. Bargh, "The Chameleon Effect: The Perception-Behavior Link and Social Interaction," *J Pers Soc Psychol* 76, no. 6 (June 1999): 893–910.

101. T. Singer and C. Lamm, "The Social Neuroscience of Empathy," *Ann NY Acad Sci* 1156 (March 2009): 81–96.

102. G. di Pellegrino, L. Fadiga, L. Fogassi, V. Gallese, and G. Rizzolatti, "Understanding Motor Events: A Neurophysiological Study," *Exp Brain Res* 91, no. 1 (1992): 176–80.

103. M. Iacoboni, "Imitation, Empathy, and Mirror Neurons," *Annu Rev Psychol* 60 (2009): 653–70.

104. M. Iacoboni and M. Dapretto, "The Mirror Neuron System and the Consequences of Its Dysfunction," *Nat Rev Neurosci* 7, no. 12 (December 2006): 942–51.

105. L. Carr, M. Iacoboni, M. C. Dubeau, J. C. Mazziotta, and G. L. Lenzi, "Neural Mechanisms of Empathy in Humans: A Relay from Neural Systems for Imitation to Limbic Areas," *Proc Natl Acad Sci USA* 100, no. 9 (April 29, 2003): 5497–5502.

106. B. Wicker, C. Keysers, J. Plailly, J. P. Royet, V. Gallese, and G. Rizzolatti, "Both of Us Disgusted in My Insula: The Common Neural Basis of Seeing and Feeling Disgust," *Neuron* 40, no. 3 (October 30, 2003): 655–64.

107. S. G. Shamay-Tsoory, J. Aharon-Peretz, and D. Perry, "Two Systems for Empathy: A Double Dissociation Between Emotional and Cognitive Empathy in Inferior Frontal Gyrus Versus Ventromedial Prefrontal Lesions," *Brain* 132, part 3 (March 2009): 617–27.

108. J. Zaki, J. Weber, N. Bolger, and K. Ochsner, "The Neural Bases of Empathic Accuracy," *Proc Natl Acad Sci USA* 106, no. 27 (July 7, 2009): 11382–87.

109. S. G. Michaud and H. Aynesworth, *Ted Bundy: Conversations with a Killer* (Irving, TX: Authorlink Press, 2000).

110. H. Cleckley, *The Mask of Sanity*, 5th ed. (Augusta, GA: Emily S. Cleckley, 1988).

111. R. J. Blair, "Responding to the Emotions of Others: Dissociating Forms of Empathy Through the Study of Typical and Psychiatric Populations," *Conscious Cogn* 14, no. 4 (December 2005): 698–718.

112. Y. Yang, A. Raine, K. L. Narr, P. Colletti, and A. W. Toga, "Localization of deformations within the amygdala in individuals with psychopathy," *Arch Gen Psychiatry* 66, no. 9 (September 2009): 986–94.

113. R. J. Blair, "The Amygdala and Ventromedial Prefrontal Cortex in Morality and Psychopathy," *Trends Cogn Sci* 11, no. 9 (September 2007): 387–92.

114. R. J. Blair, D. G. Mitchell, R. A. Richell, S. Kelly, A. Leonard, C. Newman, and S. K. Scott, "Turning a Deaf Ear to Fear: Impaired Recognition of Vocal Affect in Psychopathic Individuals," *J Abnorm Psychol* 111, no. 4 (November 2002): 682–86.

115. A. A. Marsh and R. J. Blair, "Deficits in Facial Affect Recognition Among Antisocial Populations: A Meta-Analysis," *Neurosci Biobehav Rev* 32, no. 3 (2008): 454–65.

116. A. A. Marsh, E. C. Finger, D. G. Mitchell, M. E. Reid, C. Sims, D. S. Kosson, K. E. Towbin, E. Leibenluft, D. S. Pine, and R. J. Blair, "Reduced Amygdala Response to Fearful Expressions in Children and Adolescents with Callous-Unemotional Traits and Disruptive Behavior Disorders," *Am J Psychiatry* 165, no. 6 (June 2008): 712–20.

117. R. J. Blair, "The Amygdala and Ventromedial Prefrontal Cortex: Functional Contributions and Dysfunction in Psychopathy," *Philos Trans R Soc Lond B Biol Sci* 363, no. 1503 (August 12, 2008): 2557–65.

118. S. G. Shamay-Tsoory, H. Harari, J. Aharon-Peretz, and Y. Levkovitz, "The Role of the Orbitofrontal Cortex in Affective Theory of Mind Deficits in Criminal Offenders with Psychopathic Tendencies," *Cortex* 46, no. 5 (May 2010): 668–77.

119. J. W. Buckholtz, M. T. Treadway, R. L. Cowan, N. D. Woodward, S. D. Benning, R. Li, M. S. Ansari, R. M. Baldwin, A. N. Schwartzman, E. S. Shelby, C. E. Smith, D. Cole, R. M. Kessler, and D. H. Zald, "Mesolimbic Dopamine Reward System Hypersensitivity in Individuals with Psychopathic Traits," *Nat Neurosci* 13, no. 4 (April 2010): 419–21.

120. D. M. Blonigen, B. M. Hicks, R. F. Krueger, C. J. Patrick, and W. G. Iacono, "Psychopathic Personality Traits: Heritability and Genetic Overlap with Internalizing and Externalizing Psychopathology," *Psychol Med* 35, no. 5 (May 2005): 637–48.

121. E. Viding, R. J. Blair, T. E. Moffitt, and R. Plomin, "Evidence for Substantial Genetic Risk for Psychopathy in 7-Year-Olds," *J Child Psychol Psychiatry* 46, no. 6 (June 2005): 592–97.

122. J. Coid, M. Yang, S. Ullrich, A. Roberts, and R. D. Hare, "Prevalence and Correlates of Psychopathic Traits in the Household Population of Great Britain," *Int J Law Psychiatry* 32, no. 2 (March–April 2009): 65–73.

123. B. P. Klein-Tasman and C. B. Mervis, "Distinctive Personality Characteristics of 8-, 9-, and 10-Year-Olds with Williams Syndrome," *Dev Neuropsychol* 23, no. 1–2 (2003): 269–90.

124. M. A. Martens, S. J. Wilson, and D. C. Reutens, "Research Review: Williams Syndrome: A Critical Review of the Cognitive, Behavioral, and Neuroanatomical Phenotype," *J Child Psychol Psychiatry* 49, no. 6 (June 2008): 576–608.
125. H. Tager-Flusberg and K. Sullivan. "A Componential View of Theory of Mind: Evidence from Williams Syndrome," *Cognition* 76, no. 1 (July 14, 2000): 59–90.
126. O. T. Leyfer, J. Woodruff-Borden, B. P. Klein-Tasman, J. S. Fricke, and C. B. Mervis, "Prevalence of Psychiatric Disorders in 4- to 16-Year-Olds with Williams Syndrome," *Am J Med Genet B Neuropsychiatr Genet* 141, no. 6 (September 5, 2006): 615–22.
127. M. L. Hoffman, "Empathy and Prosocial Behavior," in *Handbook of Emotions*, 3rd ed., ed. M. Lewis, J. M. Haviland-Jones, and L. Feldman Barrett (New York: Guilford Press, 2008), 440–55.
128. P. Singer, *The Life You Can Save* (New York: Random House, 2009).
129. P. Slovic, "'If I Look at the Mass I Will Never Act': Psychic Numbing and Genocide," *Judgment and Decision Making* 2 (2007): 79–95.
130. M. K. Kearney, R. B. Weininger, M. L. Vachon, R. L. Harrison, and B. M. Mount, "Self-Care of Physicians Caring for Patients at the End of Life: "'Being Connected . . . a Key to My Survival,'" *JAMA* 301, no. 11 (March 18, 2009): 1155–64, E1151.
131. J. A. Boscarino, C. R. Figley, and R. E.Adams, "Compassion Fatigue Following the September 11 Terrorist Attacks: A Study of Secondary Trauma Among New York City Social Workers," *Int J Emerg Ment Health* 6, no. 2 (Spring 2004): 57–66.

CHAPTER 5: "SOLE MATES"

1. P. J. Brunton and J. A. Russell, "The Expectant Brain: Adapting for Motherhood," *Nat Rev Neurosci* 9, no. 1 (January 2008): 11–25.
2. R. Tyzio, R. Cossart, I. Khalilov, M. Minlebaev, C. A. Hubner, A. Represa, Y. Ben-Ari, and R. Khazipov, "Maternal Oxytocin Triggers a Transient Inhibitory Switch in GABA Signaling in the Fetal Brain During Delivery," *Science* 314, no. 5806 (December 15, 2006): 1788–92.
3. C. A. Pedersen, J. A. Ascher, Y. L. Monroe, and A. J. Prange Jr., "Oxytocin Induces Maternal Behavior in Virgin Female Rats," *Science* 216, no. 4546 (May 7, 1982): 648–50.
4. J. T. Winslow and T. R. Insel, "Neuroendocrine Basis of Social Recognition," *Curr Opin Neurobiol* 14, no. 2 (April 2004): 248–53.
5. A. Gonzalez, L. Atkinson, and A. S. Fleming, "Attachment and the Comparative Psychobiology of Mothering," in *Handbook of Developmental Social Neuroscience*, ed. M. de Haan and M. R. Gunnar (New York: Guilford Press, 2009), 225–45.
6. B. J. Mattson, S. Williams, J. S. Rosenblatt, and J. I. Morrell, "Comparison of Two Positive Reinforcing Stimuli: Pups and Cocaine Throughout the Postpartum Period," *Behav Neurosci* 115, no. 3 (June 2001): 683–94.
7. R. Sprengelmeyer, D. I. Perrett, E. C. Fagan, R. E. Cornwell, J. S. Lobmaier, A. Sprengelmeyer, H. B. Aasheim, I. M. Black, L. M. Cameron, S. Crow, N. Milne, E. C. Rhodes, and A. W. Young, "The Cutest Little Baby Face: A Hormonal Link to Sensitivity to Cuteness in Infant Faces," *Psychol Sci* 20, no. 2 (February 2009): 149–54.
8. J. S. Lobmaier, R. Sprengelmeyer, B. Wiffen, and D. I. Perrett, "Female and Male Responses to Cuteness, Age and Emotion in Infant Faces," *Evol Hum Behav* 31 (2010): 16–21.
9. M. L. Glocker, D. D. Langleben, K. Ruparel, J. W. Loughead, J. N. Valdez, M. D. Griffin, N. Sachser, and R. C. Gur, "Baby Schema Modulates the Brain Reward System in Nulliparous Women," *Proc Natl Acad Sci USA* 106, no. 22 (June 2, 2009): 9115–19.
10. L. Strathearn, J. Li, P. Fonagy, and P. R. Montague, "What's in a Smile? Maternal Brain Responses to Infant Facial Cues," *Pediatrics* 122, no. 1 (July 2008): 40–51.
11. M. Noriuchi, Y. Kikuchi, and A. Senoo, "The Functional Neuroanatomy of Maternal Love: Mother's Response to Infant's Attachment Behaviors," *Biol Psychiatry* 63, no. 4 (February 15, 2008): 415–23.
12. D. G. Kleiman, "Monogamy in Mammals," *Q Rev Biol* 52, no. 1 (March 1977): 39–69.

13. Z. R. Donaldson and L. J. Young, "Oxytocin, Vasopressin, and the Neurogenetics of Sociality," *Science* 322, no. 5903 (November 7, 2008): 900–904.

14. H. E. Ross, S. M. Freeman, L. L. Spiegel, X. Ren, E. F. Terwilliger, and L. J. Young, "Variation in Oxytocin Receptor Density in the Nucleus Accumbens Has Differential Effects on Affiliative Behaviors in Monogamous and Polygamous Voles," *J Neurosci* 29, no. 5 (February 4, 2009): 1312–18.

15. H. E. Ross, C. D. Cole, Y. Smith, I. D. Neumann, R. Landgraf, A. Z. Murphy, and L. J. Young, "Characterization of the Oxytocin System Regulating Affiliative Behavior in Female Prairie Voles," *Neuroscience* 162, no. 4 (September 15, 2009): 892–903.

16. A. G. Ophir, S. M. Phelps, A. B. Sorin, and J. O. Wolff, "Social But Not Genetic Monogamy Is Associated with Greater Breeding Success in Prairie Voles," *Anim Behav* 75 (2008): 1143–54.

17. S. M. Phelps, P. Campbell, D. J. Zheng, and A. G. Ophir, "Beating the Boojum: Comparative Approaches to the Neurobiology of Social Behavior," *Neuropharmacology* 58, no. 1 (January 2010): 17–28.

18. H. Walum, L. Westberg, S. Henningsson, J. M. Neiderhiser, D. Reiss, W. Igl, J. M. Ganiban, E. L. Spotts, N. L. Pedersen, E. Eriksson, and P. Lichtenstein, "Genetic Variation in the Vasopressin Receptor 1a Gene (*AVPR1A*) Associates with Pair-Bonding Behavior in Humans," *Proc Natl Acad Sci USA* 105, no. 37 (September 16, 2008): 14153–56.

19. B. Ditzen, M. Schaer, B. Gabriel, G. Bodenmann, U. Ehlert, and M. Heinrichs, "Intranasal Oxytocin Increases Positive Communication and Reduces Cortisol Levels During Couple Conflict," *Biol Psychiatry* 65, no. 9 (May 1, 2009): 728–31.

20. H. Walum, P. Lichtenstein, J. M. Neiderhiser, D. Reiss, J. M. Ganiban, E. L. Spotts, N. L. Pedersen, H. Anckarsater, H. Larsson, L. Westberg, "Variation in the Oxytocin Receptor Gene Is Associated with Pair-Bonding and Social Behavior," *Biological Psychiatry* (October 17 2011).

21. A. Bartels and S. Zeki, "The Neural Basis of Romantic Love," *Neuroreport* 11, no. 17 (November 27, 2000): 3829–34.

22. A. Bartels and S. Zeki, "The Neural Correlates of Maternal and Romantic Love," *Neuroimage* 21, no. 3 (March 2004): 1155–66.

23. M. D. S. Ainsworth and J. Bowlby, "An Ethological Approach to Personality Development," *Am Psychol* 46, no. 4 (1991): 333–41.

24. J. Bowlby and J. Robertson, "A Two-Year Old Goes to Hospital," *Proc R Soc Med* 46, no. 6 (June 1953): 425–27.

25. H. F. Harlow and R. R. Zimmermann, "Affectional Responses in the Infant Monkey; Orphaned Baby Monkeys Develop a Strong and Persistent Attachment to Inanimate Surrogate Mothers," *Science* 130, no. 3373 (August 21, 1959): 421–32.

26. H. F. Harlow, "The Nature of Love," *Am Psychol* 13 (1958): 673–85.

27. G. A. Barr, S. Moriceau, K. Shionoya, P. Muzny, S. Gao, R. Wang, M. Sullivan. "Transitions in Infant Learning Are Modulated by Dopamine in the Amygdala," *Nat Neurosci* 12, no. 11 (November 2009): 1367–69.

28. R. M. Sullivan and P. J. Holman, "Transitions in Sensitive Period Attachment Learning in Infancy: The Role of Corticosterone," *Neurosci Biobehav Rev* 34, no. 6 (May 2010): 835–844.

29. C. Raineki, S. Moriceau, and R. M. Sullivan, "Developing a Neurobehavioral Animal Model of Infant Attachment to an Abusive Caregiver," *Biol Psychiatry* 67, no. 12 (June 15, 2010): 1137–45.

30. N. Allen, "Jaycee Lee Dugard Showed Signs of Stockholm Syndrome," *Telegraph .co.uk*, November 5, 2009.

31. J. Van Derbeken, "Jaycee Dugard's Anguished Journal Entries," *San Francisco Chronicle*, February 12, 2010.

32. L. Fitzpatrick, "A Brief History of Stockholm Syndrome," *Time*, 2009, http://www.time.com/time/nation/article/0,8599,1919757,00.html.

33. J. A. Simpson and J. Belsky, "Attachment Theory Within a Modern Evolutionary Framework," in *Handbook of Attachment,* 2nd ed., ed. J. Cassidy and P. R. Shaver (New York: Guilford Press, 2008), 131–57.

34. M. D. Ainsworth, "Infant–Mother Attachment," *Am Psychol* 34, no. 10 (October 1979): 932–37.

35. J. Solomon and C. George, "The Measurement of Attachment Security and Related Constructs in Infancy and Early Childhood," in *Handbook of Attachment,* 2nd ed., ed. C. Cassidy and P. R. Shaver (New York: Guilford Press, 2008), 383–416.

36. M. Main and J. Solomon, "Procedures for Identifying Infants as Disorganized/Disoriented During the Ainsworth Strange Situation," in *Attachment in the Preschool Years: Theory, Research and Intervention,* ed. M. T. Greenberg, D. Cichetti, and E. M. Cummings (Chicago: University of Chicago Press, 1990), 121–60.

37. N. W. Boris and C. H. Zeanah, "Practice Parameter for the Assessment and Treatment of Children and Adolescents with Reactive Attachment Disorder of Infancy and Early Childhood," *J Am Acad Child Adolesc Psychiatry* 44, no. 11 (November 2005): 1206–19.

38. J. G. Gunderson, "Borderline Personality Disorder: Ontogeny of a Diagnosis," *Am J Psychiatry* 166, no. 5 (May 2009): 530–39.

39. J. G. Gunderson, R. L. Stout, T. H. McGlashan, M. T. Shea, L. C. Morey, C. M. Grilo, M. C. Zanarini, S. Yen, J. C. Markowitz, C. Sanislow, E. Ansell, A. Pinto, and A. E. Skodol, "Ten-Year Course of Borderline Personality Disorder: Psychopathology and Function from the Collaborative Longitudinal Personality Disorders Study," *Arch Gen Psychiatry* 68, no. 6 (August 2011): 827–37.

40. K. S. Kendler, S. H. Aggen, N. Czajkowski, E. Roysamb, K. Tambs, S. Torgersen, M. C. Neale, and T. Reichborn-Kjennerud, "The Structure of Genetic and Environmental Risk Factors for DSM-IV Personality Disorders: A Multivariate Twin Study," *Arch Gen Psychiatry* 65, no. 12 (December 2008): 1438–46.

41. S. Torgersen, N. Czajkowski, K. Jacobson, T. Reichborn-Kjennerud, E. Roysamb, M. C. Neale, and K. S. Kendler, "Dimensional Representations of DSM-IV Cluster B Personality Disorders in a Population-Based Sample of Norwegian Twins: A Multivariate Study," *Psychol Med* 38, no. 11 (November 2008): 1617–25.

42. S. Torgersen, S. Lygren, P. A. Oien, I. Skre, S. Onstad, J. Edvardsen, K. Tambs, and E. Kringlen, "A Twin Study of Personality Disorders," *Compr Psychiatry* 41, no. 6 (November–December 2000): 416–25.

43. H. R. Agrawal, J. Gunderson, B. M. Holmes, K. Lyons-Ruth, "Attachment Studies with Borderline Patients: A Review," *Harv Rev Psychiatry* 12, no. 2 (March–April 2004): 94–104.

44. M. C. Zanarini and F. R. Frankenburg, "The Essential Nature of Borderline Psychopathology," *J Pers Disord* 21, no. 5 (October 2007): 518–35.

45. J. G. Gunderson and K. Lyons-Ruth, "BPD's Interpersonal Hypersensitivity Phenotype: A Gene-Environment-Developmental Model," *J Pers Disord* 22, no. 1 (February 2008): 22–41.

46. N. H. Donegan, C. A. Sanislow, H. P. Blumberg, R. K. Fulbright, C. Lacadie, P. Skudlarski, J. C. Gore, I. R. Olson, T. H. McGlashan, and B. E. Wexler, "Amygdala Hyperreactivity in Borderline Personality Disorder: Implications for Emotional Dysregulation," *Biol Psychiatry* 54, no. 11 (December 1, 2003): 1284–93.

47. M. J. Minzenberg, J. Fan, A. S. New, C. Y. Tang, and L. J. Siever, "Fronto-Limbic Dysfunction in Response to Facial Emotion in Borderline Personality Disorder: An Event-Related fMRI Study," *Psychiatry Res* 155, no. 3 (August 15, 2007): 231–43.

48. S. Baron-Cohen, S. Wheelwright, J. Hill, Y. Raste, and I. Plumb, "The 'Reading the Mind in the Eyes' Test Revised Version: A Study with Normal Adults, and Adults with Asperger Syndrome or High-Functioning Autism," *J Child Psychol Psychiatry* 42, no. 2 (February 2001): 241–51.

49. E. A. Fertuck, A. Jekal, I. Song, B. Wyman, M. C. Morris, S. T. Wilson, B. S. Brodsky, and B. Stanley, "Enhanced 'Reading the Mind in the Eyes' in Borderline Personality

Disorder Compared to Healthy Controls," *Psychol Med* 39, no. 12 (December 2009): 1979–88.

50. M. C. Zanarini, "Childhood Experiences Associated with the Development of Borderline Personality Disorder," *Psychiatr Clin North Am* 23, no. 1 (March 2000): 89–101.

51. E. H. Erikson, *Identity: Youth and Crisis* (New York: W.W. Norton, 1968).

52. M. Kosfeld, M. Heinrichs, P. J. Zak, U. Fischbacher, and E. Fehr, "Oxytocin Increases Trust in Humans," *Nature* 435, no. 7042 (June 2, 2005): 673–76.

53. J. A. Barraza and P. J. Zak, "Empathy Toward Strangers Triggers Oxytocin Release and Subsequent Generosity," *Ann NY Acad Sci* 1167 (June 2009): 182–89.

54. P. J. Zak, A. A. Stanton, and S. Ahmadi, "Oxytocin Increases Generosity in Humans," *PLoS One* 2, no. 11 (2007): e1128.

55. E. Savaskan, R. Ehrhardt, A. Schulz, M. Walter, and H. Schachinger, "Post-Learning Intranasal Oxytocin Modulates Human Memory for Facial Identity," *Psychoneuroendocrinology* 33, no. 3 (April 2008): 368–74.

56. A. J. Guastella, P. B. Mitchell, and M. R. Dadds, "Oxytocin Increases Gaze to the Eye Region of Human Faces," *Biol Psychiatry* 63, no. 1 (January 1, 2008): 3–5.

57. G. Domes, M. Heinrichs, A. Michel, C. Berger, and S. C. Herpertz, "Oxytocin Improves 'Mind-Reading' in Humans," *Biol Psychiatry* 51, no. 6 (March 15, 2007): 731–33.

58. S. M. Rodrigues, L. R. Saslow, N. Garcia, O. P. John, and D. Keltner, "Oxytocin Receptor Genetic Variation Relates to Empathy and Stress Reactivity in Humans," *Proc Natl Acad Sci USA* 106, no. 50 (December 15, 2009): 21437–41.

59. E. Hollander, J. Bartz, W. Chaplin, A. Phillips, J. Sumner, L. Soorya, E. Anagnostou, and S. Wasserman, "Oxytocin Increases Retention of Social Cognition in Autism," *Biol Psychiatry* 61, no. 4 (February 15, 2007): 498–503.

60. A. J. Guastella, S. L. Einfeld, K. M. Gray, N. J. Rinehart, B. J. Tonge, T. J. Lambert, and I. B. Hickie, "Intranasal Oxytocin Improves Emotion Recognition for Youth with Autism Spectrum Disorders," *Biol Psychiatry* 67, no. 6 (April 1, 2010): 692–94.

61. D. Huber, P. Veinante, and R. Stoop, "Vasopressin and Oxytocin Excite Distinct Neuronal Populations in the Central Amygdala," *Science* 308, no. 5719 (April 8, 2005): 245–48.

62. P. Kirsch, C. Esslinger, Q. Chen, D. Mier, S. Lis, S. Siddhanti, H. Gruppe, V. S. Mattay, B. Gallhofer, and A. Meyer-Lindenberg, "Oxytocin Modulates Neural Circuitry for Social Cognition and Fear in Humans," *J Neurosci* 25, no. 49 (December 7, 2005): 11489–93.

63. G. Domes, M. Heinrichs, J. Glascher, C. Buchel, D. F. Braus, and S. C. Herpertz, "Oxytocin Attenuates Amygdala Responses to Emotional Faces Regardless of Valence," *Biol Psychiatry* 62, no. 10 (November 15, 2007): 1187–90.

64. T. Baumgartner, M. Heinrichs, A. Vonlanthen, U. Fischbacher, and E. Fehr, "Oxytocin Shapes the Neural Circuitry of Trust and Trust Adaptation in Humans," *Neuron* 58, no. 4 (May 22, 2008): 639–50.

65. M. Di Simplicio, R. Massey-Chase, P. J. Cowen, and C. J. Harmer, "Oxytocin Enhances Processing of Positive Versus Negative Emotional Information in Healthy Male Volunteers," *J Psychopharmacol* 23, no. 3 (May 2009): 241–48.

66. A. J. Guastella, P. B. Mitchell, and F. Mathews, "Oxytocin Enhances the Encoding of Positive Social Memories in Humans," *Biol Psychiatry* 64, no. 3 (August 1 2008): 256–58.

67. S. D. Pollak and D. J. Kistler, "Early Experience Is Associated with the Development of Categorical Representations for Facial Expressions of Emotion," *Proc Natl Acad Sci USA* 99, no. 13 (June 25, 2002): 9072–76.

68. G. J. Dumont, F. C. Sweep, R. van der Steen, R. Hermsen, A. R. Donders, D. J. Touw, J. M. van Gerven, J. K. Buitelaar, and R. J. Verkes, "Increased Oxytocin Concentrations and Prosocial Feelings in Humans After Ecstasy (3,4-Methylenedioxymethamphetamine) Administration," *Soc Neurosci* 4, no. 4 (2009): 359–66.

69. M. R. Thompson, P. D. Callaghan, G. E. Hunt, J. L. Cornish, and I. S. McGregor, "A Role for Oxytocin and 5-HT(1A) Receptors in the Prosocial Effects of 3,4 Methylene-dioxymethamphetamine ('Ecstasy')," *Neuroscience* 146, no. 2 (May 11, 2007): 509–14.

70. J. S. Winston, B. A. Strange, J. O'Doherty, and R. J. Dolan, "Automatic and Intentional Brain Responses During Evaluation of Trustworthiness of Faces," *Nat Neurosci* 5, no. 3 (March 2002): 277–83.

71. R. Adolphs, D. Tranel, and A. R. Damasio, "The Human Amygdala in Social Judgment," *Nature* 393, no. 6684 (June 4, 1998): 470–74.

72. B. King-Casas, D. Tomlin, C. Anen, C. F. Camerer, S. R. Quartz, and P. R. Montague, "Getting to Know You: Reputation and Trust in a Two-Person Economic Exchange," *Science* 308, no. 5718 (April 1, 2005): 78–83.

73. E. Krumhuber, A. S. Manstead, D. Cosker, D. Marshall, P. L. Rosin, and A. Kappas, "Facial Dynamics as Indicators of Trustworthiness and Cooperative Behavior," *Emotion* 7, no. 4 (November 2007): 730–35.

74. L. Cosmides and J. Tooby, "Neurocognitive Adaptations Designed for Social Exchange," in *The Handbook of Evolutionary Psychology,* ed. D. M. Buss (Hoboken, NJ: John Wiley and Sons, 2005), 584–627.

75. P. L. Harris, M. Nunez, and C. Brett, "Let's Swap: Early Understanding of Social Exchange by British and Nepali Children," *Mem Cognit* 29, no. 5 (July 2001): 757–64.

76. L. S. Sugiyama, J. Tooby, and L. Cosmides, "Cross-Cultural Evidence of Cognitive Adaptations for Social Exchange Among the Shiwiar of Ecuadorian Amazonia," *Proc Natl Acad Sci USA* 99, no. 17 (August 20, 2002): 11537–42.

77. V. E. Stone, L. Cosmides, J. Tooby, N. Kroll, and R. T. Knight, "Selective Impairment of Reasoning About Social Exchange in a Patient with Bilateral Limbic System Damage," *Proc Natl Acad Sci USA* 99, no. 17 (August 20, 2002): 11531–36.

78. F. Krueger, K. McCabe, J. Moll, N. Kriegeskorte, R. Zahn, M. Strenziok, A. Heinecke, and J. Grafman, "Neural Correlates of Trust," *Proc Natl Acad Sci USA* 104, no. 50 (December 11, 2007): 20084–89.

79. J. Bartz, D. Simeon, H. Hamilton, S. Kim, S. Crystal, A. Braun, V. Vicens, E. Hollander, "Oxytocin Can Hinder Trust and Cooperation in Borderline Personality Disorder," *Social Cognitive and Affective Neuroscience* 6, no. 5 (October 2011): 556–63.

80. E. Andari, J. R. Duhamel, T. Zalla, E. Herbrecht, M. Leboyer, and A. Sirigu, "Promoting Social Behavior with Oxytocin in High-Functioning Autism Spectrum Disorders," *Proc Natl Acad Sci USA* 107, no. 9 (March 2, 2010): 4389–94.

CHAPTER 6: THE BRAIN OF THE BEHOLDER

1. D. Plotnikoff, "'Am I Hot or Not'" Is One Hot Site," Knight Ridder/Tribune News Service, January 30, 2001.

2. B. M. Schwartz, "Hot or Not? Website Briefly Judges Looks," *Harvard Crimson,* November 4, 2003.

3. D. Denby, *Snark* (New York: Simon & Schuster, 2009).

4. N. Wolf, *The Beauty Myth: How Images of Beauty Are Used Against Women* (New York: Harper Perennial, 2002).

5. T. Seifert, "Anthropomorphic Characteristics of Centerfold Models: Trends Towards Slender Figures Over Time," *Int J Eat Disord* 37, no. 3 (April 2005): 271–74.

6. H. G. Pope Jr., R. Olivardia, A. Gruber, and J. Borowiecki, "Evolving Ideals of Male Body Image as Seen Through Action Toys," *Int J Eat Disord* 26, no. 1 (July 1999): 65–72.

7. S. Grabe, L. M. Ward, and J. S. Hyde, "The Role of the Media in Body Image Concerns Among Women: A Meta-Analysis of Experimental and Correlational Studies," *Psychol Bull* 134, no. 3 (May 2008): 460–76.

8. D. J. Buller, *Adapting Minds: Evolutionary Psychology and the Persistent Quest for Human Nature* (Cambridge, MA: MIT Press, 2006).

9. R. L. Trivers, "Parental Investment and Sexual Selection," in *Sexual Selection and the*

Descent of Man 1871–1971, ed. B. Campbell (Chicago, IL: Aldine Publishing Company, 1972), 136–207.

10. J. L. Brown, V. Morales, and K. Summers, "A Key Ecological Trait Drove the Evolution of Biparental Care and Monogamy in an Amphibian," *American Naturalist* 175 (2010): 436–46.

11. T. Clutton-Brock, "Sexual Selection in Males and Females," *Science* 318, no. 5858 (December 21, 2007): 1882–85.

12. D. P. Schmitt, "Fundamentals of Human Mating Strategies," in *The Handbook of Evolutionary Psychology*, ed. D. M. Buss (Hoboken, NJ: John Wiley & Sons, Inc., 2005), 258–91.

13. R. D. Clark and E. Hatfield, "Gender Differences in Receptivity to Sexual Offers," *J Psychol Hum Sexuality* 2 (1989): 39–55.

14. D. P. Schmitt, "Sociosexuality from Argentina to Zimbabwe: A 48-Nation Study of sex, Culture, and Strategies of Human Mating," *Behav Brain Sci* 28, no. 2 (April 2005): 247–75; discussion 275–311.

15. G. Rhodes, "The Evolutionary Psychology of Facial Beauty," *Annu Rev Psychol* 57 (2006): 199–226.

16. J. Schwartz, "Just Average, and Therein Lay His Greatness," *New York Times*, November 16, 2004.

17. Royal Veterinary College, "Why Was the Racehorse Eclipse So Good?" *Science Daily* (June 14, 2007). Retrieved March 26, 2010, from http://www.sciencedaily.com/releases/2007/06/070611134032.htm.

18. A. Peters, "Eclipse," http://www.tbheritage.com/Portraits/Eclipse.html, accessed March 26, 2010.

19. F. Galton, *Inquiries into Human Faculty and Its Development* (London: J.M. Dent & Sons, Ltd., 1907 reprinted by Dodo Press).

20. H. C. Lie, L. W. Simmons, and G. Rhodes, "Does Genetic Diversity Predict Health in Humans?" *PLoS One* 4, no. 7 (2009): e6391.

21. H. C. Lie, G. Rhodes, and L. W. Simmons, "Genetic Diversity Revealed in Human Faces," *Evolution* 62, no. 10 (October 2008): 2473–86.

22. R. Chaix, C. Cao, and P. Donnelly, "Is Mate Choice in Humans MHC-Dependent?" *PLoS Genet* 4, no. 9 (2008): e1000184.

23. J. Havlicek and S. C. Roberts, "MHC-Correlated Mate Choice in Humans: A Review," *Psychoneuroendocrinology* 34, no. 4 (May 2009): 497–512.

24. S. C. Roberts and A. C. Little, "Good Genes, Complementary Genes and Human Mate Preferences," *Genetica* 134, no. 1 (September 2008): 31–43.

25. C. E. Garver-Apgar, S. W. Gangestad, R. Thornhill, R. D. Miller, and J. J. Olp, "Major Histocompatibility Complex Alleles, Sexual Responsivity, and Unfaithfulness in Romantic Couples," *Psychol Sci* 17, no. 10 (October 2006): 830–35.

26. W. M. Brown, M. E. Price, J. Kang, N. Pound, Y. Zhao, and H. Yu, "Fluctuating Asymmetry and Preferences for Sex-Typical Bodily Characteristics," *Proc Natl Acad Sci USA* 105, no. 35 (September 2, 2008): 12938–43.

27. S. Van Dongen, S. W. Gangestad, "Human Fluctuating Asymmetry in Relation to Health and Quality: A Meta-Analysis," *Evolution and Human Behavior* 32 (2011): 380–398.

28. I. N. Springer, B. Wannicke, P. H. Warnke, O. Zernial, J. Wiltfang, P. A. Russo, H. Terheyden, A. Reinhardt, and S. Wolfart, "Facial Attractiveness: Visual Impact of Symmetry Increases Significantly Towards the Midline," *Ann Plast Surg* 59, no. 2 (August 2007): 156–62.

29. G. Jasienska, A. Ziomkiewicz, P. T. Ellison, S. F. Lipson , and I. Thune, "Large Breasts and Narrow Waists Indicate High Reproductive Potential in Women," *Proc Biol Sci* 271, no. 1545 (June 22, 2004): 1213–17.

30. I. S. Penton-Voak, D. I. Perrett, D. L. Castles, T. Kobayashi, D. M. Burt, L. K. Murray, and R. Minamisawa, "Menstrual Cycle Alters Face Preference," *Nature* 399, no. 6738 (June 24, 1999): 741–42.

31. S. W. Gangestad, R. Thornhill, and C. E. Garver-Apgar, "Fertility in the cycle predicts women's interest in sexual opportunism," *Evol Hum Behav* 31 (2010): 400–11.

32. R. N. Pipitone and G. G. J. Gallup, "Women's Voice Attractiveness Varies Across the Menstrual Cycle," *Evol Hum Behav* 29 (2008): 268–74.

33. G. A. Bryant and M. G. Haselton, "Vocal Cues of Ovulation in Human Females," *Biol Lett* 5, no. 1 (February 23, 2009): 12–15.

34. S. C. Roberts, J. Havlicek, J. Flegr, M. Hruskova, A. C. Little, B. C. Jones, D. I. Perrett, and M. Petrie, "Female Facial Attractiveness Increases During the Fertile Phase of the Menstrual Cycle," *Proc Biol Sci* 271, suppl. 5 (August 7, 2004): S270–72.

35. K. Grammer, L. Renninger, and B. Fischer, "Disco Clothing, Female Sexual Motivation, and Relationship Status: Is She Dressed to Impress?" *J Sex Res* 41, no. 1 (February 2004): 66–74.

36. K. M. Durante, N. P. Li, and M. G. Haselton, "Changes in Women's Choice of Dress Across the Ovulatory Cycle: Naturalistic and Laboratory Task-Based Evidence," *Pers Soc Psychol Bull* 34, no. 11 (November 2008): 1451–60.

37. M. G. Haselton and S. W. Gangestad, "Conditional Expression of Women's Desires and Men's Mate Guarding Across the Ovulatory Cycle," *Horm Behav* 49, no. 4 (April 2006): 509–18.

38. M. G. Haselton, M. Mortezaie, E. G. Pillsworth, A. Bleske-Rechek, and D. A. Frederick, "Ovulatory Shifts in Human Female Ornamentation: Near Ovulation, Women Dress to Impress," *Horm Behav* 51, no. 1 (January 2007): 40–45.

39. S. W. Gangestad, R. Thornhill, and C. E. Garver, "Changes in Women's Sexual Interests and Their Partners' Mate-Retention Tactics Across the Menstrual Cycle: Evidence for Shifting Conflicts of Interest," *Proc Biol Sci* 269, no. 1494 (May 7, 2002): 975–82.

40. A. Alvergne and V. Lummaa, "Does the Contraceptive Pill Alter Mate Choice in Humans?" *Trends Ecol Evol* 25, no. 3 (March 2010): 171–79.

41. S. C. Roberts, K. Klapilova, A. C. Little, R. P. Burriss, B. C. Jones, L. M. Debruine, M. Petrie, J. Havlicek, "Relationship Satisfaction and Outcome in Women Who Meet Their Partner While Using Oral Contraception," *Proceedings. Biological Sciences / The Royal Society* (October 12 2011).

42. C. L. Apicella, A. C. Little, and F. W. Marlowe, "Facial Averageness and Attractiveness in an Isolated Population of Hunter-Gatherers," *Perception* 36, no. 12 (2007): 1813–20.

43. A. C. Little, C. L. Apicella, and F. W. Marlowe, "Preferences for Symmetry in Human Faces in Two Cultures: Data from the UK and the Hadza, an Isolated Group of Hunter-Gatherers," *Proc Biol Sci* 274, no. 1629 (December 22, 2007): 3113–17.

44. N. Tinbergen and A. C. Perdeck, "On the Stimulus Situation Releasing the Begging Response in the Newly Hatched Herring Gull Chick," *Behaviour* 3 (1950): 1–39.

45. J. E. R. Staddon, "A Note on the Evolutionary Significance of 'Supernormal' Stimuli,' *Am Nat* 109 (1975): 541–45.

46. D. Barrett, *Supernormal Stimuli: How Primal Urges Overran Their Evolutionary Purpose* (New York: W.W. Norton, 2010).

47. Angus Reid Public Opinion, *Americans Remain Divided on Allowing Gays and Lesbians to Marry,* Vision Critical, December 17, 2009.

48. R. Herrn, "On the History of Biological Theories of Homosexuality," *J Homosex* 28, no. 1–2 (1995): 31–56.

49. Family Research Council, "Homosexuality," http://www.frc.org/human-sexuality #homosexuality. Accessed April 9, 2010.

50. P. Sprigg and T. Dailey, *Getting It Straight: What the Research Shows About Homosexuality* (Washington, DC: Family Research Council, 2004).

51. R. T. Michael, J. H. Gagnon, E. O. Laumann, and G. Kolata, *Sex in America: A Definitive Survey* (New York: Warner Books, 1994).

52. W. D. Mosher, A. Chandra, and J. Jones, *Sexual Behavior and Selected Health Measures: Men and Women 15–44 Years Of Age. United States, 2002. Advance Data from*

Vital and Health Statistics: No. 362 (Hyattsville, MD: National Center for Health Statistics, 2005).

53. N. W. Bailey and M. Zuk, "Same-Sex Sexual Behavior and Evolution," *Trends Ecol Evol* 24, no. 8 (August 2009): 439–46.

54. R. Halwani, "Essentialism, Social Constructionism, and the History of Homosexuality," *J Homosex* 35, no. 1 (1998): 25–51.

55. Q. Rahman and M. S. Hull, "An Empirical Test of the Kin Selection Hypothesis for Male Homosexuality," *Arch Sex Behav* 34, no. 4 (August 2005): 461–67.

56. D. Bobrow and J. M. Bailey, "Is Male Homosexuality Maintained Via Kin Selection?" *Evol Hum Behav* 22 (2001): 361–68.

57. P. L. Vasey and N. H. Bartlett, "What Can the Samoan 'Fa'afafine' Teach Us About the Western Concept of Gender Identity Disorder in Childhood?" *Perspect Biol Med* 50, no. 4 (Autumn 2007): 481–90.

58. P. L. Vasey and D. P. VanderLaan, "An Adaptive Cognitive Dissociation Between Willingness to Help Kin and Nonkin in Samoan Fa'afafine," *Psychol Sci* [epub ahead of print].

59. D. P. Vanderlaan and P. L. Vasey, "Male Sexual Orientation in Independent Samoa: Evidence for Fraternal Birth Order and Maternal Fecundity Effects," *Arch Sex Behav* 40, no. 3 (June 2011): 495–503.

60. F. Iemmola and A. Camperio Ciani, "New Evidence of Genetic Factors Influencing Sexual Orientation in Men: Female Fecundity Increase in the Maternal Line," *Arch Sex Behav* 38, no. 3 (June 2009): 393–99.

61. Q. Rahman, A. Collins, M. Morrison, J. C. Orrells, K. Cadinouche, S. Greenfield, and S. Begum, "Maternal Inheritance and Familial Fecundity Factors in Male Homosexuality," *Arch Sex Behav* 37, no. 6 (December 2008): 962–69.

62. A. Villella and J. C. Hall, "Neurogenetics of Courtship and Mating in Drosophila," *Adv Genet* 62 (2008): 67–184.

63. E. Demir and B. J. Dickson, "Fruitless Splicing Specifies Male Courtship Behavior in Drosophila," *Cell* 121, no. 5 (June 3, 2005): 785–94.

64. D. S. Manoli, M. Foss, A. Villella, B. J. Taylor, J. C. Hall, and B. S. Baker, "Male-Specific Fruitless Specifies the Neural Substrates of Drosophila Courtship Behaviour," *Nature* 436, no. 7048 (July 21, 2005): 395–400.

65. J. M. Bailey, R. C. Pillard, K. Dawood, M. B. Miller, L. A. Farrer, S. Trivedi, and R. L. Murphy, "A Family History Study of Male Sexual Orientation Using Three Independent Samples," *Behav Genet* 29, no. 2 (March 1999): 79–86.

66. R. C. Pillard, J. Poumadere, and R. A. Carretta, "A Family Study of Sexual Orientation," *Arch Sex Behav* 11, no. 6 (December 1982): 511–20.

67. R. C. Pillard and J. D. Weinrich, "Evidence of Familial Nature of Male Homosexuality," *Arch Gen Psychiatry* 43, no. 8 (August 1986): 808–12.

68. C. J. Patterson, "Children of Lesbian and Gay Parents: Psychology, Law, and Policy," *Am Psychol* 64, no. 8 (November 2009): 727–36.

69. J. M. Bailey and R. C. Pillard, "A Genetic Study of Male Sexual Orientation," *Arch Gen Psychiatry* 48 (1991): 1089–96.

70. J. M. Bailey, R. C. Pillard, M. C. Neale, and Y. Agyei, "Heritable Factors Influence Sexual Orientation in Women," *Arch Gen Psychiatry* 50 (1993): 217–23.

71. N. Langstrom, Q. Rahman, E. Carlstrom, and P. Lichtenstein, "Genetic and Environmental Effects on Same-Sex Sexual Behavior: A Population Study of Twins in Sweden," *Arch Sex Behav* 39, no. 1 (February 2010): 75–80.

72. K. S. Kendler, L. M. Thornton, S. E. Gilman, and R. C. Kessler, "Sexual Orientation in a U.S. National Sample of Twin and Nontwin Sibling Pairs," *Am J Psychiatry* 157, no. 11 (November 2000): 1843–46.

73. B. S. Mustanski, M. G. Dupree, C. M. Nievergelt, S. Bocklandt, N. J. Schork, and D. H. Hamer, "A Genomewide Scan of Male Sexual Orientation," *Hum Genet* 116, no. 4 (March 2005): 272–78.

74. S. V. Ramagopalan, D. A. Dyment, L. Handunnetthi, G. P. Rice, and G. C. Ebers, "A

Genome-Wide Scan of Male Sexual Orientation," *J Hum Genet* 55, no. 2 (February 2010): 131–32.

75. M. G. DuPree, B. S. Mustanski, S. Bocklandt, C. Nievergelt, and D. H. Hamer, "A Candidate Gene Study of CYP19 (Aromatase) and Male Sexual Orientation," *Behav Genet* 34, no. 3 (May 2004): 243–50.

76. D. H. Hamer, S. Hu, V. L. Magnuson, N. Hu, and A. M. Pattatucci, "A Linkage Between DNA Markers on the X Chromosome and Male Sexual Orientation," *Science* 261, no. 5119 (July 16, 1993): 321–27.

77. S. Hu, A. M. Pattatucci, C. Patterson, L. Li, D. W. Fulker, S. S. Cherny, L. Kruglyak, and D. H. Hamer, "Linkage Between Sexual Orientation and Chromosome Xq28 in Males But Not in Females," *Nat Genet* 11, no. 3 (November 1995): 248–56.

78. A. F. Bogaert, "Biological Versus Nonbiological Older Brothers and Men's Sexual Orientation," *Proc Natl Acad Sci USA* 103, no. 28 (July 11, 2006): 10771–74.

79. R. Blanchard, "Quantitative and Theoretical Analyses of the Relation Between Older Brothers and Homosexuality in Men," *J Theor Biol* 230, no. 2 (September 21, 2004): 173–87.

80. P. L. Vasey and D. P. VanderLaan, "Birth Order and Male Androphilia in Samoan Fa'afafine," *Proc Biol Sci* 274, no. 1616 (June 7, 2007): 1437–42.

81. J. M. Cantor, R. Blanchard, A. D. Paterson, and A. F. Bogaert, "How Many Gay Men Owe Their Sexual Orientation to Fraternal Birth Order?" *Arch Sex Behav* 31, no. 1 (February 2002): 63–71.

82. R. Blanchard and A. F. Bogaert, "Proportion of Homosexual Men Who Owe Their Sexual Orientation to Fraternal Birth Order: An Estimate Based on Two National Probability Samples," *Am J Hum Biol* 16, no. 2 (March–April 2004): 151–57.

83. J. Balthazart, "Minireview: Hormones and Human Sexual Orientation," *Endocrinology* 52, no. 8 (August 2011): 2937–47.

84. T. Grimbos, K. Dawood, R. P. Burriss, K. J. Zucker, and D. A. Puts, "Sexual Orientation and the Second to Fourth Finger Length Ratio: A Meta-Analysis in Men and Women," *Behav Neurosci* 124, no. 2 (April 2010): 278–87.

85. I. Aharon, N. Etcoff, D. Ariely, C. F. Chabris, E. O'Connor, and H. C. Breiter, "Beautiful Faces Have Variable Reward Value: fMRI and Behavioral Evidence," *Neuron* 32, no. 3 (November 8, 2001): 537–51.

86. J. S. Winston, J. O'Doherty, J. M. Kilner, D. I. Perrett, and R. J. Dolan, "Brain Systems for Assessing Facial Attractiveness," *Neuropsychologia* 45, no. 1 (January 7, 2007): 195–206.

87. J. O'Doherty, J. Winston, H. Critchley, D. Perrett, D. M. Burt, and R. J. Dolan, "Beauty in a Smile: The Role of Medial Orbitofrontal Cortex in Facial Attractiveness," *Neuropsychologia* 41, no. 2 (2003): 147–55.

88. J. Cloutier, T. F. Heatherton, P. J. Whalen, and W. M. Kelley, "Are Attractive People Rewarding? Sex Differences in the Neural Substrates of Facial Attractiveness," *J Cogn Neurosci* 20, no. 6 (June 2008): 941–51.

89. K. C. Berridge and M. L. Kringelbach, "Affective Neuroscience of Pleasure: Reward in Humans and Animals," *Psychopharmacology (Berl)* 199, no. 3 (August 2008): 457–80.

90. S. N. Haber and B. Knutson, "The Reward Circuit: Linking Primate Anatomy and Human Imaging," *Neuropsychopharmacology* 35, no. 1 (January 2010): 4–26.

91. F. Kranz and A. Ishai, "Face Perception Is Modulated by Sexual Preference," *Curr Biol* 16, no. 1 (January 10, 2006): 63–68.

92. L. S. Sugiyama, "Physical Attractiveness in Adaptationist Perspective," in *Handbook of Evolutionary Psychology*, ed. D. M. Buss (Hoboken, NJ: John Wiley & Sons, 2005), 292–343.

93. D. Singh, "Mating Strategies of Young Women: Role of Physical Attractiveness," *J Sex Res* 41, no. 1 (February 2004): 43–54.

94. D. Singh and P. K. Randall, "Beauty Is in the Eye of the Plastic Surgeon: Waist-Hip

Ratio (WHR) and Women's Attractiveness," *Personality and Individual Differences* 43 (2007): 329–40.

95. D. Singh, "Adaptive Significance of Female Physical Attractiveness: Role of Waist-to-Hip Ratio," *J Pers Soc Psychol* 65, no. 2 (August 1993): 293–307.

96. F. Marlowe, C. Apicella, and D. Reed, "Men's Preferences for Women's Profile Waist-to-Hip Ratio in Two Societies," *Evol Hum Behav* 26 (2005): 458–68.

97. B. J. Dixson, A. F. Dixson, P. J. Bishop, and A. Parish, "Human Physique and Sexual Attractiveness in Men and Women: A New Zealand–U.S. Comparative Study," *Arch Sex Behav* 39, no. 3 (June 2010): 798–806.

98. B. J. Dixson, K. Sagata, W. L. Linklater, and A. F. Dixson, "Male Preferences for Female Waist-to-Hip Ratio and Body Mass Index in the Highlands of Papua New Guinea," *Am J Phys Anthropol* 141, no. 4 (April 2010): 620–25.

99. B. J. Dixson, G. M. Grimshaw, W. L. Linklater, and A. F. Dixson, "Eye-Tracking of Men's Preferences for Waist-to-Hip Ratio and Breast Size of Women," *Arch Sex Behav* 40, no. 1 (February 2011): 43–50.

100. J. C. Karremans, W. E. Frankenhuis, and S. Arons, "Blind Men Prefer a Low Waist-to-Hip Ratio," *Evol Hum Behav* 31 (2010): 182–86.

101. S. M. Platek and D. Singh, "Optimal Waist-to-Hip Ratios in Women Activate Neural Reward Centers in Men," *PLoS One* 5, no. 2 (February 5, 2010): e9042.

102. R. A. Lippa, "The Preferred Traits of Mates in a Cross-National Study of Heterosexual and Homosexual Men and Women: An Examination of Biological and Cultural Influences," *Arch Sex Behav* 36, no. 2 (April 2007): 193–208.

103. A. Goodman. "Sexual Addiction: Diagnosis and Treatment," *Psychiatric Times,* 15, no. 10 (October 1, 1998), http://www.psychiatrictimes.com/sexual-addiction/content/article/10168/55141.

104. American Psychiatric Association, *Diagnostic and Statistical Manual of Mental Disorders,* 4th ed., rev. ed. (Washington, DC: American Psychiatric Association, 2000).

105. C. Scorolli, S. Ghirlanda, M. Enquist, S. Zattoni, and E. A. Jannini, "Relative Prevalence of Different Fetishes," *Int J Impotence Res* 19 (2007): 432–37.

106. J. Ropelato. "Top 10 Internet Pornography Statistics," *TopTenReviews,* 2006.

107. J. S. Carroll, L. M. Padilla-Walker, L. J. Nelson, C. D. Olson, C. McNamara Barry, and S. D. Madsen, "Generation XXX: Pornography Acceptance and Use Among Emerging Adults," *J Adolesc Res* 23 (2008): 6–30.

108. "Nielsen on Internet Porn," 2003, http://www.itfacts.biz/nielsen-on-internet-porn/246.

109. S. Hamann, R. A. Herman, C. L. Nolan, and K. Wallen. "Men and Women Differ in Amygdala Response to Visual Sexual Stimuli," *Nat Neurosci* 7, no. 4 (April 2004): 411–16.

110. J. Ponseti, H. A. Bosinski, S. Wolff, M. Peller, O. Jansen, H. M. Mehdorn, C. Buchel, and H. R. Siebner, "A Functional Endophenotype for Sexual Orientation in Humans," *Neuroimage* 33, no. 3 (November 15, 2006): 825–33.

111. M. Leahy, *Porn Nation: Conquering America's #1 Addiction* (Chicago, IL: Northfield Publishing, 2008).

112. M. P. Kafka, "Hypersexual Disorder: A Proposed Diagnosis for DSM-V," *Arch Sex Behav* 39, no. 2 (April 2010): 377–400.

113. A. N. Gearhardt, S. Yokum, P. T. Orr, E. Stice, W. R. Corbin, and K. D. Brownell, "Neural Correlates of Food Addiction," *Arch Gen Psychiatry* 68, no. 8 (August 2011): 808–16.

114. P. M. Johnson and P. J. Kenny, "Dopamine D2 Receptors in Addiction-like Reward Dysfunction and Compulsive Eating in Obese Rats," *Nat Neurosci* 13, no. 5 (May 2010): 635–41.

115. A. F. Bogaert, "Asexuality: Prevalence and Associated Factors in a National Probability Sample," *J Sex Res* 41, no. 3 (August 2004): 279–87.

116. N. Prause and C. A. Graham, "Asexuality: Classification and Characterization," *Arch Sex Behav* 36, no. 3 (June 2007): 341–56.

CHAPTER 7: REMEMBERING TO FORGET

1. H. Cantril, *The Invasion from Mars: A Study in the Psychology of Panic* (Princeton, NJ: Princeton University Press, 2008).
2. B. English, "A Matter of Scales: An Ophidiophobic Tries for a New Perspective," *Boston Globe,* January 6, 2009.
3. H. S. Bracha, "Human Brain Evolution and the 'Neuroevolutionary Time-Depth Principle': Implications for the Reclassification of Fear-Circuitry-Related Traits in DSM-V and for Studying Resilience to Warzone-Related Posttraumatic Stress Disorder," *Prog Neuropsychopharmacol Biol Psychiatry* 30, no. 5 (July 2006): 827–53.
4. D. J. Stein and C. Bouwer, "A Neuro-Evolutionary Approach to the Anxiety Disorders," *J Anxiety Disord* 11, no. 4 (July–August 1997): 409–29.
5. T. Cherones, "The Pilot," *Seinfeld,* May 20, 1993
6. T. Tully, "Pavlov's Dogs," *Curr Biol* 13 (2003): R117-19.
7. J. B. Watson, "Psychology as the Behaviorist Sees It," *Psychol Rev* 20 (1920): 158–77.
8. J. B. Watson, *Behaviorism* (New Brunswick, NJ: Transaction Publishers, 1998).
9. J. B. Watson and R. Raynor, "Conditioned Emotional Reactions," *J Exper Psychol* 3 (1920): 1–14.
10. J. LeDoux, *The Emotional Brain* (New York: Simon & Schuster, 1996).
11. H. C. Pape and D. Pare, "Plastic Synaptic Networks of the Amygdala for the Acquisition, Expression, and Extinction of Conditioned Fear," *Physiol Rev* 90, no. 2 (April 2010): 419–63.
12. B. Roozendaal, B. S. McEwen, and S. Chattarji, "Stress, Memory and the Amygdala," *Nat Rev Neurosci* 10, no. 6 (June 2009): 423–33.
13. G. J. Quirk and D. Mueller, "Neural Mechanisms of Extinction Learning and Retrieval," *Neuropsychopharmacology* 33, no. 1 (January 2008): 56–72.
14. L. M. Shin and I. Liberzon, "The Neurocircuitry of Fear, Stress, and Anxiety Disorders," *Neuropsychopharmacology* 35, no. 1 (January 2010): 169–91.
15. C. Herry, S. Ciocchi, V. Senn, L. Demmou, C. Muller, and A. Luthi, "Switching On and Off Fear by Distinct Neuronal Circuits," *Nature* 454, no. 7204 (July 31, 2008): 600–606.
16. T. Amano, C. T. Unal, and D. Pare, "Synaptic Correlates of Fear Extinction in the Amygdala," *Nat Neurosci* 13, no. 4 (April 2010): 489–94.
17. M. Davis, D. L. Walker, L. Miles, and C. Grillon, "Phasic Vs Sustained Fear in Rats and Humans: Role of the Extended Amygdala in Fear Vs Anxiety," *Neuropsychopharmacology* 35, no. 1 (January 2010): 105–35.
18. W. B. Cannon, *Bodily Changes in Pain, Hunger, Fear, and Rage* (New York: D. Appleton and Company, 1915).
19. W. B. Cannon, "Voodoo Death," *Psychosom Med* 19, no 3 (May–June 1957): 182–90.
20. M. A. Samuels, "'Voodoo' Death Revisited: The Modern Lessons of Neurocardiology," *Cleve Clin J Med* 74, suppl. 1 (February 2007): S8–16.
21. C. M. Albert, C. U. Chae, K. M. Rexrode, J. E. Manson, and I. Kawachi, "Phobic Anxiety and Risk of Coronary Heart Disease and Sudden Cardiac Death Among Women," *Circulation* 111, no. 4 (February 1, 2005): 480–87.
22. L. D. Kubzansky, I. Kawachi, A. Spiro III, S. T. Weiss, P. S. Vokonas, and D. Sparrow, "Is Worrying Bad for Your Heart? A Prospective Study of Worry and Coronary Heart Disease in the Normative Aging Study," *Circulation* 95, no. 4 (1997): 818–24.
23. J. W. Smoller, M. H. Pollack, S. Wassertheil-Smoller, R. D. Jackson, A. Oberman, N. D. Wong, and D. Sheps, "Panic Attacks and Risk of Incident Cardiovascular Events Among Postmenopausal Women in the Women's Health Initiative Observational Study," *Arch Gen Psychiatry* 64, no. 10 (October 2007): 1153–60.
24. G. Berrios, "Anxiety Disorders: A Conceptual History," *J Affect Disord* 56, no. 2–3 (December 1999): 83–94.
25. S. Freud, "The Justification for Detaching from Neurasthenia a Particular Syndrome:

The Anxiety-Neurosis (1894)," in *Early Psychoanalytic Writings*, ed. P. Rieff (New York: Macmillan Publishing Company, 1963): 93–117.

26. S. Freud, "Heredity and the Aetiology of the Neuroses (1896)," in *Early Psychoanalytic Writings*, ed. P. Rieff (New York: Macmillan Publishing Company, 1963: 137–50.

27. E. Jones. *The Life and Work of Sigmund Freud*, Vol. 3 (New York: Basic Books, 1957).

28. R. C. Kessler, P. Berglund, O. Demler, R. Jin, K. R. Merikangas, and E. E. Walters. "Lifetime Prevalence and Age-of-Onset Distributions of DSM-IV Disorders in the National Comorbidity Survey Replication," *Arch Gen Psychiatry* 62, no. 6 (June 2005): 593–602.

29. K. R. Merikangas, M. Ames, L. Cui, P. E. Stang, T. B. Ustun, M. Von Korff, and R. C. Kessler, "The Impact of Comorbidity of Mental and Physical Conditions on Role Disability in the US Adult Household Population," *Arch Gen Psychiatry* 64, no. 10 (October 2007): 1180–88.

30. S. Lissek, A. S. Powers, E. B. McClure, E. A. Phelps, G. Woldehawariat, C. Grillon, and D. S. Pine, "Classical Fear Conditioning in the Anxiety Disorders: A Meta-Analysis," *Behav Res Ther* 43, no. 11 (November 2005): 1391–1424.

31. S. Lissek, S. Rabin, R. E. Heller, D. Lukenbaugh, M. Geraci, D. S. Pine, and C. Grillon, "Overgeneralization of Conditioned Fear as a Pathogenic Marker of Panic Disorder," *Am J Psychiatry* 167, no. 1 (January 2010): 47–55.

32. A. Etkin and T. D. Wager, "Functional Neuroimaging of Anxiety: A Meta-Analysis of Emotional Processing in PTSD, Social Anxiety Disorder, and Specific Phobia," *Am J Psychiatry* 164, no. 10 (October 2007): 1476–88.

33. M. R. Milad, B. T. Quinn, R. K. Pitman, S. P. Orr, B. Fischl, and S. L. Rauch, "Thickness of Ventromedial Prefrontal Cortex in Humans Is Correlated with Extinction Memory," *Proc Natl Acad Sci USA* 102, no. 30 (July 26, 2005): 10706–11.

34. M. W. Gilbertson, M. E. Shenton, A. Ciszewski, K. Kasai, N. B. Lasko, S. P. Orr, and R. K. Pitman, "Smaller Hippocampal Volume Predicts Pathologic Vulnerability to Psychological Trauma," *Nat Neurosci* 5, no. 11 (November 2002): 1242–47.

35. S. L. Rauch, L. M. Shin, E. Segal, R. K. Pitman, M. A. Carson, K. McMullin, P. J. Whalen, and N. Makris, "Selectively Reduced Regional Cortical Volumes in Post-Traumatic Stress Disorder," *Neuroreport* 14, no. 7 (May 23, 2003): 913–16.

36. S. Mineka and A. Ohman, "Phobias and Preparedness: The Selective, Automatic, and Encapsulated Nature of Fear," *Biol Psychiatry* 52, no. 10 (November 15, 2002): 927–37.

37. M. Cook and S. Mineka, "Observational Conditioning of Fear to Fear-Relevant Versus Fear-Irrelevant Stimuli in Rhesus Monkeys," *J Abnorm Psychol* 98, no. 4 (November 1989): 448–59.

38. M. Cook and S. Mineka, "Selective Associations in the Observational Conditioning of Fear in Rhesus Monkeys," *J Exp Psychol Anim Behav Process* 16, no. 4 (October 1990): 372–89.

39. J. S. DeLoache and V. LoBue, "The Narrow Fellow in the Grass: Human Infants Associate Snakes with Fear," *Dev Sci* 12 (2009): 201–207.

40. J. Da Costa, "On Irritable Heart: A Clinical Study of a Form of Functional Cardiac Disorder and Its Consequences," *Am J Med Sci* 61 (1871): 17–52.

41. W. H. R. Rivers, "An Address on the Repression of War Experience," *Lancet* 191 (1918): 173–77.

42. B. P. Gersons and I. V. Carlier, "Post-Traumatic Stress Disorder: The History of a Recent Concept," *Br J Psychiatry* 161 (December 1992): 742–48.

43. J. A. Bodkin, H. G. Pope, M. J. Detke, and J. I. Hudson, "Is PTSD Caused by Traumatic Stress?" *J Anxiety Disord* 21, no. 2 (2007): 176–82.

44. R. L. Spitzer, M. B. First, and J. C. Wakefield, "Saving PTSD from Itself in DSM-V," *J Anxiety Disord* 21, no. 2 (2007): 233–41.

45. American Psychiatric Association, *Diagnostic and Statistical Manual of Mental Disorders*, 4th ed., rev. ed. (Washington, DC: American Psychiatric Association, 2000).

46. R. J. McNally, "Progress and Controversy in the Study of Posttraumatic Stress Disorder," *Annu Rev Psychol* 54 (2003): 229–52.

47. T. Jovanovic and K. J. Ressler, "How the Neurocircuitry and Genetics of Fear Inhibition May Inform Our Understanding of PTSD," *Am J Psychiatry* 167, no. 6 (June 2010): 648–662.

48. M. R. Milad and G. J. Quirk, "Neurons in Medial Prefrontal Cortex Signal Memory for Fear Extinction," *Nature* 420, no. 6911 (November 7, 2002): 70–74.

49. M. R. Milad, C. I. Wright, S. P. Orr, R. K. Pitman, G. J. Quirk, and S. L. Rauch, "Recall of Fear Extinction in Humans Activates the Ventromedial Prefrontal Cortex and Hippocampus in Concert," *Biol Psychiatry* 62, no. 5 (September 1, 2007): 446–54.

50. R. C. Kessler, "Posttraumatic Stress Disorder: The Burden to the Individual and to Society," *J Clin Psychiatry* 61, suppl. 5 (2000): 4–12; discussion 13–14.

51. J. W. Smoller, "Genetics of Mood and Anxiety Disorders," in *Psychiatric Genetics: Applications in Clinical Practice,* ed. J. W. Smoller, B. Rosen-Sheidley, and M. T. Tsuang (Washington, DC: American Psychiatric Press, 2008), 131–76.

52. J. W. Smoller, E. Gardner-Schuster, and M. Misiaszek, "Genetics of Anxiety: Would the Genome Recognize the DSM?" *Depress Anxiety* 25, no. 5 (April 14, 2008): 368–77.

53. S. Mineka and R. Zinbarg, "A Contemporary Learning Theory Perspective on the Etiology of Anxiety Disorders: It's Not What You Thought It Was," *Am Psychol* 61, no. 1 (January 2006): 10–26.

54. D. S. Pine, "Research Review: A Neuroscience Framework for Pediatric Anxiety Disorders," *J Child Psychol Psychiatry* 48, no. 7 (July 2007): 631–48.

55. A. Simmons, I. Strigo, S. C. Matthews, M. P. Paulus, and M. B. Stein, "Anticipation of Aversive Visual Stimuli Is Associated with Increased Insula Activation in Anxiety-Prone Subjects," *Biol Psychiatry* 60, no. 4 (August 15, 2006): 402–409.

56. M. B. Stein, A. N. Simmons, J. S. Feinstein, and M. P. Paulus, "Increased Amygdala and Insula Activation During Emotion Processing in Anxiety-Prone Subjects," *Am J Psychiatry* 164, no. 2 (February 2007): 318–27.

57. H. R. Cremers, L. R. Demenescu, A. Aleman, R. Renken, M. J. van Tol, N. J. van der Wee, D. J. Veltman, and K. Roelofs, "Neuroticism Modulates Amygdala-Prefrontal Connectivity in Response to Negative Emotional Facial Expressions," *Neuroimage* 49, no. 1 (January 1, 2010): 963–70.

58. E. M. Drabant, K. McRae, S. B. Manuck, A. R. Hariri, and J. J. Gross, "Individual Differences in Typical Reappraisal Use Predict Amygdala and Prefrontal Responses," *Biol Psychiatry* 65, no. 5 (March 1, 2009): 367–73.

59. B. C. Reeb-Sutherland, R. E. Vanderwert, K. A. Degnan, P. J. Marshall, K. Perez-Edgar, A. Chronis-Tuscano, D. S. Pine, and N. A. Fox, "Attention to Novelty in Behaviorally Inhibited Adolescents Moderates Risk for Anxiety," *J Child Psychol Psychiatry* 50, no. 11 (November 2009): 1365–72.

60. C. E. Schwartz, P. S. Kunwar, D. N. Greve, L. R. Moran, J. C. Viner, J. M. Covino, J. Kagan, S. E. Stewart, N. C. Snidman, M. G. Vangel, and S. R. Wallace, "Structural Differences in Adult Orbital and Ventromedial Prefrontal Cortex Predicted by Infant Temperament at 4 Months of Age," *Arch Gen Psychiatry* 67, no. 1 (January 2010): 78–84.

61. C. E. Schwartz, C. I. Wright, L. M. Shin, J. Kagan, and S. L. Rauch, "Inhibited and Uninhibited Infants 'Grown Up': Adult Amygdalar Response to Novelty," *Science* 300, no. 5627 (June 20, 2003): 1952–53.

62. S. Y. Hill, K. Tessner, S. Wang, H. Carter, and M. McDermott, "Temperament at 5 Years of Age Predicts Amygdala and Orbitofrontal Volume in the Right Hemisphere in Adolescence," *Psychiatry Res* 182, no. 1 (April 30, 2010): 14–21.

63. J. W. Smoller, M. P. Paulus, J. A. Fagerness, S. Purcell, L. Yamaki, D. Hirshfeld-Becker, J. Biederman, J. R. Rosenbaum, J. Gelernter, and M. B. Stein, "Influence of RGS2 on Anxiety-Related Temperament, Personality, and Brain Function," *Arch Gen Psychiatry* 65 (2008): 298–308.

64. N. H. Kalin, S. E. Shelton, A. S. Fox, J. Rogers, T. R. Oakes, and R. J. Davidson. "The Serotonin Transporter Genotype Is Associated with Intermediate Brain Phenotypes

That Depend on the Context of Eliciting Stressor," *Mol Psychiatry* 13, no. 11 (November 2008): 1021–27.

65. A. Holmes, Q. Lit, D. L. Murphy, E. Gold, and J. N. Crawley, "Abnormal Anxiety-Related Behavior in Serotonin Transporter Null Mutant Mice: The Influence of Genetic Background," *Genes Brain Behav* 2, no. 6 (December 2003): 365–80.

66. C. S. Barr, T. K. Newman, C. Shannon, C. Parker, R. L. Dvoskin, M. L. Becker, M. Schwandt, M. Champoux, K. P. Lesch, D. Goldman, S. J. Suomi, and J. D. Higley, "Rearing Condition and rh5-HTTLPR Interact to Influence Limbic-Hypothalamic-Pituitary-Adrenal Axis Response to Stress in Infant Macaques," *Biol Psychiatry* 55, no. 7 (April 1, 2004): 733–38.

67. T. Canli and K. P. Lesch. "Long Story Short: The Serotonin Transporter in Emotion Regulation and Social Cognition," *Nat Neurosci* 10, no. 9 (August 28, 2007): 1103–09.

68. M. R. Munafo, S. M. Brown, and A. R. Hariri, "Serotonin Transporter (5-HTTLPR) Genotype and Amygdala Activation: A Meta-Analysis," *Biol Psychiatry* 63, no. 9 (May 1, 2008): 852–57.

69. L. Pezawas, A. Meyer-Lindenberg, E. M. Drabant, B. A. Verchinski, K. E. Munoz, B. S. Kolachana, M. F. Egan, V. S. Mattay, A. R. Hariri, and D. R. Weinberger, "5-HTTLPR Polymorphism Impacts Human Cingulate-Amygdala Interactions: A Genetic Susceptibility Mechanism for Depression," *Nat Neurosci* 8, no. 6 (June 2005): 828–34.

70. V. Krishnan, M. H. Han, D. L. Graham, O. Berton, W. Renthal, S. J. Russo, Q. Laplant, A. Graham, M. Lutter, D. C. Lagace, S. Ghose, R. Reister, P. Tannous, et al., "Molecular Adaptations Underlying Susceptibility and Resistance to Social Defeat in Brain Reward Regions," *Cell* 131, no. 2 (October 19, 2007): 391–404.

71. J. P. Chhatwal, L. Stanek-Rattiner, M. Davis, and K. J. Ressler, "Amygdala BDNF Signaling Is Required for Consolidation But Not Encoding of Extinction," *Nat Neurosci* 9, no. 7 (July 2006): 870–72.

72. D. C. Choi, K. A. Maguschak, K. Ye, S. W. Jang, K. M. Myers, and K. J. Ressler, "Prelimbic Cortical BDNF Is Required for Memory of Learned Fear But Not Extinction or Innate Fear," *Proc Natl Acad Sci USA* 107, no. 6 (February 9, 2010): 2675–80.

73. M. F. Egan, M. Kojima, J. H. Callicott, T. E. Goldberg, B. S. Kolachana, A. Bertolino, E. Zaitsev, B. Gold, D. Goldman, M. Dean, B. Lu, D. R. Weinberger, "The BDNF val66met Polymorphism Affects Activity-Dependent Secretion of BDNF and Human Memory and Hippocampal Function," *Cell* 112, no. 2 (January 24, 2003): 257–69.

74. J. Peters, L. M. Dieppa-Perea, L. M. Melendez, and G. J. Quirk, "Induction of Fear Extinction with Hippocampal-Infralimbic BDNF," *Science* 328, no. 5983 (June 4, 2010): 1288–90.

75. Z. Y. Chen, D. Jing, K. G. Bath, A. Ieraci, T. Khan, C. J. Siao, D. G. Herrera, M. Toth, C. Yang, B. S. McEwen, B. L. Hempstead, and F. S. Lee,"Genetic Variant BDNF (Val66Met) Polymorphism Alters Anxiety-Related Behavior," *Science* 314, no. 5796 (October 6, 2006): 140–43.

76. F. Soliman, C. E. Glatt, K. G. Bath, L. Levita, R. M. Jones, S. S. Pattwell, D. Jing, N. Tottenham, D. Amso, L. H. Somerville, H. U. Voss, G. Glover, D. J. Ballon, et al., "A Genetic Variant BDNF Polymorphism Alters Extinction Learning in Both Mouse and Human," *Science* 327, no. 5967 (February 12, 2010): 863–66.

77. T. W. Bredy, H. Wu, C. Crego, J. Zellhoefer, Y. E. Sun, and M. Barad, "Histone Modifications Around Individual BDNF Gene Promoters in Prefrontal Cortex Are Associated with Extinction of Conditioned Fear," *Learn Mem* 14, no. 4 (April 2007): 268–76.

78. M. Lyons, J. Goldberg, S. Eisen, M. True, M. Tsuang, J. Meyer, and W. Henderson, "Do Genes Influence Exposure to Trauma? A Twin Study of Combat," *Am J Med Genet* 48 (1993): 22–27.

79. M. B. Stein, K. L. Jang, S. Taylor, P. A. Vernon, and W. J. Livesley, "Genetic and Environmental Influences on Trauma Exposure and Posttraumatic Stress Disorder Symptoms: A Twin Study," *Am J Psychiatry* 159, no. 10 (October 2002): 1675–81.

80. K. L. Jang, M. B. Stein, S. Taylor, G. J. Asmundson, and W. J. Livesley, "Exposure to

Traumatic Events and Experiences: Aetiological Relationships with Personality Function," *Psychiatry Res* 120, no. 1 (August 30, 2003): 61–69.

81. J. M. Hettema, M. C. Neale, J. M. Myers, C. A. Prescott, and K. S. Kendler, "A Population-Based Twin Study of the Relationship Between Neuroticism and Internalizing Disorders," *Am J Psychiatry* 163, no. 5 (May 2006): 857–64.

82. K. Tambs, N. Czajkowsky, E. Roysamb, M. C. Neale, T. Reichborn-Kjennerud, S. H. Aggen, J. R. Harris, R. E. Orstavik, and K. S. Kendler, "Structure of Genetic and Environmental Risk Factors for Dimensional Representations of DSM-IV Anxiety Disorders," *Br J Psychiatry* 195, no. 4 (October 2009): 301–307.

83. O. J. Bienvenu, J. M. Hettema, M. C. Neale, C. A. Prescott, and K. S. Kendler, "Low Extraversion and High Neuroticism as Indices of Genetic and Environmental Risk for Social Phobia, Agoraphobia, and Animal Phobia," *Am J Psychiatry* 164, no. 11 (November 2007): 1714–21.

84. C. E. Schwartz, C. I. Wright, L. M. Shin, J. Kagan, P. J. Whalen, K. G. McMullin, and S. L. Rauch, "Differential Amygdalar Response to Novel Versus Newly Familiar Neutral Faces: A Functional MRI Probe Developed for Studying Inhibited Temperament," *Biol Psychiatry* 53, no. 10 (May 15, 2003): 854–62.

85. H. Hu, E. Real, K. Takamiya, M. G. Kang, J. Ledoux, R. L. Huganir, and R. Malinow, "Emotion Enhances Learning Via Norepinephrine Regulation of AMPA-Receptor Trafficking," *Cell* 131, no. 1 (October 5, 2007): 160–73.

86. J. L. McGaugh and B. Roozendaal, "Drug Enhancement of Memory Consolidation: Historical Perspective and Neurobiological Implications," *Psychopharmacology (Berl)* 202, no. 1–3 (January 2009): 3–14.

87. R. Hurlemann, H. Walter, A. K. Rehme, J. Kukolja, S. C. Santoro, C. Schmidt, K. Schnell, F. Musshoff, C. Keysers, W. Maier, K. M. Kendrick, and O. A. Onur, "Human Amygdala Reactivity Is Diminished by the Beta-Noradrenergic Antagonist Propranolol," *Psychol Med* 40, no. 11 (November 2010): 1839–48.

88. O. A. Onur, H. Walter, T. E. Schlaepfer, A. K. Rehme, C. Schmidt, C. Keysers, W. Maier, and R. Hurlemann, "Noradrenergic Enhancement of Amygdala Responses to Fear," *Soc Cogn Affect Neurosci* 4, no. 2 (June 2009): 119–26.

89. L. Cahill, B. Prins, M. Weber, and J. L. McGaugh, "Beta-Adrenergic Activation and Memory for Emotional Events," *Nature* 371, no. 6499 (October 20, 1994): 702–704.

90. R. K. Pitman, K. M. Sanders, R. M. Zusman, A. R. Healy, F. Cheema, N. B. Lasko, L. Cahill, and S. P. Orr, "Pilot Study of Secondary Prevention of Posttraumatic Stress Disorder with Propranolol," *Biol Psychiatry* 51, no. 2 (January 15, 2002): 189–92.

91. T. L. Holbrook, M. R. Galarneau, J. L. Dye, K. Quinn, and A. L. Dougherty, "Morphine Use After Combat Injury in Iraq and Post-Traumatic Stress Disorder," *N Engl J Med* 362, no. 2 (January 14, 2010): 110–17.

92. M. C. Jones, "A Laboratory Study of Fear: The Case of Peter," *Pedagog Sem* 31 (1924): 308–15.

93. L. Goossens, S. Sunaert, R. Peeters, E. J. Griez, and K. R. Schruers, "Amygdala Hyperfunction in Phobic Fear Normalizes After Exposure," *Biol Psychiatry* 62, no. 10 (November 15, 2007): 1119–25.

94. D. L. Walker, K. J. Ressler, K. T. Lu, and M. Davis, "Facilitation of Conditioned Fear Extinction by Systemic Administration or Intra-Amygdala Infusions of D-Cycloserine as Assessed with Fear-Potentiated Startle in Rats," *J Neurosci* 22, no. 6 (March 15, 2002): 2343–51.

95. K. J. Ressler, B. O. Rothbaum, L. Tannenbaum, P. Anderson, K. Graap, E. Zimand, L. Hodges, and M. Davis, "Cognitive Enhancers as Adjuncts to Psychotherapy: Use of D-Cycloserine in Phobic Individuals to Facilitate Extinction of Fear," *Arch Gen Psychiatry* 61, no. 11 (November 2004): 1136–44.

96. A. J. Guastella, M. R. Dadds, P. F. Lovibond, P. Mitchell, and R. Richardson, "A Randomized Controlled Trial of the Effect of D-Cycloserine on Exposure Therapy for Spider Fear," *J Psychiatr Res* 41, no. 6 (September 2007): 466–71.

97. A. J. Guastella, R. Richardson, P. F. Lovibond, R. M. Rapee, J. E. Gaston, P. Mitchell, and M. R. Dadds, "A Randomized Controlled Trial of D-Cycloserine Enhancement of Exposure Therapy for Social Anxiety Disorder," *Biol Psychiatry* 63, no. 6 (March 15, 2008): 544–49.

98. S. G. Hofmann, A. E. Meuret, J. A. Smits, N. M. Simon, M. H. Pollack, K. Eisenmenger, M. Shiekh, and M. W. Otto, "Augmentation of Exposure Therapy with D-Cycloserine for Social Anxiety Disorder," *Arch Gen Psychiatry* 63, no. 3 (March 2006): 298–304.

99. M. G. Kushner, S. W. Kim, C. Donahue, P. Thuras, D Adson, M. Kotlyar, J. McCabe, J. Peterson, and E. B. Foa, "D-Cycloserine Augmented Exposure Therapy for Obsessive-Compulsive Disorder," *Biol Psychiatry* 62, no. 8 (October 15, 2007): 835–38.

100. M. M. Norberg, J. H. Krystal, and D. F. Tolin, "A Meta-Analysis of D-Cycloserine and the Facilitation of Fear Extinction and Exposure Therapy," *Biol Psychiatry* 63, no. 12 (June 15, 2008): 1118–26.

101. M. W. Otto, D. F. Tolin, N. M. Simon, G. D. Pearlson, S. Basden, S. A. Meunier, S. G. Hofmann, K. Eisenmenger, J. H. Krystal, and M. H. Pollack, "Efficacy of D-Cycloserine for Enhancing Response to Cognitive-Behavior Therapy for Panic Disorder," *Biol Psychiatry* 67, no. 4 (February 15, 2010): 365–70.

102. S. Wilhelm, U. Buhlmann, D. F. Tolin, S. A. Meunier, G. D. Pearlson, H. E. Reese, P. Cannistraro, M. A. Jenike, and S. L. Rauch, "Augmentation of Behavior Therapy with D-Cycloserine for Obsessive-Compulsive Disorder," *Am J Psychiatry* 165, no. 3 (March 2008): 335–41; quiz 409.

103. J. H. Han, S. A. Kushner, A. P. Yiu, H. L. Hsiang, T. Buch, A. Waisman, B. Bontempi, R. L. Neve, P. W. Frankland, and S. A. Josselyn, "Selective Erasure of a Fear Memory," *Science* 323, no. 5920 (March 13, 2009): 1492–96.

104. M. H. Monfils, K. K. Cowansage, E. Klann, and J. E. LeDoux, "Extinction-Reconsolidation Boundaries: Key to Persistent Attenuation of Fear Memories," *Science* 324, no. 5929 (May 15, 2009): 951–55.

105. D. Schiller, M. H. Monfils, C. M. Raio, D. C. Johnson, J. E. Ledoux, and E. A. Phelps, "Preventing the Return of Fear in Humans Using Reconsolidation Update Mechanisms," *Nature* 463, no. 7277 (January 7, 2010): 49–53.

106. M. Kindt, M. Soeter, and B. Vervliet, "Beyond Extinction: Erasing Human Fear Responses and Preventing the Return of Fear," *Nat Neurosci* 12, no. 3 (March 2009): 256–58.

107. M. Soeter and M. Kindt, "Dissociating Response Systems: Erasing Fear from Memory," *Neurobiol Learn Mem* 94, no. 1 (July 2010): 30–41.

108. H. P. Beck, S. Levinson, and G. Irons, "Finding Little Albert: A Journey to John B. Watson's Infant Laboratory," *Am Psychol* 64, no. 7 (October 2009): 605–14.

CHAPTER 8: THE NEW NORMAL

1. M. Ridley, *Nature via Nurture* (New York: HarperCollins Publishers, 2003).

2. A. Petronis, "Epigenetics as a Unifying Principle in the Aetiology of Complex Traits and Diseases," *Nature* 465, no. 7299 (June 10, 2010): 721–27.

3. A. Raj, S. A. Rifkin, E. Andersen, and A. van Oudenaarden, "Variability in Gene Expression Underlies Incomplete Penetrance," *Nature* 463, no. 7283 (February 18, 2010): 913–18.

4. G. Will, "Handbook Suggests That Deviations from 'Normality' Are Disorders," *Washington Post*, February 28, 2010.

5. E. Shorter, "Why Psychiatry Needs Therapy," *Wall Street Journal*, February 27, 2010.

6. C. A. Boyle, S. Boulet, L. A. Schieve, R. A. Cohen, S. J. Blumberg, M. Yeargin-Allsopp, S. Visser, and M. D. Kogan, "Trends in the Prevalence of Developmental Disabilities in US Children, 1997–2008," *Pediatrics* 127, no 6 (June 2011): 1034–42.

7. Cross-National Collaborative Group, "The Changing Rate of Major Depression. Cross-National Comparisons," *JAMA* 268, no. 21 (December 2, 1992): 3098–3105.

8. CDC, "Prevalence of Autism Spectrum Disorders: Autism and Developmental Disabilities Monitoring Network, United States, 2006," *MMWR* 58, SS–10 (2009).

9. R. Mojtabai. "Increase in Antidepressant Medication in the US Adult Population Between 1990 and 2003," *Psychother Psychosom* 77, no. 2 (2008): 83–92.

10. World Health Organization, *The World Health Report 2001: Mental Health: New Understanding, New Hope* (Geneva, Switzerland: The World Health Organization, 2001).

11. International Schizophrenia Consortium, "Rare Chromosomal Deletions and Duplications Increase Risk of Schizophrenia," *Nature* 455, no. 7210 (September 11, 2008): 237–41.

12. International Schizophrenia Consortium, S. M. Purcell, N. R. Wray, J. L. Stone, P. M. Visscher, M. C. O'Donovan, P. F. Sullivan, and P. Sklar, "Common Polygenic Variation Contributes to Risk of Schizophrenia and Bipolar Disorder," *Nature* 460, no 7256 (August 6, 2009): 748–52.

13. M. A. Ferreira, M. C. O'Donovan, Y. A. Meng, I. R. Jones, D. M. Ruderfer, L. Jones, J. Fan, G. Kirov, R. H. Perlis, E. K. Green, J. W. Smoller, D. Grozeva, J. Stone, et al., "Collaborative Genome-Wide Association Analysis Supports a Role for ANK3 and CACNA1C in Bipolar Disorder," *Nat Genet* 40, no. 9 (September 2008): 1056–1058.

14. D. Pinto, A. T. Pagnamenta, L. Klei, R. Anney, D. Merico, R. Regan, J. Conroy, T. R. Magalhaes, C. Correia, B. S. Abrahams, J. Almeida, E. Bacchelli, G. D. Bader, et al., "Functional Impact of Global Rare Copy Number Variation in Autism Spectrum Disorders," *Nature* 466, no. 7304 (July 15, 2010): 368–72.

15. K. S. Kendler, P. Zachar, and C. Craver, "What Kinds of Things Are Psychiatric Disorders?" *Psychol Med* (September 22, 2010): 1–8.

16. P. Zachar, "The Practical Kinds Model as a Pragmatist Theory of Classification," *Philosophy, Psychiatry, and Psychology* 9 (2002): 219–27.

17. P. Zachar and K. S. Kendler, "Psychiatric Disorders: A Conceptual Taxonomy," *Am J Psychiatry* 164, no. 4 (April 2007): 557–65.

18. E. E. Southard, "Psychopathology and Neuropathology: The Problems of Teaching and Research Contrasted," *JAMA* 58 (1912): 914–16.

19. J. C. Wakefield, "The Concept of Mental Disorder: Diagnostic Implications of the Harmful Dysfunction Analysis," *World Psychiatry* 6, no. 3 (October 2007): 149–56.

20. N. Craddock, K. Kendler, M. Neale, J. Nurnberger, S. Purcell, M. Rietschel, R. Perlis, S. L. Santangelo, T. Schulze, J. W. Smoller, and A. Thapar, "Dissecting the Phenotype in Genome-Wide Association Studies of Psychiatric Illness," *Br J Psychiatry* 195, no. 2 (August 2009): 97–99.

21. J. W. Smoller, E. Gardner-Schuster, and M. Misiaszek, "Genetics of Anxiety: Would the Genome Recognize the DSM?" *Depress Anxiety* 25, no. 4 (April 14, 2008): 368–77.

22. J. Huang, R. H. Perlis, P. H. Lee, A. J. Rush, M. Fava, G. S. Sachs, J. Lieberman, S. P. Hamilton, P. Sullivan, P. Sklar, S. Purcell, and J. W. Smoller, "Cross-Disorder Genomewide Analysis of Schizophrenia, Bipolar Disorder, and Depression," *Am J Psychiatry*, 167, no. 10 (October 2010): 1254–63.

23. S. E. McCarthy, V. Makarov, G. Kirov, A. M. Addington, J. McClellan, S. Yoon, D. O. Perkins, D. E. Dickel, M. Kusenda, O. Krastoshevsky, V. Krause, R. A. Kumar, D. Grozeva, et al., "Microduplications of 16p11.2 Are Associated with Schizophrenia," *Nat Genet* 41, no. 11 (November 2009): 1223–27.

24. F. J. McMahon, N. Akula, T. G. Schulze, P. Muglia, F. Tozzi, S. D. Detera-Wadleigh, C. J. Steele, R. Breuer, J. Strohmaier, J. R. Wendland, M. Mattheisen, T. W. Muhleisen, W. Maier, et al., "Meta-Analysis of Genome-Wide Association Data Identifies a Risk Locus for Major Mood Disorders on 3p21.1," *Nat Genet* 42, no. 2 (February 2010): 128–31.

25. A. Guilmatre, C. Dubourg, A. L. Mosca, S. Legallic, A. Goldenberg, V. Drouin-Garraud, V. Layet, A. Rosier, S. Briault, F. Bonnet-Brilhault, F. Laumonnier, S. Odent, G. Le Vacon, et al., "Recurrent Rearrangements in Synaptic and Neurodevelopmental Genes and Shared Biologic Pathways in Schizophrenia, Autism, and Mental Retardation," *Arch Gen Psychiatry* 66, no. 9 (September 2009): 947–56.

26. J. Sebat, D. L. Levy, and S. E. McCarthy, "Rare Structural Variants in Schizophrenia: One Disorder, Multiple Mutations; One Mutation, Multiple Disorders," *Trends Genet* 25, no. 12 (December 2009): 528–35.

27. N. M. Williams, I. Zaharieva, A. Martin, K. Langley, K. Mantripragada, R. Fossdal, H. Stefansson, K. Stefansson, P. Magnusson, O. O. Gudmundsson, O. Gustafsson, P. Holmans, M. J. Owen, et al., "Rare Chromosomal Deletions and Duplications in Attention-Deficit Hyperactivity Disorder: A Genome-Wide Analysis," *Lancet* 376, no. 9750 (October 23, 2010): 1401–1408.

28. D. F. Levinson, J. Duan, S. Oh, K. Wang, A. R. Sanders, J. Shi, N. Zhang, B. J. Mowry, A. Olincy, F. Amin, C. R. Cloninger, J. M. Silverman, N. G. Buccola, et al., "Copy Number Variants in Schizophrenia: Confirmation of Five Previous Findings and New Evidence for 3q29 Microdeletions and VIPR2 Duplications," *Am J Psychiatry* 168, no. 3 (March 2011): 302–16.

29. V. Vacic, S. McCarthy, D. Malhotra, F. Murray, H. H. Chou, A. Peoples, V. Makarov, S. Yoon, A. Bhandari, R. Corominas, L. M. Iakoucheva, O. Krastoshevsky, V. Krause, et al., "Duplications of the Neuropeptide Receptor Gene VIPR2 Confer Significant Risk for Schizophrenia," *Nature* 471, no. 7339 (March 24, 2011): 499–503.

30. R. Plomin, C. M. Haworth, and O. S. Davis, "Common Disorders Are Quantitative Traits," *Nat Rev Genet* 10, no. 12 (December 2009): 872–78.

31. R. Poulton, A. Caspi, T. E. Moffitt, M. Cannon, R. Murray, and H. Harrington, "Children's Self-Reported Psychotic Symptoms and Adult Schizophreniform Disorder: A 15-Year Longitudinal Study," *Arch Gen Psychiatry* 57, no. 11 (November 2000): 1053–58.

32. R. J. Linscott and J. van Os, "Systematic Reviews of Categorical Versus Continuum Models in Psychosis: Evidence for Discontinuous Subpopulations Underlying a Psychometric Continuum. Implications for DSM-V, DSM-VI, and DSM-VII," *Annu Rev Clin Psychol* 6 (April 27, 2010): 391–419.

33. T. Insel, "Research Domain Criteria (RDoC)," http://www.nimh.nih.gov/research-funding/rdoc.shtml. Accessed May 1, 2010.

34. T. R. Insel and B. N. Cuthbert, "Endophenotypes: Bridging Genomic Complexity and Disorder Heterogeneity," *Biol Psychiatry* 66, no. 11 (December 1, 2009): 988–89.

INDEX

abnormal: definition of, 14, 330. *See also* mental disorders/illness; normal-abnormal

abortion, 105–6

abuse/adversity, childhood: anxiety and, 308; attachment and, 199–202, 206–7, 326, 327; biology of normal and, 326–27, 328; BPD and, 213; brain development and, 103–10, 116–17; fear and, 308; memory and, 19–20; mental disorders/illness and, 18, 19–20; nature and nurture and, 94, 103–10, 116–17, 119–21, 122, 328; normal-abnormal and, 118; oxytocin and, 217; PTSD and, 300; in Romania, 105–8, 131, 135; sensitive periods and, 110, 326; trust and, 222

acrophobia, 316

adaptation: biology of normal and, 325, 327, 337; biology of nurture and, 132, 135–36; of brain, 96, 135–36, 143–44; Darwin's theory about, 7; mind reading and, 143–44; normal-abnormal debate, 118

addiction, 269, 270–73, 313

adrenaline. *See* epinephrine

aggression, 58, 68, 103, 153, 154, 178, 180, 327, 336–37

agoraphobia, 294, 295, 296, 305, 309

agreeableness, 60–62

Aigner-Clark, Julie, 84, 85, 88

Ainsworth, Mary, 196, 203

airplane flight (fear example), 284–87, 288, 321

Albert B. *See* "Little Albert"

Allport, Gordon, 59

Am I Hot or Not (web site), 227–28

American Academy of Pediatrics, 86

American Kennel Club, 63

American Psychiatric Association, 15, 28, 329

amnesia, dissociative, 21

amygdala: anxiety and, 291, 296, 305–6, 307, 309; attachment and, 189; BDNF gene and, 306, 307; biology of normal and, 327; BPD and, 213; damage to, 179; emotional memory and, 310, 313, 315, 320; emotional reactions and, 179, 180, 181; facial recognition and, 101, 141; fear and, 216–17, 284, 285–86, 288, 289, 290, 296, 304, 305–6, 307, 310; functions of, 55, 76; mind reading and, 141; oxytocin and, 216–17; PTSD and, 304; serotonin and, 70, 72, 73, 80, 112; temperament and, 54–56, 58, 66–67, 70, 72, 73, 80, 327; trust and, 216–17, 218, 220; vasopressin in, 216–17

Amytal, 19

anger, 13, 104, 118, 122, 154, 172, 213, 327, 336–37

animals: Big Five and, 62; children raised by, 93–94; domestication of, 150–54; interspecies communication and, 150; mind reading and, 142–43, 150–54; mother-child bonding among, 190–91, 197; and quest to understand mind and brain, 2; same-sex behavior among, 253, 260; temperament of, 154; theory of mind of, 148–49, 150, 154, 154*n*. *See also* cats; chimpanzees; dogs; monkeys

anorexia, 24

Anthony, Helen, 274

antidepressants. *See* drugs/medications

antipsychiatry movement, 28

antipsychotics. *See* drugs/medications

anxiety: benefits of, 291–92; biology of normal and, 336; childhood abuse/adversity and, 103, 108, 109, 118, 308; cognition and, 305–6; compassion fatigue and, 183; definition of, 291, 331, 336; disorders, 4, 279, 294–97, 301–10, 312–15, 316, 331, 336, 337; environment and, 305; experiences and, 305, 308; fear and, 294–97; genes/genetics and, 73, 79–81, 305–8, 306*n*, 309; harm-avoidance mechanism and, 33, 70; health and, 293–94; nature and nurture and, 103, 108, 109, 112, 118, 128, 304–8; neurotransmitters and, 72; normal-abnormal boundary and, 166–67, 309; prevalence of, 295–96; and quest for understanding of mind and brain, 4; resilience and, 128; separation, 202; serotonin and, 68, 70, 80, 112; shrinking penis case and, 24–25, 26–27; stranger, 202, 204; stress and, 305, 308; temperament and, 57, 306;